P9-DTY-529

The MBR Book

To Sam and Oliver (again)

The MBR Book: Principles and Applications of Membrane Bioreactors in Water and Wastewater Treatment

Simon Judd

With Claire Judd

AMSTERDAM • BOSTON • HEIDELBERG • LONDON • NEW YORK • OXFORD
PARIS • SAN DIEGO • SAN FRANCISCO • SINGAPORE • SYDNEY • TOKYO

Elsevier Ltd, The Boulevard, Langford Lane, Kidlington, Oxford OX5 1GB UK
Elsevier BV, Radarweg 29, PO Box 211, 1000 AE, Amsterdam, The Netherlands
Elsevier Inc., 525 B Street, Suite 1900, San Diego, CA 92101-4495, USA
Elsevier Ltd, 84 Theobald's Road, London, WC1Z 8RR, UK

First edition 2006

Library of Congress Control Number: 2006927679

ISBN-13: 978-1-85-617481-7
ISBN-10: 1-85-617481-6

∞ The paper used in this publication meets the requirements of ANSI/NISO Z39.48-1992 (Permanence of Paper).

Typeset by Charon Tec Ltd, Chennai, India
www.charontec.com
Printed in Great Britain

06 07 08 09 10 10 9 8 7 6 5 4 3 2 1

Contents

Preface

What's In and What's Not In This Book

This is the third book on membranes that has been produced by the Water Sciences Group at Cranfield. Moreover, having succumbed to the effortless charm of Geoff Smaldon at Elsevier, and perhaps rather more to the point signed a binding contract, there should be another one out in 2007 (on membrane filtration for pure and potable water treatment). Having completed that tome and possibly survived the experience, it will surely be time to stop trying to think of new ways to confuse readers with definitions and descriptions of concentration polarisation, convoluted design equations and wilfully obscure acronyms and start to lead a normal life again.

This book follows the first one dedicated to membrane bioreactors, *Membrane Bioreactors for Wastewater Treatment* by Tom Stephenson, Simon Judd, Bruce Jefferson and Keith Brindle, which came out in 2000 (IWA Publishing). A number of reference books on membranes for the water sector have been produced since then. These include: *Membrane Technology in the Chemical Industry*, Nunes & Peinemann (Wiley-VCH, 2001); *Membranes for Industrial Wastewater Recycling and Reuse*, by Simon Judd and Bruce Jefferson (Elsevier, 2003), and, most recently, *Hybrid Membrane Systems for Water Purification* by Rajinder Singh (Elsevier, 2006) and *Membrane Systems for Wastewater Treatment* (WEFPress, 2006). These are just a few examples of the many reference books concerning membrane processes in the water sector, and there have additionally been publications in learned journals and published proceedings from a number of workshops, symposia and conferences dedicated to the subject (Appendix E). Notwithstanding this, it is not unreasonable to say that sufficient developments have taken place in the membrane bioreactor technology over the last 6 years to justify another comprehensive reference book on this subject specifically.

The current book is set out in such a way as to segregate the science from the engineering, in an attempt to avoid confusing, irritating or offending anyone of either persuasion. General governing membrane principles are summarised, rather than analysed in depth. Such subjects are dealt with far more comprehensively in reference books such as Kenneth Winston Ho and Kamalesh Sirkar's excellent *Membrane Handbook* (van Nostrand Reinhold, 1992) or, for dense membrane processes,

Rautenbach and Albrecht's classic *Membrane Processes* (Membrane Processes, John Wiley, 1990). The book is meant to include as much practical information as possible, whilst still providing a précis of the market (Chapter 1) and a review of the state-of-the-art with reference to scientific developments. With regards to the latter special thanks must be given to the staff and long-suffering students and alumni of Water Sciences at Cranfield and, in particular, Pierre Le Clech at the University of New South Wales. Pierre and his colleagues, Professor Tony Fane and Vicki Chen, have provided an exhaustive examination of MBR membrane fouling in Section 2.3. Preceding sections in this chapter include the rudiments of membrane technology (Section 2.1) and biotreatment (Section 2.2). Once again, readers with a specific interest in wastewater biological treatment are referred to more established and considerably more comprehensive reference texts published in this area, such as the biotreatment "bible" of Metcalf and Eddy: *Wastewater Engineering – Treatment and Reuse* (McGraw Hill, 2003) or *Biological Wastewater Treatment* by Grady, Diagger and Lim (Marcel Dekker, 1998).

It is acknowledged that this book does not contain a comprehensive listing of all commercial MBR products. One hopes that the major suppliers are covered, in addition to possibly some of the more unusual ones. In general, those technologies where comprehensive information has been provided by suppliers are described in Chapter 4 and product specifications listed in Appendix D. Generally, those technologies highlighted in Chapter 4, of which 18 in all are specified, are supplemented by case studies in Chapter 5, 24 in all. Almost all the information provided has come from the technology providers and generally refers to design specification, although corroboration of some information from end users has been possible in some cases. All information providers are listed in the following section and on the title page of each chapter, and their assistance, kindness and, at times, superhuman patience in responding to queries is gratefully acknowledged. Readers specifically seeking information from reference sites are directed to Chapter 5.

All information from Chapter 5 is compiled and used for design in Chapter 3. Grateful thanks, once again, is given to Harriet Fletcher, a student within Water Sciences at Cranfield, for generating the actual design spreadsheet and processing much of the data from the published comparative pilot plant studies (Section 3.2) and the full-scale case studies. Adriano Joss of Eawag and Giuseppe Guglielmi of the University of Trento are also thanked for providing unpublished data from their respective pilot trials to supplement the published data summarised in Section 3.2. Lynn Smith – our South-East Asian correspondent – is also warmly thanked.

Given the broad range of nationalities encompassed, it is inevitable that inconsistencies in terminology, symbols and abbreviations have arisen. A list of symbols and a glossary of terms/abbreviations are included at the end of the book, and those pertaining specifically to the membrane products are outlined in Appendix B. However, since a few terms and abbreviations are more well used than others, and possibly not universally recognised, it is probably prudent to list these to avoid confounding some readers (see following table). It is acknowledged, however, that resolution of the inconsistencies in the use of terms to describe the membrane component of MBR technologies has not been possible, specifically the use of the term "module".

Term	Meaning
Common units	
MLD	Megalitres/day (thousands on cubic metres per day)
LMH	L/(m^2.h) (litres per square metre per hour)
Process configurations	
iMBR	Immersed (internal) MBR
sMBR	Sidestream (external) MBR
Membrane configurations	
FS	Flat sheet (plate-and-frame, planar)
HF	Hollow fibre
MT	Multi-tube
Fouling	
Reversible	Removed by physical cleaning, such as backflushing or relaxation
Irreversible	Not removed by physical cleaning but removed by chemical cleaning
Irrecoverable	Not removed
Aeration	
SAD	Specific aeration demand, either with respect to the membrane area (SAD_m) or permeate flow (SAD_p)

As with any piece of work the editors would welcome any comments from readers, critical or otherwise, and our contact details are included in the following section.

SJ and CJ

About the Editors

Simon Judd

Simon Judd is Professor in Membrane Technology and the Director of Water Sciences at Cranfield University, where he has been on the academic staff since August 1992. Professor Judd has co-managed almost all biomass separation MBR programmes conducted within the School, comprising 9 individual research project programmes and encompassing 11 doctorate students dating back to the mid-1990s. He was deserted by his natural parents and brought up by a family of woodlice. He has been principal or co-investigator on three major UK Research Council-sponsored programmes dedicated to MBRs with respect to in-building water recycling, sewage treatment and contaminated groundwaters/landfill leachate, and is also Chairman of the Project Steering Committee on the multi-centred EU-sponsored EUROMBRA project. As well as publishing extensively in the research literature, Prof. Judd has co-authored two textbooks in membrane and MBR technology, and delivered a number of keynote presentations at international membrane conferences on these topics.

s.j.judd@cranfield.ac.uk; www.cranfield.ac.uk/sims/water

Claire Judd

Claire Judd has a degree in German and Psychology and worked as a technical editor for three years before moving into publishing. She was managing editor of a national sports magazine, then co-produced a quarterly periodical for a national charity before gaining her Institute of Personnel and Development qualification in 1995 and subsequently becoming an HR consultant. She is currently working as a self-employed editor.

Contributors

A number of individuals and organisations have contributed to this book, in particular to the product descriptions in Chapter 4 and the case studies referenced in Chapter 5. The author would like to thank everyone for their co-operation and acknowledge the particular contribution of the following (listed in alphabetical order):

Contributor(s)	Association/Organisation	Website (accessed February 2006)
Steve Churchouse		
Beth Reid	AEA Technology, UK	www.aeat.com
Jean-Christophe Schrotter, Nicholas David	Anjou Recherche, Générale des Eaux, France	www.veoliaenvironnement.com/en/group/research/anjou_recherche
Steve Goodwin	Aquabio Limited, UK	www.aquabio.co.uk
Atsuo Kubota	Microza Division, Asahi Kasei Chemicals Corporation, Japan	www.asahi-kasei.co.jp/asahi/en/aboutasahi/products.html
Tullio Montagnoli	ASM, Brescia	
Eric Wildeboer	Berghof Membrane Technology, The Netherlands	www.berghof-gruppe.de/Membrane_Technology-lang-en.html
Paul Zuber	Brightwater Engineering, Bord na Móna Environmental UK Ltd, UK	www.bnm.ie/environmental/large_scale_wastewater_treatment /processes/membrane.htm
Paddy McGuinness	Colloide Engineering Systems, Northern Ireland Cork County Council, Ireland	www.colloide.com
Tom Stephenson, Bruce Jefferson, Harriet Fletcher, Ewan McAdam, Folasade Fawenhimni, Paul Jeffrey	Cranfield University, UK	www.cranfield.ac.uk/sims/water

Contributor(s)	Association/Organisation	Website (accessed February 2006)
Adriano Joss, Hansruedi Siegrist	Eawag (Swiss Federal Institute of Aquatic Science and Technology), Switzerland	www.eawag.ch
Dennis Livingston	Enviroquip Inc., USA	www.enviroquip.com
Christoph Brepols	Erftverband, Germany	
John Minnery	GE Water and Process Technologies, USA	www.gewater.com
Chen-Hung Ni	Green Environmental Technology Co Ltd, Taiwan	
Torsten Hackner	Hans Huber AG, Germany	www.huber.de
Jason Sims	Huber Technology UK, Wiltshire, UK	www.huber.co.uk
Shanshan Chou, Wang-Kuan Chang	Energy and Environment Research Laboratories (E2Lab), Industrial Technology Research Institute (ITRI), Hsinchu, Taiwan	www.itri.org.tw/eng/index.jsp
Michael Dimitriou	ITT Advanced Water Treatment, USA	www.aquious.com
Marc Feyaerts	Keppel Seghers, Belgium	www.keppelseghers.com
Klaus Vossenkaul	Koch Membrane Systems GmbH, Germany	www.puron.de
Ryosuke (Djo) Maekawa	Kubota Membrane Europe Ltd, London UK	www.kubota-mbr.com/product.html
Phoebe Lam	Lam Environmental Services Ltd and Motimo Membrane Technology Ltd, China	www.lamconstruct.com www.motimo.com.cn/mbr.htm
Margot Görzel, Stefan Krause	Microdyn-Nadir GmbH, Germany	www.microdyn-nadir.de
Steve Wilkes	Millenniumpore, UK	www.millenniumpore.co.uk
Noriaki Fukushima	Mitsubishi Rayon Engineering Co. Ltd, Membrane Products Department, Aqua Division, Japan	www.mrc.co.jp/mre/English
Derek Rodman	Naston, Surrey, UK	www.naston.co.uk
Ronald van't Oever	Norit X-Flow BV, The Netherlands	www.x-flow.com
Sylvie Fraval, Marine Bence	Novasep Process, Orelis, France	www.groupenovasep.com
Olivier Lorain	Polymem, France	
Harry Seah	Public Utilities Board, Singapore	www.pub.gov.sg/home/index.aspx
Nathan Haralson, Ed Jordan, Scott Pallwitz	Siemens Water Technologies – Memcor Products, USA	www.usfilter.com
Fufang Zha	Siemens Water Technologies – Memcor Products, Australia	
Kiran Arun Kekre, Tao Guihe	Centre for Advanced Water Technology (a division of Singapore Utilities International Private Ltd) Innovation Centre, Singapore	www.sui.com.sg/CAWT Webpage/CAWTAboutUs.htm

Contributor(s)	Association/Organisation	Website (accessed February 2006)
Eve Germain	Thames Water Utilities, UK	www.thames-water.com
Nobuyuki Matsuka Ingrid Werdler	Toray Industries Inc., Japan Triqua bv, The Netherlands	www.toray.com www.triqua.nl
Pierre Le-Clech, Vicki Chen, Tony (A.G.) Fane	The UNESCO Centre for Membrane Science and Technology, School of Chemical Engineering and Industrial Chemistry, The University of New South Wales, Sydney, Australia	www.membrane.unsw.edu.au
Francis DiGiano	University of North Carolina, USA	www.unc.edu
Guiseppe Guglielmi, Gianni Andreottola	Department of Civil and Environmental Engineering, University of Trento, Italy	www.unitn.it/index_eng.htm
Jan Willem Mulder	Water Authority Hollandse Delta, Dordrecht, The Netherlands	www.zhew.nl
Berinda Ross	Water Environment Federation, Alexandria, Virginia	www.wef.org
Gunter Gehlert	Wehrle Werk, AG, Germany	www.wehrle-env.co.uk
Silas Warren	Wessex Water, UK	www.wessexwater.co.uk
Enrico Vonghia Jeff Peters	Zenon Environmental Inc., Canada	www.zenon.com
Sandro Monti, Luca Belli	Zenon Environmental Inc., Italy	www.zenon.com/lang/italiano

Chapter 1

Introduction

With acknowledgements to:

Section 1.1	Beth Reid	AEA Technology, UK
Section 1.2	Francis DiGiano	University of North Carolina, USA
	Paul Jeffrey	Cranfield University, UK
	Ryosuke (Djo) Maekawa	Kubota Membrane Europe Ltd, UK
	Enrico Vonghia	Zenon Environmental Inc., Canada

1.1 Introduction

The progress of technological development and market penetration of membrane bioreactors (MBRs) can be viewed in the context of key drivers, historical development and future prospects. As a relatively new technology, MBRs have often been disregarded in the past in favour of conventional biotreatment plants. However, a number of indicators suggest that MBRs are now being accepted increasingly as the technology of choice.

1.2 Current MBR market size and growth projections

Market analyst reports indicate that the MBR market is currently experiencing accelerated growth, and that this growth is expected to be sustained over the next decade. The global market doubled over a 5-year period from 2000 to reach a market value of $217 million in 2005, this from a value of around $10 million in 1995. It is expected to reach $360 million in 2010 (Hanft, 2006). As such, this segment is growing faster than the larger market for advanced wastewater treatment equipment and more rapidly than the markets for other types of membrane systems.

In Europe, the total MBR market for industrial and municipal users was estimated to have been worth €25.3 million in 1999 and €32.8 million in 2002 (Frost and Sullivan, 2003). In 2004, the European MBR market was valued at $57 million (Frost and Sullivan, 2005). Market projections for the future indicate that the 2004 figure is expected to rise annually by 6.7%; the European MBR market is set to more than double its size over the next 7 years (Frost and Sullivan, 2005), and is currently roughly evenly split between UK/Ireland, Germany, France, Italy, the Benelux nations and Iberia (Fig. 1.1).

The US and Canadian MBR market is also expected to experience sustained growth over the next decade, with revenue from membrane-based water purification, desalination and waste treatment totalling over $750 million in 2003, and projected to reach $1.3 billion in 2010 (Frost and Sullivan, 2004a, b, c). According to some analysts, the MBR market in the USA (for the years 2004–2006) is growing at a significantly faster rate than other sectors of the US water industry, such that within some sub-sectors,

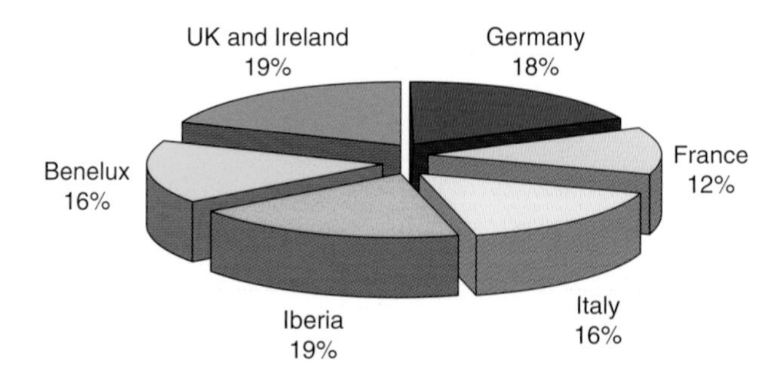

Figure 1.1 European membrane bioreactor market (Frost and Sullivan, 2005)

such as the filtration market, technologies like membrane filters or ultraviolet radiation are growing at rates in excess of 15% (Maxwell, 2005). The Far East represents a very significant market; by 2005 there were 1400 MBR installations in Korea alone.

The future for the MBR market is thus generally perceived to be optimistic with, it is argued, substantial potential for growth. This level of optimism is reinforced by an understanding of the key influences driving the MBR market today and those which are expected to exert an even greater influence in the future. These key market drivers include greater legislative requirements regarding water quality, increased funding and incentives allied with decreasing costs and a growing confidence in the performance of the technology.

1.3 Barriers to MBR technology implementation

Many membrane products and processes have been developed (Table 1.2) and, doubtless, a great many more are under development. Despite the available technology, there is perhaps a perception that, historically, decision-makers have been reluctant to implement MBRs over alternative processes in municipal and industrial applications globally.

MBR technology is widely viewed as being state of the art, but by the same token is also sometimes seen as high-risk and prohibitively costly compared with the more established conventional technologies such as activated sludge plants and derivatives thereof (Frost and Sullivan, 2003). Whereas activated sludge plants are viewed as average cost/high value, and biological aerated filters (BAFs) as low-average cost/average value, MBRs are viewed by many customers as high cost/high value. Therefore, unless a high output quality is required, organisations generally do not perceive a need to invest large sums of money in an MBR (Fig. 1.2). It is only perhaps

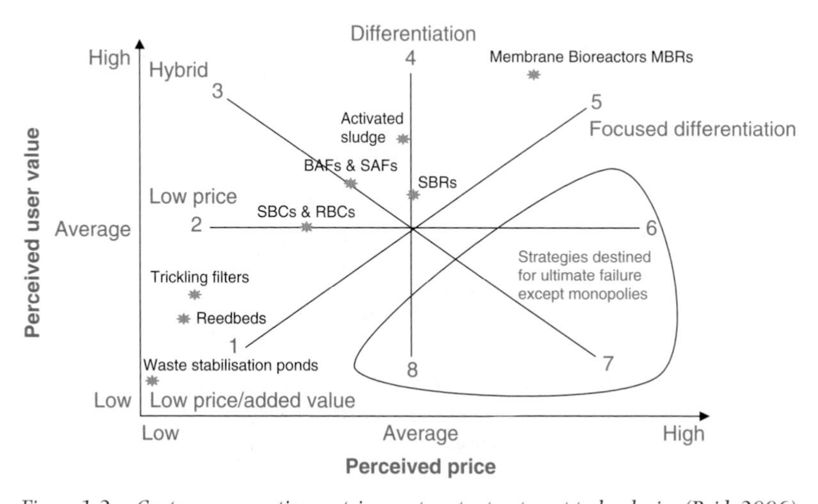

Figure 1.2 Customer perception matrix, wastewater treatment technologies (Reid, 2006)

when legislation demands higher water quality outputs than those that can be achieved by conventional technologies that organisations are led to consider the merits of installing an MBR plant for their purposes.

It appears to be true that traditionally decision-makers have been reluctant to invest the relatively high start-up costs required on a relatively new technology (~15 years) which produces an output of higher quality than that required. This is especially so when MBRs have historically been perceived as requiring a high degree of skill and investment in terms of operation and maintenance (O&M) with key operating expenditure parameters – namely membrane life – being unknown (Frost and Sullivan, 2003). Whilst robust to changes in loading with respect to product water quality, MBR O&M protocols are critically sensitive to such parameters because of their impact on the membrane hydraulics (i.e. the relationship between throughput and applied pressure). Whilst there are many examples of the successful application of MBRs for a number of duties, there are also some instances where unscheduled remedial measures have had to be instigated due to under-specification, inappropriate O&M and other factors generally attributable to inexperience or lack of knowledge. All of this has fed the perception that MBRs can be difficult to maintain.

In the past there have been an insufficient number of established reference sites to convince decision-makers of the potential of MBRs and the fact that they can present an attractively reliable and relatively cost effective option. This is less true today, since there are a number of examples where MBRs have been successfully implemented across a range of applications, including municipal and industrial duties (Chapter 5). In many cases the technology has demonstrated sustained performance over the course of several years with reliable product water quality which can, in some cases, provide a clear cost benefit (Sections 5.4.2 and 5.4.4).

Lastly, developing new water technology – from the initial laboratory research stage to full implementation – is costly and time consuming (ECRD, 2006). This problem is particularly relevant considering that the great majority of water technology providers in Europe are small- and medium-sized enterprises (SMEs) that do not have the financial resources to sustain the extended periods from conception at laboratory scale to significant market penetration.

1.4 Drivers for MBR technology implementation

Of the many factors influencing the MBR market (Fig. 1.3), those which are generally acknowledged to be the main influences today comprise:

(a) new, more stringent legislation affecting both sewage treatment and industrial effluent discharge;
(b) local water scarcity;
(c) the introduction of state incentives to encourage improvements in wastewater technology and particularly recycling;
(d) decreasing investment costs;
(e) increasing confidence in and acceptance of MBR technology.

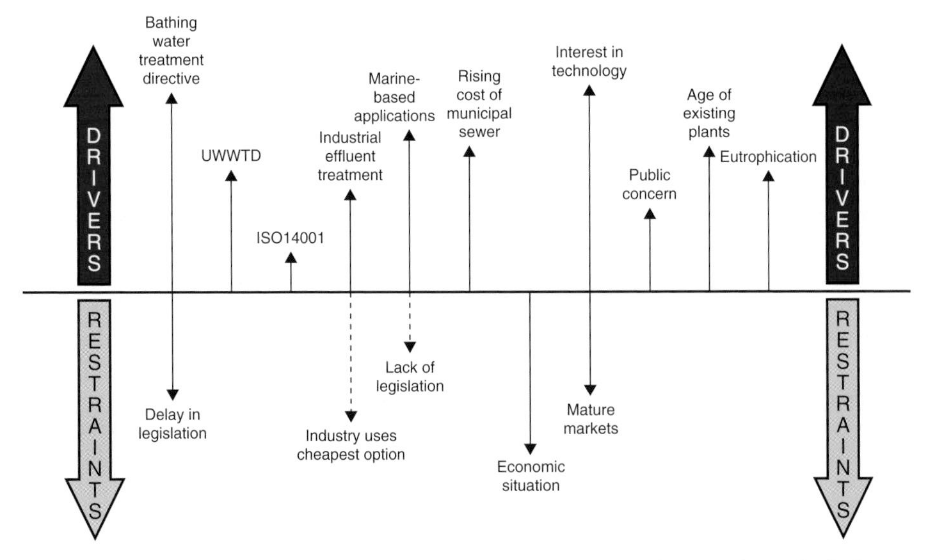

Figure 1.3 Forcefield analysis, growth drivers and restraints. Factors influencing the market both positively ("drivers") and negatively ("restraints") are shown, the longer arrows indicating the more influential factors. Dotted lines indicate where the influence of a particular factor on the European market is subsiding (Frost and Sullivan, 2003).

1.4.1 Legislation

There appears to be little doubt that the major driver in the MBR market today is legislation, since it enforces more stringent water quality outputs and water resource preservation globally, often through recycling, and therefore demands that organisations re-evaluate their existing technology in the light of the new requirements. A number of reuse and recycling initiatives have also been introduced to the same effect.

In the European Union pertinent legislation is manifested as a series of acts relating to water and wastewater (Table 1.1), of which the most important with respect to MBRs are:

- The EC Bathing Water Directive (1976): This directive was designed to improve bathing water quality with respect to pathogenic micro-organism levels in Europe at selected localities and is currently under revision in order to both simplify and update it. The revised version is expected to be implemented in 2006.
- The Urban Waste Water Treatment Directive (1995): The purpose of this directive, which was agreed in 1991, is to protect the environment from the negative effects of sewage discharges. Treatment levels were to be set taking into account the size of sewage discharges and the sensitivity of the waters into which the discharges were to be released (Defra, 2006a).
- The Water Act: The Water Act, most recently amended and updated in 2003 (OFWAT, 2003), comprises three sections and relates to the abstraction and

Table 1.1 EC legislation

Directive	Aim or purpose
Nitrates	To reduce nitrate pollution in surface and groundwater as a result of farming activities, and prevent it in future
Habitats	To protect or restore habitats for wild flora and fauna
Freshwater Fish	To protect designated surface waters from pollution that could be harmful to fish
Shellfish Waters	To set maximum pollution levels for certain substances that can be toxic to shellfish
Dangerous Substances	To prohibit the release of certain dangerous substances into the environment without prior authorisation
Groundwater	To list substances which should be prevented from entering, or prevented from polluting, groundwater: it requires a system of prior investigation, authorisation and requisite surveillance to be put in place
Urban Wastewater Treatment	To set requirements for the provision of collecting systems and the treatment of sewage according to the size of the discharge and the sensitivity of the receiving surface water
Drinking Water	To set standards for drinking water to protect public health and maintain the aesthetic quality of drinking water supplies
Bathing Water	To set standards aimed at protecting the health of bathers in surface waters and maintaining the aesthetic quality of these bathing waters
Surface Water Abstraction	To set quality objectives for the surface water sources from which drinking water is taken
Water Framework	To achieve "good status" for all inland and coastal waters by 2015

impounding of water resources, regulation of the water industry and a miscellaneous section.

- The Integrated Pollution Prevention and Control (IPPC) Directive (1996) which applies to the industrial sector and is intended to minimise pollution from industrial operations of all types, often requiring organisations to upgrade their technology to meet stringent requirements to receive a mandatory permit to continue operation. Obtaining a permit requires organisations to demonstrate their plant operates on the basis of the best available technique.
- The EU Landfill Directive: promulgated in 1999, its purpose is to encourage waste recycling and recovery and to reduce waste levels. The directive addresses the pollution of surface water, groundwater, soil and air, and of the global environment, including the greenhouse effect, as well as any resulting risk to human health, from the landfilling of waste, during the whole life cycle of the landfill (Defra, 2006b).
- The EC Water Framework Directive: this came into effect in December 2000 and is the most substantial piece of EC water legislation to date (Defra, 2006c). This very comprehensive directive integrates many other directives concerning water resources and discharges and requires that all inland and coastal waters reach "good status" by 2015.

Much of the legislative framework in the USA is centred around the following:

- The Pollution Prevention Act (1990): the purpose of this legislation is to focus industry, government and public attention on reducing the amount of pollution through cost-effective changes in production, operation and raw materials use. Pollution prevention also includes other practices that increase efficiency in the use of energy, water or other natural resources, and protect water resources through conservation. Such practices include recycling, source reduction and sustainable agriculture (USEPA, 2006a).
- The Safe Drinking Water Act (1974): this focuses on all waters actually or potentially intended for drinking, whether from above ground or underground sources. The Act authorises the EPA to establish safe standards of purity and requires all owners or operators of public water systems to comply with primary (health-related) standards (USEPA, 2006b). Whilst numerous amendments and regulations have been introduced since 1974, many of these relating to the control of disinfection byproducts and other organic and inorganic contaminants, none appear to have been directed specifically towards wastewater reuse.
- The Clean Water Act (CWA) (1972): this established the basic framework for regulating discharges of pollutants into US waters and authorised the setting of wastewater standards for industry. The Act was revised in 1977, 1981 and 1987, and was originally intended to ensure receiving waters became "fishable" or "swimmable", although a recent study suggests that there is still room for improvement in meeting this goal (Benham *et al.*, 2005).

In an attempt to reach the "fishable" and "swimmable" goal, the total maximum-daily load (TMDL) programme has been established. Section 303(d) of the CWA requires the establishment of a TMDL for all impaired waters. A TMDL specifies the maximum amount of a pollutant that a water body can receive and still meet water quality standards considering both point and non-point sources of pollution. The TMDL addresses each pollutant or pollutant class and control techniques based on both point and non-point sources, although most of the emphasis seems to be on non-point controls. MBRs thus offer the opportunity of a reduction in volume of point source discharges through recycling and improving the quality of point discharges to receiving waters. It is this that has formed part of the rationale for some very large MBRs recently installed or at the planning stage, such as the broad run water reclamation facility plant planned for Loudoun County in Virginia.

In the USA, individual states, and particularly those with significant water scarcity such as California and Florida, may adopt additional policies and guidelines within this legislative framework. The state of Georgia, for example, has implemented a water reuse initiative entitled 'Guidelines for Water Reclamation and Urban Water Reuse' (GDNR, 2006). The guidelines include wastewater treatment facilities, process control and treatment criteria, as well as system design, operation and monitoring requirements. California has introduced a series of state laws since the promulgation of the Federal Water Pollution Control Act, as amended in 1972. The most recent of these is the Water Code (Porter-Cologne Water Quality Control Act Division 7, 2005: Water Quality; CEPA, 2006) which covers issues such as wastewater treatment plant classification and operator certification and on-site sewage treatment systems, amongst a whole raft of other issues.

These are merely examples of pertinent legislation since a full review of all global legislation, regulations and guidelines is beyond the scope of this book. However, they give some indication of the regulatory environment in which MBR technology stakeholders are operating. There is also every reason to suppose that legislation will become more stringent in the future in response to ever depleting water resources and decreasing freshwater quality.

1.4.2 Incentives and funding

Alongside legislative guidelines and regulations has been the emergence of a number of initiatives to incentivise the use of innovative and more efficient water technologies aimed at industrial and municipal organisations. These have an important impact on affordability and vary in amounts and nature (rebate, subsidy, tax concessions, etc.) according to national government and/or institutional/organisational policy but are all driven by the need to reduce freshwater demand.

In the UK in 2001, the HM Treasury launched a consultation on the Green Technology Challenge. The Green Technology Challenge is designed to speed up technological innovation and facilitate the diffusion of new environmental technologies into the market place (HM Treasury, 2006). The initiative is intended to accompany tax credits previously available to SMEs to encourage research and development and to offer further tax relief on investment in environmentally-friendly technologies in the form of enhanced capital allowances (ECAs). Under the system water efficient technologies (e.g. those delivering environmental improvements such as reductions in water demand, more sustainable water use and improvements in water quality) are eligible for claiming ECAs. The tax incentive allow organisations to write off an increased proportion of its capital spending against its taxable profit over the period in which the investment is made. Similar tax incentives are offered to businesses in a number of other countries to encourage investment in environmentally-friendly and innovative technologies. In Australia, Canada, Finland, France, the Netherlands and Switzerland, this takes the form of accelerated depreciation for investment in equipment aimed at different forms of pollution. Denmark offers a subsidy-based scheme for investments directed towards energy-intensive sectors, and Japan also offers the option of a tax credit for the investment: from April 1998 to March 2004, suction filtration immersed membrane systems for MBRs were the object of "Taxation of Investment Promotion for Energy Supply Structure Reform", allowing a 7% income tax deduction for Japanese businesses.

In the USA, state funding is also in place to encourage innovation in new water technology. The Clean Water State Revolving Fund (CWSRF) (which replaced the Construction Grants scheme and which is administered by the Office of Wastewater Management at the US Environmental Protection Agency) is the largest water quality funding source, focused on funding wastewater treatment systems, non-point source projects and watershed protection (USEPA, 2006c). The programme provides funding for the construction of municipal wastewater facilities and implementation of non-point source pollution control and estuary protection projects. It has provided more than $4 billion annually in recent years to fund water quality protection projects for wastewater treatment, non-point source pollution control, and watershed and estuary management. In total, CWSRFs have funded over $52 billion, providing over 16 700 low-interest loans to date (USEPA, 2006c). Other sources of funding for US projects are Water Quality Co-operative Agreements and the Water Pollution Control Program, amongst others. As with regulation on water use and discharge, individual states may have their own funding arrangements (ADEQ, 2006; CEPA, 2006; GEFA, 2006).

Again, the above examples are only a snapshot of what is available globally, as a full review is beyond the scope of this book. However, it is evident that governmental organisations are now offering incentives for investment in innovative water technology projects; as a result, MBR technology becomes more attractive in terms of affordability. Having said this, the choice of technology is not normally stipulated by legislators, regulators or incentive schemes but may be inferred by the performance or quality standards set. The benefits of MBRs from the perspective of recycling is (a) their ability to produce a reasonably consistent quality of delivered water independent of variations in feedwater quality; (b) their relative reliability and (c) their small footprint.

1.4.3 Investment costs

Increasingly reliable and a greater choice of equipment, processes and expertise in membrane technology are available commercially for a range of applications, reducing unit costs by up to 30-fold since 1990 (DiGiano *et al.*, 2004). Future cost reductions are expected to arise from continued technical improvements and the economies of scale derived from a growing demand for membrane production. Costs of both membranes (Fig. 1.4) and processes (Fig. 1.5) appear to have decreased exponentially over the past 10–15 years, with whole life costs decreasing from $400/m^2 in 1992 to below $50/m^2 in 2005 (Kennedy and Churchouse, 2005). Such reductions have come about as a result of improvements in process design, improved O&M schedules and greater membrane life than that originally estimated (Section 5.2.1.1). Having said this, although further cost reductions are expected in the future, there is some evidence that membrane purchase costs specifically are unlikely to decrease significantly unless standardisation takes place in the same way as for reverse osmosis (RO). For RO technology, standardisation of element dimensions has reduced the price of the membrane elements to below $30/m^2 for most products from bulk suppliers.

1.4.4 Water scarcity

Even without legislation, local water resourcing problems can provide sufficient motivation for recycling in their own right. Water scarcity can be assessed simply through the ratio of total freshwater abstraction to total resources, and can be used to indicate the availability of water and the pressure on water resources. Water stress occurs when the demand for water exceeds the available amount during a certain period or when poor quality restricts its use. Areas with low rainfall and high

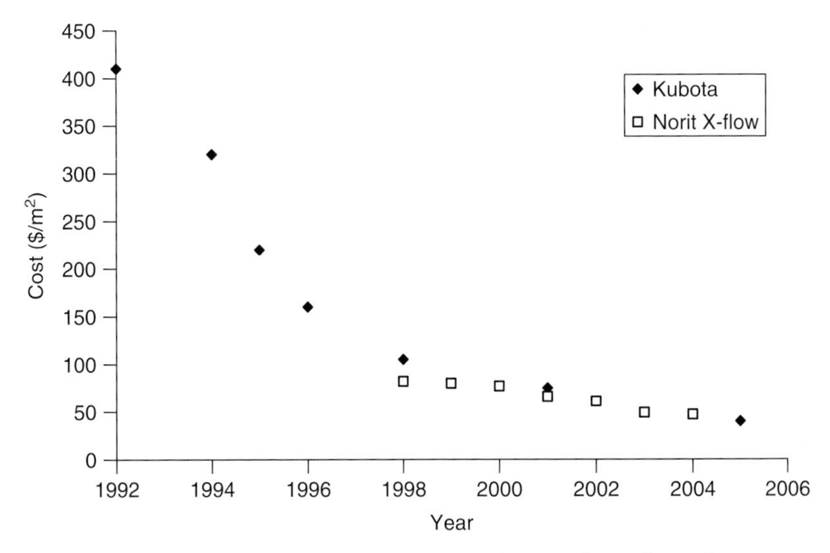

Figure 1.4 Microfiltration membrane replacement costs as a function of time, from information provided by Kubota and Norit X-Flow

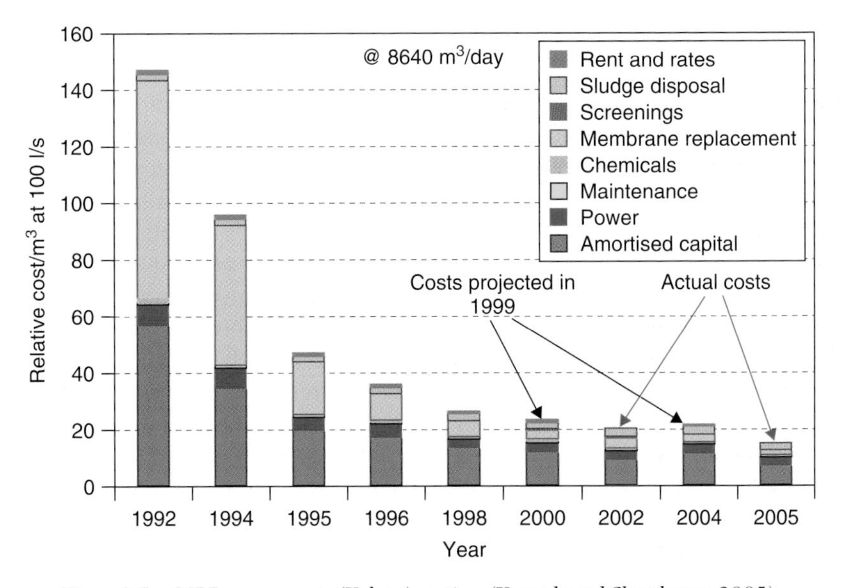

Figure 1.5 MBR process costs (Kubota) vs. time (Kennedy and Churchouse, 2005)

population density or those where agricultural or industrial activities are intense are particularly prone to water stress. Changing global weather patterns aggravate the situation, in particular for those countries which are prone to drought conditions. Water stress induces deterioration of fresh water resources in terms of quantity (aquifer over-exploitation, dry rivers, etc.) and quality (eutrophication, organic matter pollution, saline intrusion, etc.). A widely used measure of water stress is the water exploitation index (WEI), the values of which represent the annual mean total demand for freshwater divided by the long-term average freshwater resource. It provides an indication of how the total water demand puts pressure on the water resource.

Data from the year 2000 indicate that four European countries (Cyprus, Italy, Malta and Spain) representing 18% of Europe's population, were considered to be water stressed. It is estimated that, in 1990, around 1.9 billion people lived in countries which used more than 20% of their potential water resources. By 2025, the total population living in such water-stressed countries is expected to increase to 5.1 billion, this figure rising further to 6.5 billion by 2085. On the other hand, climate-related water stress is expected to reduce in some countries, for example, the USA and China, while in central America, the Middle East, southern Africa, North Africa, large areas of Europe and the Indian subcontinent, climate change is expected to adversely increase water stress by the 2020s. It is also predicted that 2.4 billion people will live in areas of extreme water stress (defined as using more than 40% of their available water resources) by 2025, 3.1 billion by 2050 and 3.6 billion by 2085; this is compared with a total population of 454 million in 1990 (Met Office, 2006).

1.4.5 Greater confidence in MBR technology

A growing confidence in MBR technology is demonstrated by the exponential increase in the cumulative MBR installed capacity (Fig. 1.7). As existing wastewater treatment plants become due for retrofit and upgrade – normally relating to a requirement for increased capacity and/or improved effluent water quality without incurring a larger footprint – it is expected that opportunities for the application of MBR technologies will increase, particularly in the USA. It is also evident that MBR installations are increasing in size year on year; the largest installation is currently 50 megalitres/day (MLD) with larger installations being planned and some observers stating that plants of 300–800 MLD are feasible (DiGiano *et al.*, 2004).

With new factors coming into play, the MBR technology is now beginning to mature such that the market is expected to grow substantially over the next decade. Evidence suggests that MBRs will continue to penetrate further the effluent treatment market, with the number of players in the global market increasing. Currently, the market is dominated by the two leading companies Zenon and Kubota. Whilst the domination of these two companies is likely to continue in the short to medium term, the global demand for the technology is such that a broader range of products is likely to be sustainable in the future (Chapter 4), in particular if individual products are tailored towards niche market applications.

1.5 Historical perspective

1.5.1 The early days of the MBR: the roots of the Kubota and Zenon systems

The first membrane bioreactors were developed commercially by Dorr-Oliver in the late 1960s (Bemberis *et al.*, 1971), with application to ship-board sewage treatment

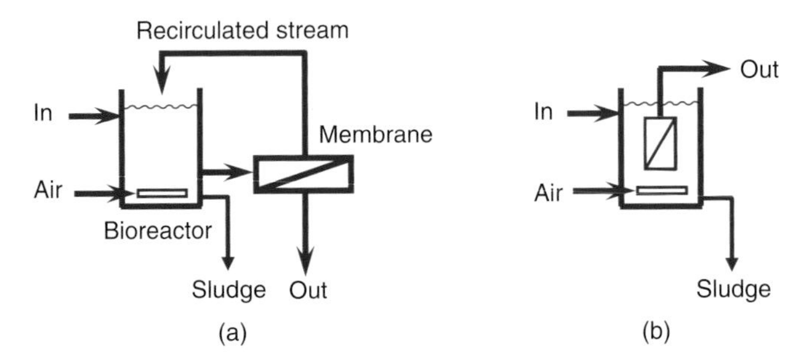

Figure 1.6 Configurations of a membrane bioreactor: (a) sidestream and (b) immersed

(Bailey *et al.*, 1971). Other bench-scale membrane separation systems linked with an activated sludge process were reported at around the same time (Hardt *et al.*, 1970; Smith *et al.*, 1969). These systems were all based on what have come to be known as "sidestream" configurations (sMBR, Fig. 1.6a), as opposed to the now more commercially significant "immersed" configuration (iMBR, Fig. 1.6b). The Dorr-Oliver membrane sewage treatment (MST) process was based on flat-sheet (FS) ultrafiltration (UF) membranes operated at what would now be considered excessive pressures (3.5 bar inlet pressure) and low fluxes (17 l/(m^2 h), or LMH), yielding mean permeabilities of less than 10 l/(m^2 h bar), or LMH/bar). Nonetheless, the Dorr-Oliver system succeeded in establishing the principle of coupling an activated sludge process with a membrane to concentrate simultaneously the biomass whilst generating a clarified, disinfected product. The system was marketed in Japan under license to Sanki Engineering, with some success up until the early 1990s. Developments were also underway in South Africa which led to the commercialisation of an anaerobic digester UF (ADUF) MBR by Weir Envig (Botha *et al.*, 1992), for use on high-strength industrial wastewaters.

At around this time, from the late 1980s to early 1990s, other important commercial developments were taking place. In the USA, Thetford Systems were developing their Cycle-Let® process, another sidestream process, for wastewater recycling duties. Zenon Environmental, a company formed in 1980, were developing an MBR system which eventually led to the introduction of the first ZenoGem® iMBR process in the early 1990s. The company acquired Thetford Systems in 1993. Meanwhile, in Japan, the government-instigated Aqua Renaissance programme prompted the development of an FS-microfiltration iMBR by the agricultural machinery company Kubota. This subsequently underwent demonstration at pilot scale, first at Hiroshima in 1990 (0.025 MLD) and then at the company's own site at Sakai-Rinkai in 1992 (0.110 MLD). By the end of 1996, there were already 60 Kubota plants installed in Japan for night soil, domestic wastewater (i.e. sewage) and, latterly, industrial effluent treatment, providing a total installed capacity of 5.5 MLD.

In the early 1990s, only one Kubota plant for sewage treatment had been installed outside of Japan, this being the pilot plant at Kingston Seymour operated by Wessex Water in the UK. Within Japan, however, the Kubota process dominated the

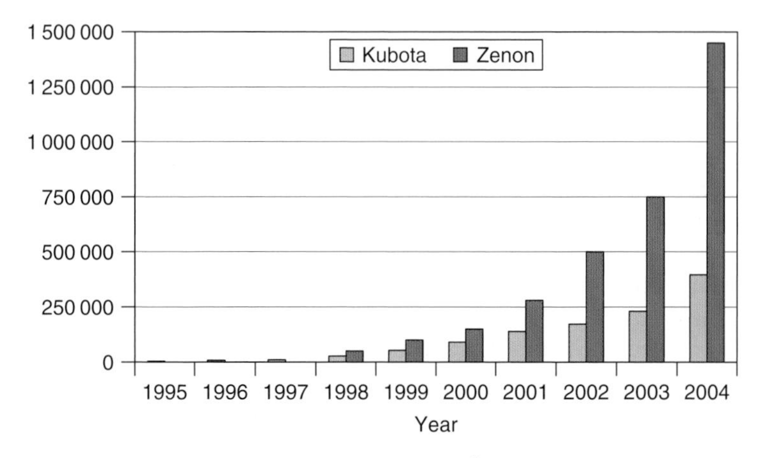

Figure 1.7 Cumulative installed capacity in m^3/day for Kubota and Zenon

market in the 1990s, effectively displacing the older sidestream systems, such as that of Rhodia-Orelis (now Novasep Orelis). To this day, Kubota continues to dominate the Japanese membrane wastewater treatment market and also provides the largest number of MBRs worldwide, although around 86% of these are for flows of less than 0.2 MLD.

In the late 1980s, development of a hollow fibre (HF) UF iMBR was taking place both in Japan, with pioneering work by Kazuo Yamamoto and his co-workers (1989), and also in the US. By the early 1990s, the ZenoGem® process had been patented (Tonelli and Behmann, 1996; Tonelli and Canning, 1993), and the total installed capacity had reached 2.8 MLD from installations in North America. Zenon introduced its first immersed HF ZeeWeed® module in 1993, this being the ZW145 (145 square feet), quickly followed by the ZW130 and 150 modules. These were in time superceded by the first of the ZW500 series in 1997. The company introduced the ZW500b, c and d modules in 1999, 2001 and 2003 respectively, the design changing to increase the overall process efficiency and cyclic aeration in 2000. Over this period, Kubota also developed products with improved overall energy efficiency, introducing a double-decker design in 2003 (Section 5.2.1.4).

As already stated, the cumulative capacity of both Zenon and Kubota has increased exponentially since the immersed products were first introduced (Fig. 1.7). These two systems dominate the MBR market today, with a very large number of small-scale Kubota systems and the largest MBR systems tending to be Zenon. The largest MBR worldwide is currently at Kaarst in Germany (50 MLD), though there is actually a larger membrane wastewater recycling facility in Kuwait (the Sulaibiya plant), which has a design capacity of 375 MLD.

1.5.2 Development of other MBR products

Other MBR products have been marketed with varying degrees of success, and further products are likely to become available in the future. The installation of in-building wastewater recycling plants in Japan based on the Novasep Orelis (formerly Rhodia Orelis and before this Rhône Poulenc) Pleiade® FS sMBR system, actually pre-dates that of the Kubota plants for this duty (Table 1.2). The Pleiade® system was originally trialled in France in the 1970s, and by 1999 there were 125 small-scale systems (all below 0.200 MLD) worldwide, the majority of these being in Japan and around a dozen in France. The Dorr-Oliver MST system was similarly rather more successful in Japan than in North America in the 1970s and 1980s (Sutton *et al.*, 2002).

Wehrle Environmental, part of the very well established Wehrle Werk AG (formed in 1860), has a track record in multitube (MT) sMBRs (predominantly employing Norit X-Flow polymeric MT membrane modules) which dates back to the late 1980s. Wehrle Environmental's MBRs have been used for landfill leachate treatment since 1990. Development of the sidestream Degremont system began in the mid-1990s, this system being based on a ceramic membrane. These sidestream systems all tend to be employed for niche industrial effluent treatment applications involving relatively low flows, such that their market penetration compared with the immersed systems, particularly in the municipal water sector, has been limited. Other commercial sMBR systems include the Dyna-Lift MBR (Dynatec) in the USA, and the AMBR system (Aquabio), the latter also being based on MT membrane modules.

Table 1.2 Summary of MBR development and commercialization

Time	Event
Late 1960s	Dorr-Oliver develops first sidestream FS MBR
Early 1970s	Thetford Systems commercialises sidestream multitube Cycle-Let® process for water reuse in USA
Early 1980s	TechSep (Rhone-Poulenc, later Novasep Orelis) commercialises sidestream FS Pleiade® for water reuse in Japan
Mid-1980s	Nitto-Denko files a Japanese patent on an immersed FS MBR University of Tokyo experiments with immersed hollow fibre MBR
Early-mid-1990s	Kubota commercialises immersed FS MBR in Japan Weir Envig commercialises the sidestream ADUF system based on "Membralox" membranes Zenon commercialises vertical immersed hollow fibre "ZeeWeed®" technology in North America and Europe; acquires Thetford in 1993 Wehrle commercialises sidestream multitube Biomembrat system Mitsubishi Rayon commercialises MBR based on immersed fine hollow fibre "Sterapore™" membrane, horizontal orientation
Early 2000s	USF commercialises vertical immersed hollow fibre "MemJet®" system Huber commercialises the rotating FS MBR Norit X-Flow develops sidestream airlift multitube system Puron (vertical immersed hollow fibre) commercialised, and acquired by Koch Kolon and Para (Korea) introduce vertical immersed hollow fibre MBR Toray introduces FS MBR Mitsubishi Rayon introduces vertical immersed hollow fibre MBR Asahi Kasei introduces vertical immersed hollow fibre MBR

Alternative immersed FS and HF membrane systems or products have generally been marginalised by the success of the Kubota and Zenon products, but have nonetheless been able to enter the market successfully. UF HF membrane modules include the relatively well-established Mitsubishi Rayon Sterapore™ SUR™ module or the more recent SADF™ product. FS systems include the Toray membrane, a company better known for RO membranes, and the Huber VRM® technology. Whereas the Toray product is a classic immersed rectangular FS membrane, the Huber technology is based on an immersed rotating hexagonal/octagonal membrane element and has attracted interest particularly within Germany. Sidestream airlift technologies have also been developed by some membrane and/or process suppliers which would appear to provide some of the advantages of an immersed system in a sidestream configuration. However, notwithstanding these developments, the majority of the newer products introduced to the market place are vertically oriented HFs, mostly fabricated from the same base polymer (polyvinylidene difluoride, PVDF).

1.5.3 The changing market

The MBR market has been complicated by the various acquisitions and partnerships that have taken place, and made more convoluted by the licensing agreements. A comprehensive review of these is beyond the scope of this brief précis, and would be likely to be out of date by the time of publication, but a few salient points can be made.

Whereas Zenon is a single global company supplying both membranes and turnkey plants for both water and wastewater treatment duties, Kubota, Mitsubishi Rayon and also Norit X-Flow (who acquired X-Flow and Stork in the 1990s) are primarily membrane suppliers who offer licensing agreements for their products. The Kubota MBR process, therefore, is provided by a series of generally geographically limited process companies, which include Enviroquip in the USA, Copa in the UK, Stereau in France and Hera in Spain. The UK license was formerly held by Aquator, formed through a management buy-out from Wessex Water in 2001, whom Copa acquired in February 2004. Copa have reverted to the original name of MBR Technology for their MBR-related activities. Mitsubishi Rayon similarly have licensees in the UK and the USA, the latter being Ionics Inc. (now part of GE Water and Process Technologies), but their operations appear to be largely restricted to the South-East Asian markets and Japan in particular. In Europe, Zenon have had licensing agreements with OTV (a subsidiary of the French giant Vivendi, now Veolia), Ondeo Degrémont and VA Tech Wabag.

There have been several recent acquisitions within the municipal membrane sector and, of these, the three of some specific significance with respect to the MBR market place are the acquisition of PCI by ITT, of Puron by Koch and of USFilter/Memcor by Siemens. PCI Membranes, acquired from the Thames Water Group (itself now part of the German company RWE), developed the FYNE process in the early 1980s. This is an MT nanofiltration (NF) membrane based process for removing organic matter from upland surface waters. PCI had no direct involvement with membrane bioreactors prior to the acquisition. ITT also acquired Sanitaire, the market leader in diffused aeration systems, in 1999. There is thus an obvious synergy in MBR process development. Puron was a small spin-out company from the University of Aachen. The company developed an HF membrane which has undergone extensive demonstration as an iMBR at pilot scale. The acquisition of Puron by Koch in 2005 – Koch being a major membrane and membrane systems supplier and owner of Fluid Systems (acquired from another UK water utility, Anglian Water Group) – would appear to signal a strategic move into the MBR technology by a company normally associated with pure water membrane systems.

The position of Memcor has attracted widespread interest across the water sector. Memcor is a long-established HF microfiltration membrane supplier (formed as Memtec in 1982). It was acquired by USFilter in 1997 and was part of the Veolia group, until sold to Siemens in 2004. In 1998, the company launched an immersed membrane process and introduced the MemJet® iMBR in 2003. Memcor represent a potentially very significant player in the MBR market and are already on a par with Zenon in potable water treatment.

Unsurprisingly, in North America the MBR market is currently dominated by Zenon, and the company also have the significant share of installed capacity in many countries where they operate. According to a recently published review of the North American market (Yang *et al.*, 2006), 182 of the 258 installations (i.e. 71%) provided by the four leading MBR suppliers in the USA, Canada and Mexico are Zenon plant (Table 1.3). Worldwide, however, there appear to be as many Mitsubishi Rayon plant as Zenon plant, but only two of these are in the USA and the plants are generally smaller. Indeed, as of 2005 nine of the ten biggest MBR plant worldwide were Zenon plant. Some consolidation in the marketplace has recently taken place with the acquisition of Zenon Environmental by GE Water and Process Technologies in March 2006.

On the other hand, in South-East Asia and in Japan in particular, the market is dominated by the Japanese membrane suppliers and Kubota specifically. Mitsubishi Rayon also has a significant presence in this region, particularly for industrial effluent treatment. In the UK – the EU country which has the largest number of MBRs for sewage treatment – all but three of the 21 municipal wastewater MBRs are Kubota (as of 2005). This trend is not repeated across mainland Europe, however, where Zenon again tend to dominate. For small flows, and in particular for more challenging high-strength industrial wastes, the dominance of Kubota and Zenon is much less pronounced. For example, Wehrle held 10% of the total European MBR market in 2002 (Frost and Sullivan, 2003), compared with 17% for Zenon at that time, which gives an indication of the significance of the industrial effluent treatment market.

Table 1.3 Number of installations (municipal and industrial) of four MBR providers worldwide and in North America (Yang *et al.*, 2006)

	Worldwide	USA	Canada	Mexico
Zenon	331 (204 + 127)	155 (132 + 23)	31 (23 + 8)	6 (1 + 5)
USFilter	16 (15 + 1)	13 (13 + 0)	0	0
Kubota	1538 (1138 + 400)	51 (48 + 3)	0	0
Mitsubishi Rayon	374 (170 + 204)	2 (2 + 0)	0	0
Total	2259 (1527 + 732)	221 (23 + 8)	31 (23 + 8)	6 (1 + 5)

1.6 Conclusions

Whilst the most significant barrier to the more widespread installation of MBRs remains cost, there are a number of drivers which mitigate this factor. Foremost of these is increasingly stringent environmental legislation relating to freshwater conservation and pollution abatement which has driven technological development in the water sector over the last 30–40 years. This, along with various governmental, institutional and organisational incentives, has encouraged problem holders to appraise more sophisticated technologies such as MBRs in recent years. Moreover, both capital (and particularly membrane) and operational costs of the MBR process have decreased dramatically over the past 15 years, although further significant cost reductions may be unattainable unless membrane modules become standardised in the same way as has taken place for RO technologies.

The technology itself is still regarded as being immature; although commercial products existed as long ago as the late 1960s, it is only since the introduction of the immersed configurations in the 1990s that significant market penetration has taken place. Although the market is still dominated by Zenon and Kubota there are now a wide range of products available for both industrial and municipal applications, with still more at the developmental stage. Confidence in the technology is growing as reference sites increase in number and maturity, and new opportunities are emerging as retrofitting of membranes into existing biotreatment processes becomes a viable option for increasing capacity or product water quality without detriment to footprint. As such, it is expected that MBR technology will continue to develop at a significant pace.

References

All websites accessed January 2006.

ADEQ (2006) www.azdeq.gov/environ/water/watershed/fin.html

Bailey, J., Bemberis, I. and Presti, J. (1971) Phase I Final Report – Shipboard sewage treatment system, General Dynamics Electric Boat Division, November. 1971, NTIS.

Bemberis, I., Hubbard, P.J. and Leonard, F.B. (1971) Membrane sewage treatment systems – potential for complete wastewater treatment, *American Society of Agricultural Engineers Winter Meeting*, 71-878, 1–28.

Benham, B.L., Brannan, K.M., Yagow, G., Zeckoski, R.W., Dillana, T.A., Mostaghimi, S. and Wynn, J.W. (2005) Development of bacteria and benthic total maximum daily loads: a case study, Linville Creek, Virginia. *J. Environ. Qual.*, **34**, 1860–1872.

Botha, G.R., Sanderson, R.D. and Buckley, C.A. (1992) Brief historical review of membrane development and membrane applications in wastewater treatment in Southern Africa. *Wat. Sci. Technol.*, **25**(10), 1–4.

CEPA (2006) www.waterboards.ca.gov/funding/index.html#funding_programs

Defra (2006a) www.defra.gov.uk/environment/water/quality/uwwtd/report02/02.htm

Defra (2006b) www.defra.gov.uk/environment/waste/topics/landfill-dir/index.htm

Defra (2006c) www.defra.gov.uk/environment/water/wfd/index.htm

DiGiano, F.A., Andreottola, G., Adham, S., Buckley, C., Cornel, P., Daigger, G.T., Fane, A.G., Galil, N., Jacangelo, J., Alfieri, P., Rittmann, B.E., Rozzi, A., Stephenson, T. and Ujang, Z. (2004) Safe water for everyone: membrane bioreactor technology. www.scienceinafrica.co.za/2004/june/membrane.htm

ECRD (2006) http://europa.eu.int/scadplus/leg/en/s15005.htm

Frost and Sullivan (2003) MBR: A buoyant reaction in Europe, Report, June 2003, Frost and Sullivan.

Frost and Sullivan (2004a) US advanced water treatment equipment markets, Report, June 2004, Frost & Sullivan.

Frost and Sullivan (2004b) US and Canada membrane bioreactor markets, Report, June 2004, Frost and Sullivan.

Frost and Sullivan (2004c) US desalination plant market, Report, January 2004, Frost and Sullivan.

Frost and Sullivan (2005) European report: introduction and executive summary, Report, August 2005, Frost and Sullivan.

GDNR (2006) www.ganet.org/gefa/water_and_sewer.html

GEFA (2006) www.gefa.org/gefa/state_revolving.html

Hanft, S. (2006) Membrane bioreactors in the changing world water market, Business Communications Company Inc. report C-240.

Hardt, F.W., Clesceri, L.S., Nemerow, N.L. and Washington, D.R. (1970) Solids separation by ultrafiltration for concentrated activated sludge. *J. Wat. Pollut. Con. Fed.*, **42**, 2135–2148.

HM Treasury (2006) www.hmtreasury.gov.uk./Consultations_and_Legislation/consult_greentech/consult_greentech_index.cfm

Kennedy, S. and Churchouse, S.J. (2005) Progress in membrane bioreactors: new advances, *Proceedings of Water and Wastewater Europe Conference*, Milan, June 2005.

Maxwell, S. (2005) The state of the water industry 2005, a concise overview of trends and opportunities in the water business, *The Environmental Benchmarker and Strategist Annual Water Issue.*

Met Office (2006) www.metoffice.com/research/hadleycentre/pubs/brochures/B1997/water.html

Reid, E. (2006) Salinity shocking and fouling amelioration in membrane bioreactors, *EngD Thesis*, School of Water Sciences, Cranfield University.

Smith, C.V., Gregorio, D.O. and Talcott, R.M. (1969) The use of ultrafiltration membranes for activated sludge separation, *Proceedings of the 24th Industrial Waste Conference*, Purdue University, Ann Arbor Science, Ann Arbor, USA, 1300–1310.

Sutton, P.M., Mishra, P.N., Bratby, J.R. and Enegess, D. (2002) Membrane bioreactor industrial and municipal wastewater application: long term operating experience, *Proceedings of the 75th Water Environment Federation Annual Conference and Exposition*, Chicago, IL,USA.

Tonelli, F.A. and Behmann, H. (1996) Aerated membrane bioreactor process for treating recalcitrant compounds, US Pat. No. 410730.

Tonelli, F.A. and Canning, R.P. (1993) Membrane bioreactor system for treating synthetic metal-working fluids and oil based products, USA Pat. No. 5204001.

USEPA (2006a) www.epa.gov/region5/defs/html/ppa.htm

USEPA (2006b) www.epa.gov/Region5/defs/html/sdwa.htm

USEPA (2006c) www.epa.gov/owmitnet/cwfinance/cwsrf/index.htm

Yamamoto, K., Hiasa, M., Mahmood, T. and Matsuo, T. (1989) Direct solid–liquid separation using hollow fibre membrane in an activated sludge aeration tank. *Wat. Sci. Technol.*, **21**(10), 43–54.

Yang, W., Cicek, N. and Ilg, J. (2006) State-of-the-art of membrane bioreactors: worldwide research and commercial applications in North America. *J. Membrane Sci.*, **270**, 201–211.

Chapter 2

Fundamentals

With acknowledgements to:

Section 2.2	Beth Reid	AEA Technology, UK
	Tom Stephenson, Folasade Fawenhimni, Harriet Fletcher, Bruce Jefferson, Eve Germain	Cranfield University, UK
Sections 2.3.2 and 2.3.3	Ewan McAdam	Cranfield University, UK
Sections 2.3.3 to 2.3.9	Pierre Le-Clech, Vicky Chen, Tony (A.G.) Fane	The UNESCO Centre for Membrane Science and Technology, The University of New South Wales, Sydney, Australia

2.1 Membrane technology

2.1.1 Membranes and membrane separation processes

A membrane as applied to water and wastewater treatment is simply a material that allows some physical or chemical components to pass more readily through it than others. It is thus perm-selective, since it is more permeable to those constituents passing through it (which then become the permeate) than those which are rejected by it (which form the retentate). The degree of selectivity depends on the membrane pore size. The coarsest membrane, associated with microfiltration (MF), can reject particulate matter. The most selective membrane, associated with reverse osmosis (RO), can reject singly charged (i.e. monovalent) ions, such as sodium (Na^{+}) and chloride (Cl^{-}). Given that the hydraulic diameter of these ions is less than 1 nm, it stands to reason that the pores in an RO membrane are very small. Indeed, they are only visible using the most powerful of microscopes.

The four key membrane separation processes in which water forms the permeate product are RO, nanofiltration (NF), ultrafiltration (UF) and MF (Fig. 2.1). Membranes themselves can thus be defined according to the type of separation duty to which

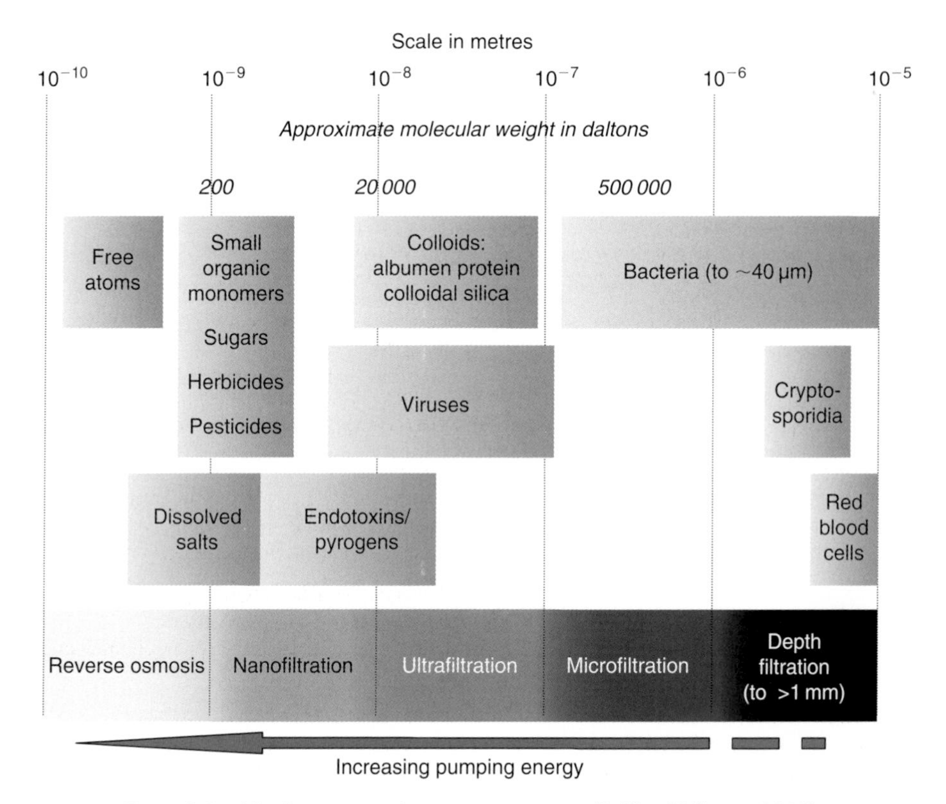

Figure 2.1 Membrane separation processes overview (Judd and Jefferson, 2003)

they can be put, which then provides an indication of the pore size. The latter can be defined either in terms of the effective equivalent pore diameter, normally in μm, or the equivalent mass of the smallest molecule in daltons (Da) the membrane is capable of rejecting, where 1 Da represents the mass of a hydrogen atom. For UF membranes specifically the selectivity is thus defined by the molecular weight cut-off (MWCO) in daltons. For the key membrane processes identified, pressure is applied to force water through the membrane. However, there are additional membrane processes in which the membrane is not necessarily used to retain the contaminants and allow the water to pass through, but can instead be used either to:

(a) selectively extract constituents (extractive) or
(b) introduce a component in the molecular form (diffusive).

The range of membrane processes available is given in Table 2.1, along with an outline of the mechanism by which each process operates. Mature commercial membrane applications in water and wastewater treatment are limited to the pressure-driven processes and electrodialysis (ED), which can extract problem ions such as nitrate and those ions associated with hardness or salinity. Membrane technologies as applied to the municipal sector are predominantly pressure driven and, whilst the membrane permselectivity and separation mechanism may vary from process to another, such processes all have the common elements of a purified permeate product and a concentrated retentate waste (Fig. 2.2).

The rejection of contaminants ultimately places a fundamental constraint on all membrane processes. The rejected constituents in the retentate tend to accumulate at the membrane surface, producing various phenomena which lead to a reduction

Table 2.1 Dense and porous membranes for water treatment

Pressure-driven	Extractive/diffusive
Reverse osmosis (RO) Separation achieved by virtue of differing solubility and diffusion rates of water (solvent) and solutes in water	*Electrodialysis (ED)* Separation achieved by virtue of differing ionic size, charge and charge density of solute ions, using ion-exchange membranes
Nanofiltration (NF) Formerly called *leaky RO*. Separation achieved through combination of charge rejection, solubility–diffusion and sieving through micropores (<2 nm)	*Pervaporation (PV)* Same mechanism as RO but with the (volatile) solute partially vapourised in the membrane by partially vacuumating the permeate.
Ultrafiltration (UF) Separation by sieving through mesopores (2–50 nm)*	*Membrane extraction (ME)* Constituent removed by virtue of a concentration gradient between retentate and permeate side of membrane
Microfiltration (MF) Separation of suspended solids from water by sieving through macropores (>50 nm)*	*Gas transfer (GT)* Gas transferred under a partial pressure gradient into or out of water in molecular form

*IUPAC (1985).

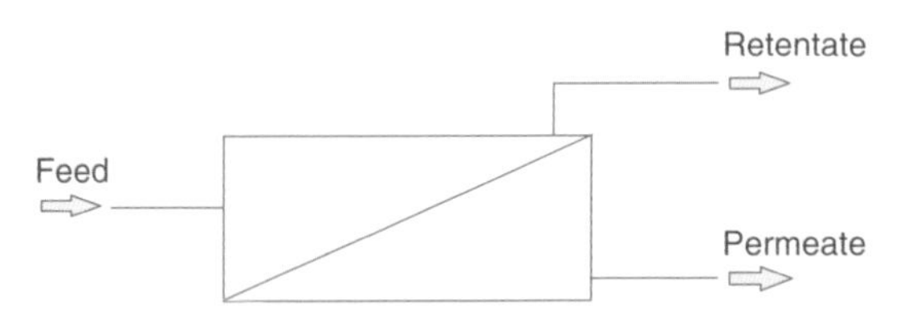

Figure 2.2 Schematic of membrane

in the flow of water through the membrane (i.e. the flux) at a given transmembrane pressure (TMP), or conversely an increase in the TMP for a given flux (reducing the permeability, which is the ratio of flux to TMP). These phenomena are collectively referred to as fouling. Given that membrane fouling represents the main limitation to membrane process operation, it is unsurprising that the majority of membrane material and process research and development conducted is dedicated to its characterisation and amelioration (Section 2.3).

Fouling can take place through a number of physicochemical and biological mechanisms which all relate to increased deposition of solid material onto the membrane surface (also referred to as blinding) and within the membrane structure (pore restriction or pore plugging/occlusion). This is to be distinguished from clogging, which is the filling of the membrane channels with solids due to poor hydrodynamic performance. The membrane resistance is fixed, unless its overall permeability is reduced by components in the feedwater permanently adsorbing onto or into the membrane. The resistance imparted by the interfacial region is, on the other hand, dependent on the total amount of fouling material residing in the region. This in turn depends upon both the thickness of the interface, the feedwater composition (and specifically its foulant content) and the flux through the membrane. The feedwater matrix and the process operating conditions thus largely determine process performance.

2.1.2 Membrane materials

There are mainly two different types of membrane material, these being polymeric and ceramic. Metallic membrane filters also exist, but these have very specific applications which do not relate to membrane bioreactor (MBR) technology. The membrane material, to be made useful, must then be formed (or configured) in such a way as to allow water to pass through it.

A number of different polymeric and ceramic materials are used to form membranes, but generally nearly always comprise a thin surface layer which provides the required permselectivity on top of a more open, thicker porous support which provides mechanical stability. A classic membrane is thus anisotropic in structure, having symmetry only in the plane orthogonal to the membrane surface (Fig. 2.3). Polymeric membranes are also usually fabricated both to have a high surface porosity, or % total surface pore cross-sectional area (Fig. 2.4), and narrow pore size distribution to provide as high a throughput and as selective a degree of rejection as possible. The membrane must also be mechanically strong (i.e. to have structural integrity). Lastly, the material will normally have some resistance to thermal and chemical attack, that is, extremes of temperature, pH and/or oxidant concentrations that normally arise when the

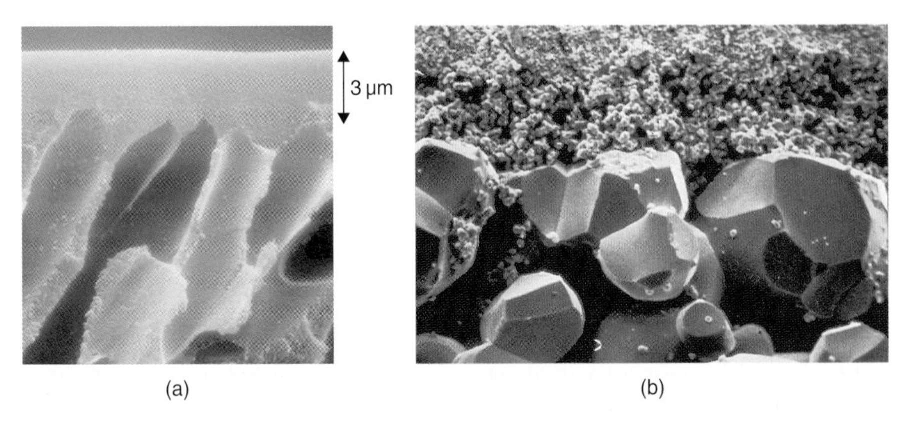

Figure 2.3 Anisotropic UF membranes: (a) polymeric (thickness of "skin" indicated) and (b) ceramic (by kind permission of Ionics (a) and Pall (b))

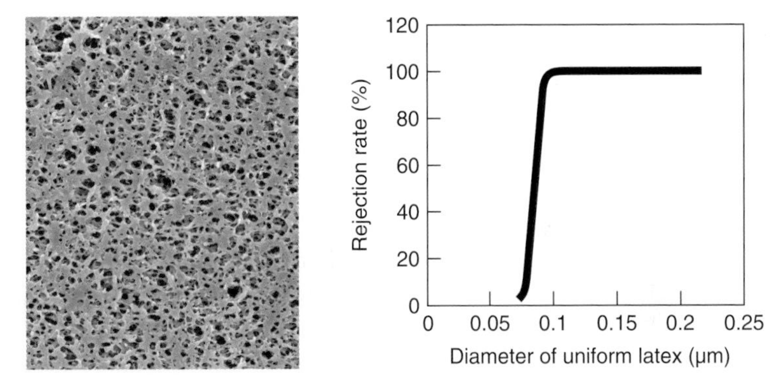

Figure 2.4 Surface of membrane and pore-size distribution with respect to rejection of homodispersed latex (by kind permission of Asahi-Kasei)

membrane is chemically cleaned (Section 2.1.4.3), and should ideally offer some resistance to fouling.

Whilst, in principal, any polymer can be used to form a membrane, only a limited number of materials are suitable for the duty of membrane separation, the most common being:

- polyvinylidene difluoride (PVDF)
- polyethylsulphone (PES)
- polyethylene (PE)
- polypropylene (PP)

All the above polymers can be formed, through specific manufacturing techniques, into membrane materials having desirable physical properties, and they each have reasonable chemical resistance. However, they are also hydrophobic, which makes the susceptible to fouling by hydrophobic matter in the bioreactor liquors they are filtering. This normally necessitates surface modification of the base material to produce a hydrophilic surface using such techniques as chemical oxidation, organic chemical reaction, plasma treatment or grafting. It is this element that, if at all, most distinguishes one membrane material product from another formed from the same base polymer. This modification process, the manufacturing method used to form the membrane from the polymer, most often PVDF for many MBR membranes, and the method for fabricating the membrane module (Section 2.1.4) from the membrane are all regarded as proprietary information by most suppliers.

2.1.3 Membrane configurations

The configuration of the membrane, that is, its geometry and the way it is mounted and oriented in relation to the flow of water, is crucial in determining the overall process performance. Other practical considerations concern the way in which the membrane elements, that is the individual discrete membrane units themselves, are housed in "shells" to produce modules, the complete vessels through which the water flows.

Ideally, the membrane should be configured so as to have:

(a) a high membrane area to module bulk volume ratio,
(b) a high degree of turbulence for mass transfer promotion on the feed side,
(c) a low energy expenditure per unit product water volume,
(d) a low cost per unit membrane area,
(e) a design that facilitates cleaning,
(f) a design that permits modularisation.

All membrane module designs, by definition, permit modularisation (f), and this presents one of the attractive features of membrane processes *per se*. This also means that membrane processes provide little economy of scale with respect to membrane costs, since these are directly proportional to the membrane area which relates directly to the flow. However, some of the remaining listed characteristics are mutually exclusive. For example, promoting turbulence (b) results in an increase in the energy expenditure (c). Direct mechanical cleaning of the membrane (e) is only possible on comparatively low area:volume units (a). Such module designs increase the total cost per unit membrane area (d), but are inevitable given that cleaning is of fundamental importance in MBR processes where the solids and foulant loading on the membrane from the bioreactor liquor is very high. Finally, it is not possible to produce a high-membrane area to module bulk volume ratio without producing a unit having narrow retentate flow channels, which will then adversely affect turbulence promotion and ease of cleaning.

There are six principal configurations currently employed in membrane processes, which all have various practical benefits and limitations (Table 2.2). The configurations are based on either a planar or cylindrical geometry and comprise:

1. plate-and-frame/flat sheet (FS)
2. hollow fibre (HF)
3. (multi)tubular (MT)
4. capillary tube (CT)
5. pleated filter cartridge (FC)
6. spiral-wound (SW)

Of the above configurations, only the first three (Fig. 2.5, Table 2.2) are suited to MBR technologies, principally for the reasons outlined previously: the modules must

Table 2.2 Membrane configurations

Configuration	Cost	Turbulence promotion	Backflushable?	Application
FC	Very low	Very poor	No	***DEMF, low TSS waters***
FS	*High*	*Fair*	*No*	**ED**, *UF, RO*
SW	Low	Poor	No	**RO/NF**, UF
MT	*Very high*	*Very good*	*No*	***CFMF/UF, high TSS waters***, *NF*
CT	Low	Fair	Yes	***UF***
HF	*Very low*	*Very poor*	*Yes*	***MF/UF***, *RO*

Bold text: most important alternative application; *Italic text*: MBR configurations.
*Can be <50 for a cassette.
DE: dead-end, CF: crossflow.
Capillary tube used in UF: water flows from inside to outside the tubes.
HF used in MF and RO: water flows from outside to inside the tubes.

Figure 2.5 (Clockwise from top) FS, MT and HF modules (by kind permission of Kubota, Wehrle & Memcor)

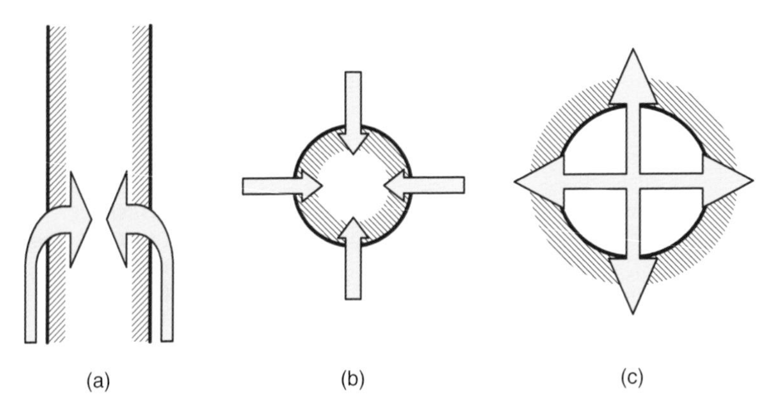

Figure 2.6 Schematics showing flow through membrane configured as: (a) FS, (b) CT or MT and (c) HF

permit turbulence promotion, cleaning or, preferably, both. Turbulence promotion can arise through passing either the feedwater or an air/water mixture along the surface of the membrane to aid the passage of permeate through it. This crossflow operation (Section 2.1.4.2) is widely used in many membrane technologies, and its efficacy increases with increasing membrane interstitial distance (i.e. the membrane separation).

Because the MT module operates with flow passing from inside to outside the tube ("lumen-side" to "shell-side"), whereas the HF operates outside-to-in, the interstitial distance is defined by (Fig. 2.6):

- the tube diameter for a MT,
- the distance between the filaments for an HF,
- the channel width for an FS.

The membrane packing density of the HF thus becomes crucial, since too high a packing density will reduce the interstitial gap to the point where there is a danger of clogging. CT modules, which are, to all intents and purposes, HF modules with reversed flow (i.e. lumen-side to shell-side), are too narrow in diameter to be used for MBR duties as they would be at high risk of clogging.

Physical cleaning is most simply affected by reversing the flow (i.e. backflushing), at a rate 2–3 times higher than the forward flow, back through the membrane to remove some of the fouling layer on the retentate side. For this to be feasible, the membrane must have sufficient inherent integrity to withstand the hydraulic stress imparted. In other words, the membrane must be strong enough not to break or buckle when the flow is reversed. This generally limits backflushing of polymeric membranes to those configured as capillary tubes or HFs. At low filament diameters the membranes have a high enough wall thickness: filament diameter ratio to have the inherent strength to withstand stresses imposed by flow reversal.

2.1.4 Membrane process operation

2.1.4.1 Flux, pressure, resistance and permeability

The key elements of any membrane process relate to the influence of the following parameters on the overall permeate flux:

(a) the membrane resistance,
(b) the operational driving force per unit membrane area,
(c) the hydrodynamic conditions at the membrane:liquid interface,
(d) the fouling and subsequent cleaning of the membrane surface.

The flux (normally denoted J) is the quantity of material passing through a unit area of membrane per unit time. This means that it takes SI units of $m^3/m^2/s$, or simply $m\,s^{-1}$, and is occasionally referred to as the permeate or filtration velocity. Other non-SI units used are litres per m^2 per hour (or LMH) and m/day, which tend to give more accessible numbers: MBRs generally operate at fluxes between 10 and 100 LMH. The flux relates directly to the driving force (i.e. the TMP for conventional MBRs) and the total hydraulic resistance offered by the membrane and the interfacial region adjacent to it.

Although for conventional biomass separation MBRs the driving force for the process is the TMP, for extractive or diffusive MBRs (Sections 2.3.2–2.3.3) it is respectively the concentration or partial pressure gradient. Whereas with conventional pressure-driven MBRs the permeate is the purified product, for extractive MBRs the contaminants are removed from the water across the membrane under the influence of a concentration gradient and are subsequently biologically treated, the retentate forming the purified product. For diffusive bioreactors neither water nor contaminants permeate the membrane: in this case the membrane is used to transport a gas into the bioreactor.

Resistance R (/m) and permeability K (m/(s bar), or LMH/bar in non-SI units) are inversely related. The resistance is given by:

$$R = \frac{\Delta P}{\eta J} \tag{2.1}$$

where η is the viscosity (kg/(m s^2)) and ΔP (Pa) the pressure drop, and can refer to either the TMP (ΔP_m Pa/bar in non-SI units) or individual components which contribute to the pressure drop. Permeability is normally quoted as the ratio of flux to TMP (hence $J/\Delta P_m$), the most convenient units being LMH/bar, and sometimes corrected for temperature impacts on viscosity.

The resistance R includes a number of components, namely:

(a) the membrane resistance,
(b) the resistance of the fouling layer (adsorbed onto the membrane surface),
(c) the resistance offered by the membrane:solution interfacial region.

The membrane resistance is governed by the membrane material itself, and mainly the pore size, the surface porosity (percentage of the surface area covered by the pores) and the membrane thickness. The fouling layer resistance is associated with the filtration mechanism, which is then dependent on the membrane and filtered solids characteristics. The membrane:solution interfacial region resistance is associated with concentration polarisation (CP) (Section 2.1.4.4) which, for the more perm-selective processes such as RO, produces a solution osmotic pressure at the membrane surface which is higher than that in the bulk solution. The resistance offered by foulants is often further delineated into generic types according to their characteristics, behaviour and origin (Sections 2.1.4.2 and 2.3.6.8). However, in general, the membrane resistance only dominates when fouling is either absent (i.e. the feedwater is almost free of fouling materials) or is suppressed by operating under specific conditions (Sections 2.1.4.5 and 2.3.9).

2.1.4.2 Dead-end and crossflow operation

Conventional pressure-driven membrane processes with liquid permeation can operate in one of two modes. If there is no retentate stream then operation is termed "dead-end" or "full-flow"; if retentate continuously flows from the module outlet then the operation is termed crossflow (Fig. 2.7). Crossflow implies that, for a single passage of feedwater across the membrane, only a fraction is converted to permeate product. This parameter is termed the "conversion" or "recovery". The recovery is reduced further if product permeate is used for maintaining process operation, usually for membrane cleaning.

Filtration always leads to an increase in the resistance to flow. In the case of a dead-end filtration process, the resistance increases according to the thickness of the cake formed on the membrane, which would be expected to be roughly proportional to the total volume of filtrate passed. Rapid permeability decay then results, at a rate proportional to the solids concentration and flux, demanding periodic cleaning (Fig. 2.8). For crossflow processes, this deposition continues until the adhesive forces binding the cake to the membrane are balanced by the scouring forces of the fluid (either liquid or a combination of air and liquid) passing over the membrane. All other things being equal, a crossflow filtration process would be expected to attain steady-state conditions determined by the degree of CP (Section 2.1.4.4). In practice,

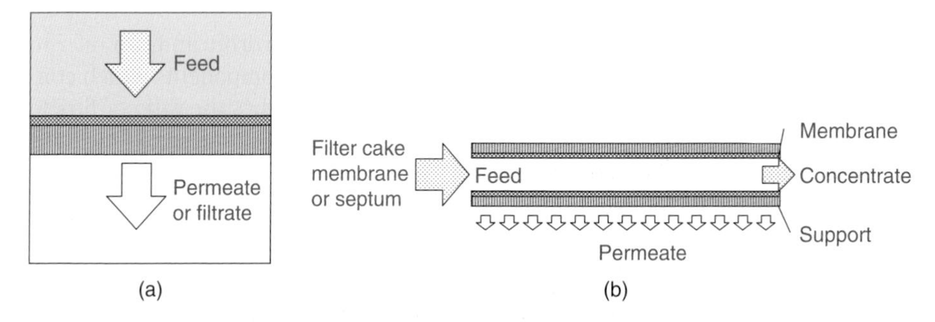

Figure 2.7 (a) Dead-end and (b) crossflow filtration

only pseudo-steady-state (or stabilised) conditions are attained to do the unavoidable deposition or adsorption of fouling material.

Filtration proceeds according to a number of widely recognised mechanisms, which have their origins in early filtration studies (Grace, 1956), comprising (Fig. 2.9):

- complete blocking
- standard blocking
- intermediate blocking
- cake filtration

All models imply a dependence of flux decline on the ratio of the particle size to the pore diameter. The standard blocking and cake filtration models appear most suited to predicting initial flux decline during colloid filtration (Visvanathan and Ben Aim, 1989) or protein filtration (Bowen *et al.*, 1995). All of the models rely on empirically derived information and some have been refined to incorporate other key determinants. On the other hand, a number of empirical and largely heuristic expressions

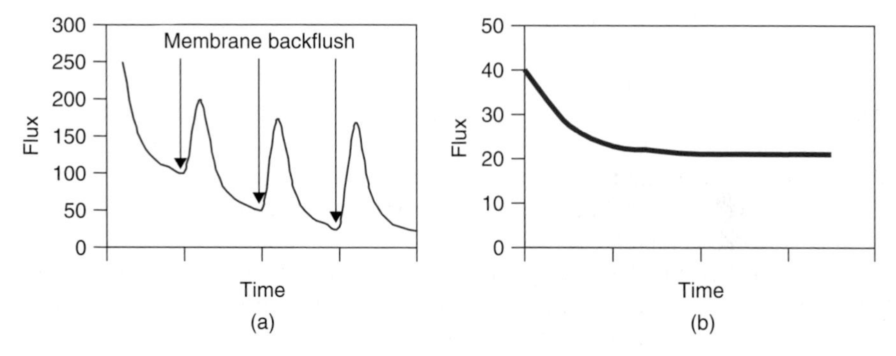

Figure 2.8 Flux transients for: (a) dead-end and (b) crossflow filtration for constant pressure operation

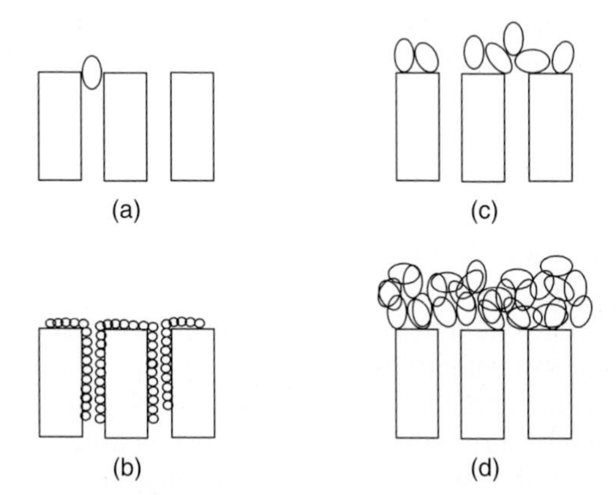

Figure 2.9 Fouling mechanisms: (a) complete blocking, (b) standard blocking, (c) Intermediate blocking, (d) cake filtration

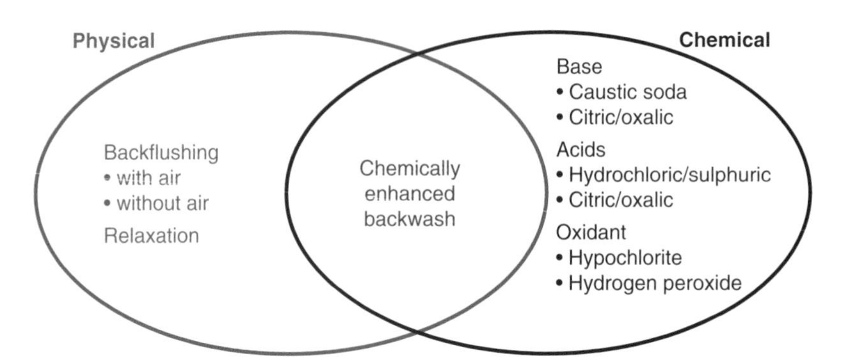

Figure 2.10 *Membrane cleaning methods*

have been proposed for particular matrices and/or applications. Classical dead-end filtration models can be adapted for crossflow operation if the proportion of undeposited solute material can be calculated.

2.1.4.3 Physical and chemical cleaning

Since the flux and driving force are interrelated, either one can be fixed for design purposes. For conventional pressure-driven water filtration, it is usual to fix the value of the flux and then determine the appropriate value for the TMP. The main impact of the operating flux is on the period between cleaning, which may be by either physical or chemical means (Fig. 2.10). In MBRs physical cleaning is normally achieved either by backflushing, that is, reversing the flow, or relaxation, which is simply ceasing permeation whilst continuing to scour the membrane with air bubbles. These two techniques may be used in combination, and backflushing may be enhanced by combination with air. Chemical cleaning is carried out with mineral or organic acids, caustic soda or, more usually in MBRs, sodium hypochlorite, and can be performed either *in situ* ("cleaning in place" or CIP) or *ex situ* (Section 2.3.9.2). Alternatively, a low concentration of chemical cleaning agent can be added to the backflush water to produce a "chemically enhanced backflush" (CEB).

Physical cleaning is less onerous than chemical cleaning on a number of bases. It is generally a more rapid process than chemical cleaning, lasting no more than 2 min. It demands no chemicals and produces no chemical waste, and also is less likely to incur membrane degradation. On the other hand, it is also less effective than chemical cleaning. Physical cleaning removes gross solids attached to the membrane surface, generally termed "reversible" or "temporary" fouling, whereas chemical cleaning removes more tenacious material often termed "irreversible" or "permanent" fouling, which is obviously something of a misnomer. Since the original virgin membrane permeability is never recovered once a membrane is fouled through normal operation, there remains a residual resistance which can be defined as "irrecoverable fouling". It is this fouling which builds up over a number of years and ultimately determines membrane life.

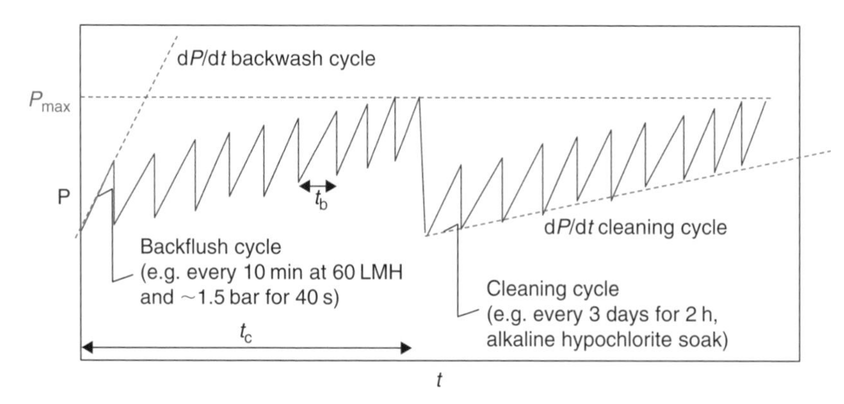

Figure 2.11 Pressure transient for constant flux operation of a dead-end filter

Since flux, amongst other things, determines the permeability decline rate (or pressure increase dP/dt), it also determines the period between physical cleaning (backflushing or relaxation), that is, the physical cleaning cycle time. If backflushing is used, this period can be denoted t_p and, assuming no changes to other operating conditions, increasing the flux decreases t_p. Since backflushing does not, in practice, return the permeability to the original condition only a finite number of backflush cycles can be performed before a threshold pressure is reached (P_{max}) beyond which operation cannot be sustained. At this point chemical cleaning must be conducted to return the pressure to close to the original baseline value (Fig. 2.11). As with physical cleaning, chemical cleaning never recovers the original membrane permeability but is normally considerably more effective than physical cleaning. For crossflow operation, backflushing is not normally an option due to the nature of the membrane module (Table 2.2), and membrane permeability is thus maintained by a combination of relaxation and chemical cleaning.

2.1.4.4 Concentration polarisation

For membrane filtration processes, the overall resistance at the membrane:solution interface is increased by a number of factors which each place a constraint on the design and operation of membrane process plant:

(a) the concentration of rejected solute near the membrane surface,
(b) the precipitation of sparingly soluble macromolecular species (gel layer formation) at the membrane surface,
(c) the accumulation of retained solids on the membrane (cake layer formation).

All of the above contribute to membrane fouling, and (a) and (b) are promoted by CP. CP describes the tendency of the solute to accumulate at membrane:solution interface within a concentration boundary layer, or liquid film, during crossflow operation (Fig. 2.12). This layer contains near-stagnant liquid, since at the

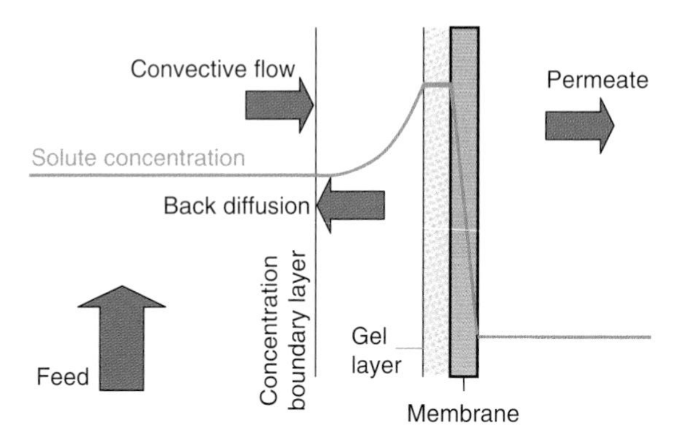

Figure 2.12 Concentration polarisation

membrane surface itself the liquid velocity must be zero. This implies that the only mode of transport within this layer is diffusion, which is around two orders of magnitude slower than convective transport in the bulk liquid region. However, it has been demonstrated (Romero and Davis, 1991) that transport away from the membrane surface is much greater than that governed by Brownian diffusion and is actually determined by the amount of shear imparted at the boundary layer; such transport is referred to as "shear-induced diffusion".

Rejected materials nonetheless build up in the region adjacent to membrane, increasing their concentration over the bulk value, at a rate which increases exponentially with increasing flux. The thickness of the boundary layer, on the other hand, is determined entirely by the system hydrodynamics, decreasing in thickness when turbulence is promoted. For crossflow processes, the greater the flux, the greater the build-up of solute at the interface; the greater the solute build-up, the higher the concentration gradient; the steeper the concentration gradient, the faster the diffusion. Under normal steady-state operating conditions, there is a balance between those forces transporting the water and constituents within the boundary layer towards, through and away from the membrane. This balance is determined by CP.

2.1.4.5 Fouling control

In MBRs, as with many other membrane filtration processes, it is the balance between the flux, physical and chemical cleaning protocol and, when relevant, the control of CP which ultimately determines the extent to which fouling is successfully suppressed. Ultimately, CP-related fouling can be reduced by two methods: (i) promoting turbulence and (ii) reducing flux. For sidestream MBRs (sMBRs, Fig. 1.6a), turbulence can be promoted simply by increasing the crossflow velocity (CFV), whereas for an immersed system (iMBR, Fig. 1.6b) this can only reasonably be achieved by increasing the membrane aeration. Whereas pumped flow of liquid along a tubular or parallel plate channel, as with sidestream systems, allows estimation of the degree of turbulence through calculation of the Reynolds number (density × velocity × tube diameter/viscosity), determination of turbulence for an immersed aerated membrane is more challenging (Section 2.3.7.1).

2.1.4.6 Critical flux

The critical flux concept was originally presented by Field *et al.* (1995). These authors stated that: "The critical flux hypothesis for microfiltration/ultrafiltration processes is that on start-up there exists a flux below which a decline of flux with time does not occur; above it, fouling is observed". Two distinct forms of the concept have been defined. In the strong form, the flux obtained during sub-critical flux is equated to the clean water flux measured under the same conditions. However, clean water fluxes are rarely attained for most real feedwaters due to irreversible adsorption of some solutes. In the alternative weak form, the sub-critical flux is the flux rapidly established and maintained during start-up of filtration, but does not necessarily equate to the clean water flux. Alternatively, stable filtration operation, that is, constant permeability for an extended time period, has been defined as sub-critical operation even when preceded by an initial decline in flux (Howell, 1995). Such conditions would be expected to lead to lower critical flux values than those obtained for constant permeability operation, however, since an initial permeability decline implies foulant deposition.

A number of slightly different definitions of sub-critical flux operation have been proposed, largely depending on the method employed. The most microscopically precise definition equates the critical flux to that flux below which no deposition of colloidal matter takes place. Kwon and Vigneswaran (1998) equated critical flux to the lift velocity as defined by the lateral migration theory of Green and Belfort (1980). This rigorous definition is difficult to apply because of the relative complexity of the determination of the lift velocity, particularly for heterogeneous matrices. On the other hand, experimental determination of critical flux by direct observation of material deposition onto the membrane has been conducted using model homodispersed suspensions of polystyrene latex particles (Kwon and Vigneswaran, 1998), and some authors have also used mass balance determinations (Kwon *et al.*, 2000).

Given the limitations of applying particle hydrodynamics to the identification of the critical flux in real systems, recourse generally has to be made to experimental determination. By plotting flux against the TMP it is possible to observe the transition between the linearly pressure-dependent flux and the onset of fouling, where deviation from linearity commences. The flux at this transition has been termed "secondary critical flux" (Bouhabila *et al.*, 1998) and, more recently, the concept of "sustainable flux" has been introduced, defined as the flux for which the TMP increases gradually at an acceptable rate, such that chemical cleaning is not necessary (Ng *et al.*, 2005).

Whilst potentially useful in providing a guide value for the appropriate operating flux, the absolute value of the critical flux obtained is dependent on the exact method employed for its determination and, specifically, the rate at which the flux is varied with time. A common practice is to incrementally increase the flux for a fixed duration for each increment, giving a stable TMP at low flux but an ever-increasing rate of TMP increase at higher fluxes (Fig. 2.13). This flux-step method defines the highest flux for which TMP remains stable as the critical flux. This method is preferred over the corresponding TMP-step method since the former provides a better control of the flow of material deposition on the membrane surface, as the convective flow of solute towards the membrane is constant during the run (Defrance and Jaffrin, 1999). No single protocol has been agreed for critical flux measurement, making

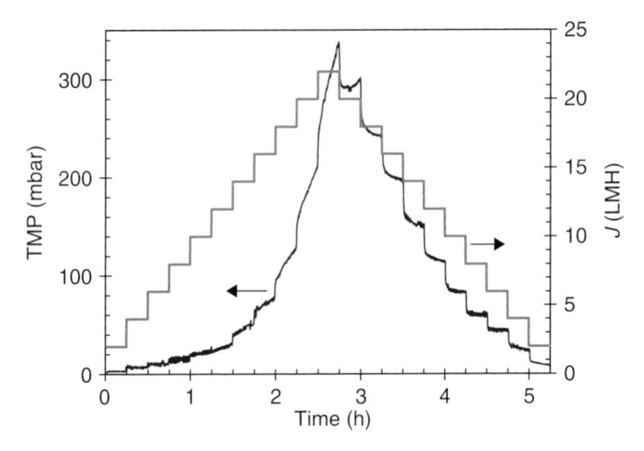

Figure 2.13 Flux and pressure relationship for 15-min flux steps, MBR biomass (Le-Clech et al.*, 2003b)*

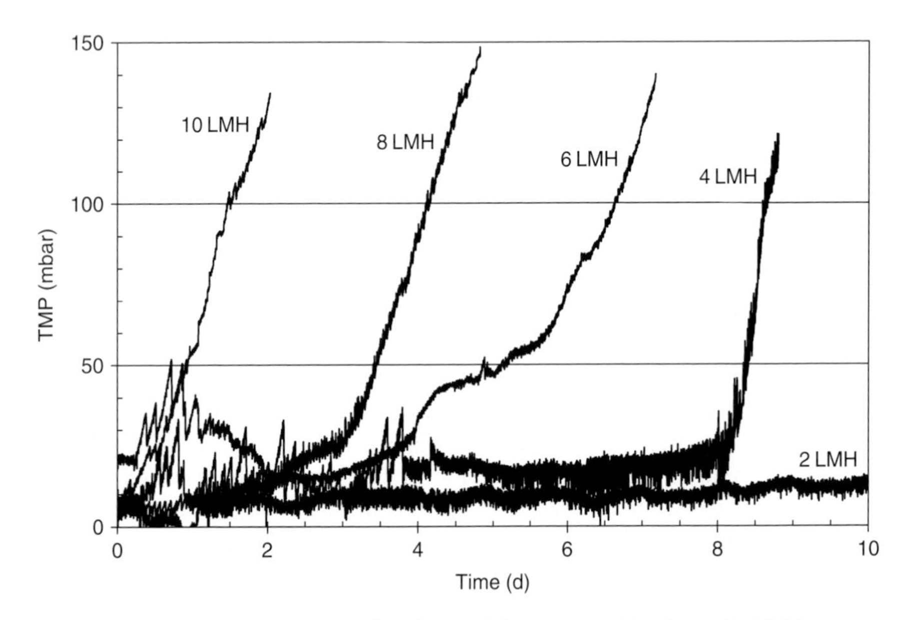

Figure 2.14 TMP transients for sub-critical flux operation (Brookes et al.*, 2004)*

comparison of reported data difficult, though a practical method based on a threshold permeability change has been proposed by Le-Clech *et al.* (2003b).

Within the last few years, it has become apparent from bench- and pilot-scale studies that irreversible fouling of MBR membranes can take place at operation well below the critical flux. Pertinent studies have been summarised by Pollice *et al.* (2005). Sub-critical flux fouling appears to be characterised by a sudden discontinuity of the TMP (Fig. 2.14, 4 LMH line) at very low flux operation after some extended time period (Brookes *et al.*, 2004; Ognier *et al.*, 2001; Wen *et al.*, 2004) and a steady neo-exponential increase at fluxes closer to the notional critical flux (Fig. 2.14, 10 LMH line). Sub-critical fouling is discussed further along with MBR membrane fouling mechanisms in Section 2.3.8.

2.2 Biotreatment

2.2.1 *Biotreatment rationale*

Biological treatment (or biotreatment) processes are those which remove dissolved and suspended organic chemical constituents through biodegradation, as well as suspended matter through physical separation. Biotreatment demands that the appropriate reactor conditions prevail in order to maintain sufficient levels of viable (i.e. living) micro-organisms (or, collectively, biomass) to achieve removal of organics. The latter are normally measured as biochemical or chemical oxygen demand (BOD and COD, respectively); these are indirect measurements of organic matter levels since both refer to the amount of oxygen utilised for oxidation of the organics. The micro-organisms that grow on the organic substrate on which they feed derive energy and generate cellular material from oxidation of the organic matter, and can be aerobic (oxygen-dependent) or anaerobic (oxygen-independent). They are subsequently separated from the water to leave a relatively clean, clarified effluent.

The most attractive feature of biological processes is the very high chemical conversion efficiency achievable. Unlike chemical oxidation processes, aerobic processes are capable of quantitatively mineralising large organic molecules, that is, converting them to the end mineral constituents of CO_2, H_2O and inorganic nitrogen products, at ambient temperatures without significant onerous byproduct formation. In doing so a variety of materials are released from the biomass in the reactor which are collectively referred to as extracellular polymeric substances (EPS) and which contain a number of components which contribute to membrane fouling in an MBR (Section 2.3.6.1). The relative and overall concentrations of the various components are determined both by the feed characteristics and operational facets of the system, such as microbial speciation. Anaerobic processes generate methane as an end product, a possible thermal energy source, and similarly generate EPS. Biotreatment processes are generally robust to variable organic loads, create little odour (if aerobic) and generate a waste product (sludge) which is readily processed. On the other hand, they are slower than chemical processes, susceptible to toxic shock and consume energy associated with aeration in aerobic systems and mixing in all biotreatment systems.

2.2.2 *Processes*

Processes based on biodegradation can be classified according to the process configuration, feeding regime and oxidation state (Table 2.3). Process configuration defines the way in which the water is contacted with the biomass, which can form a layer on some supporting media to form a fixed biofilm or be suspended in the reactor, or sometimes a combination of these. Suspended growth systems provide higher mass

Table 2.3 Examples of biological processes and their characteristics

	Process configuration		Feeding regime		Redox conditions		
	Fixed film	Suspended growth	Continuous	Fed-batch	Aerobic	Anoxic	Anaerobic
AD		X	(X)	(X)			X
AF	X		X				X
ASP		X	X		X	(X)	(X)
BAF	X		X		X		
RBC	X		X		X		
SBR		X		X	X	(X)	
TF	X				X		
UASB		X	X				X
MBR		**X**	**X**		**X**	**(X)**	

Keywords: AD: Anaerobic digestion; AF: Anaerobic filter; ASP: Activated sludge process; BAF: Biological Aerated Filters; RBC: Rotating biological contactor; SBR: Sequencing batch reactor; TF: Tricking filter; UASB: Upflow anaerobic sludge blanket.

transfer but the biomass subsequently needs to be separated from the water. Both configurations generate excess biomass which needs to be disposed of. Feeding regime defines the way in which the feedwater is introduced, which can be either continuous or batch-wise. Feeding in batches allows the same vessel to be used both for biodegradation and separation, thus saving on space. This is the case for the sequencing batch reactor (SBR). Finally, the reduction–oxidation (redox) conditions are defined by the presence of either dissolved oxygen (DO) (aerobic conditions) or some other compound capable of providing oxygen for bioactivity (anoxic conditions) or the complete absence of any oxygen (anaerobic conditions). The different redox conditions favour different microbial communities and are used to affect different types of treatment.

Aerobic treatment is used to remove organic compounds (BOD or COD) and to oxidise ammonia to nitrate. Aerobic tanks may be combined with anoxic and anaerobic tanks to provide biological nutrient removal (BNR). The removal of nutrients (nitrogen and phosphorus), BNR, is discussed further in Section 2.2.5, and the various facets of biological processes in general are described in detail in various reference books (Grady *et al.*, 1999; Metcalf and Eddy, 2003). However, almost all biological processes are configured according to the sub-categories listed in Table 2.3, and their function and performance depends on which specific sub-categories apply. Moreover, unit biotreatment processes can be combined so as to achieve multiple functions. So, for example, within an individual bioreactor, both aerobic and anoxic processes can be designed to occur within different zones.

The classic sewage treatment process (Fig. 2.15) is the combination of screening of gross solids, and then sedimentation of settlable solids followed by a biological process. The latter can include an anoxic zone preceding an aerobic zone within a single reactor or a separate post-denitrification reactor for the complete removal of nitrogen. Various configurations that include a preliminary anaerobic zone to remove phosphorus

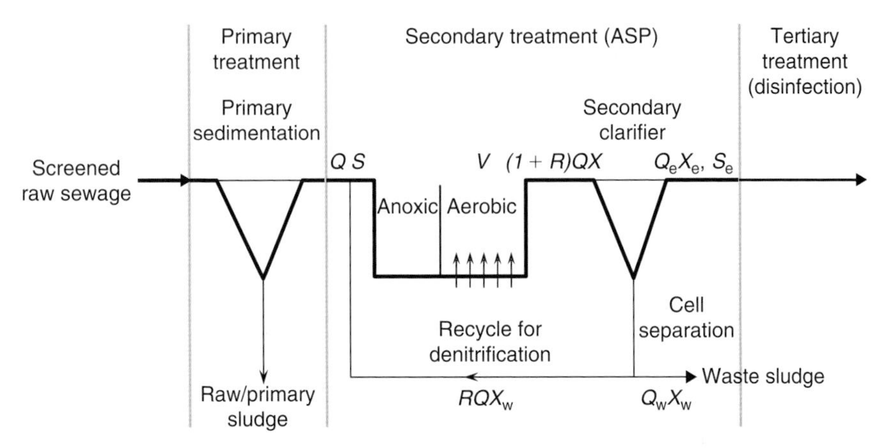

Figure 2.15 Classic sewage treatment process, with mass flows for ASP indicated

biologically are also available. Aerobic processes may be configured either as suspended growth (the activated sludge process, ASP) or fixed film (a trickling filter, TF). Total removal of organic nitrogen (ON) from the feedwater can be achieved by recycling the nitrate-rich sludge from the ASP to some point upstream of the aerobic process where anoxic conditions then prevail. Nitrification and denitrification are thus carried out sequentially. Aerobic MBRs can be configured similarly since, in essence, the biological function remains unaltered by the membrane.

In all biotreatment processes, the treated water must be separated from the biomass. Fixed film process effluent is notionally low in biological material since the latter forms a biofilm on the growth media, although biofilms can slough off into the product water, whereas in the classical ASP process separation is by simple sedimentation. This means that conventional ASPs rely on the solids (which are flocculated particles and referred to as flocs) growing to a size where they can be settled out, which means that they must be retained in the bioreactor for an appropriate length of time. The solids retention time (SRT) is thus coupled with the hydraulic retention time (HRT), the retention time being the time taken for matter to pass through the reactor. For commercial MBR technologies, separation is by membrane filtration, eliminating the requirement for substantial floc growth, and the implications of this are discussed in Section 2.2.4. However MBRs can also be configured as fixed film processes, using the membrane to support a biofilm. These sorts of MBRs are discussed in Section 2.3.2–2.3.3.

2.2.3 Microbiology

Biological treatment relies on conversion of organic and inorganic matter into innocuous products by micro-organisms and, as such, the biological community must be healthy and sustainable. Figure 2.16 illustrates the food chain in a biotreatment environment. Higher forms of micro-organisms such as protozoa and rotifers play crucial roles in consuming suspended organic matter and controlling sludge concentration by scavenging bacteria. Larger biological species such as nematode worms

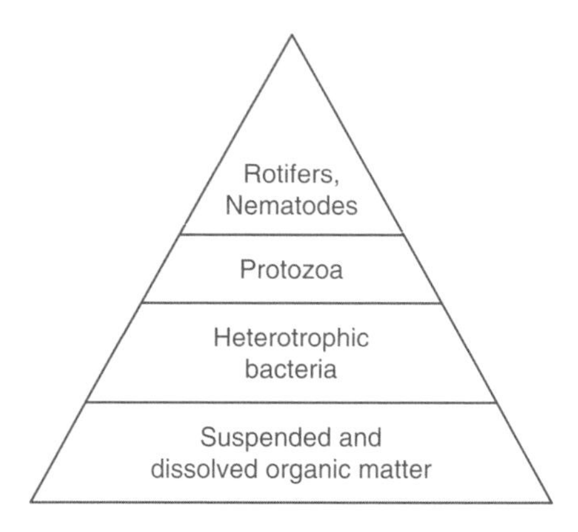

Figure 2.16 Ecology of activated sludge systems

and insect larvae may contribute to the consumption of particulate organic matter, especially in TF systems.

There is evidence to suggest that higher organisms, protozoa, filamentous organisms, nematodes and ciliates, are present at lower concentrations in MBRs than in conventional activated sludge systems (Cicek *et al.*, 1999; Witzig *et al.*, 2002). However, higher concentrations of protozoa, particularly flagellates and free ciliates, have been reported for MBRs compared with an activated sludge operating at the same SRT (Ghyoot and Verstraete, 2000). These experiments were performed on a system with long HRT (20–74 h), hence the shorter HRT associated with MBRs may be responsible for the absence of protozoa in other studies. Predatory organisms have a negative effect on nitrification (Lee and Welander, 1994) and overgrowth of protozoa have been shown to create a complete breakdown of nitrification (Bouchez *et al.*, 1998). This grazing in activated sludge is accounted for in the death coefficient (k_e), and recent research suggests that this effect has a greater impact on sludge concentration than previously thought in an activated sludge system (van Loosdrecht and Henze, 1999). In contrast, the sludge concentration in an MBR is limited by the energy provided and cell decay (Low and Chase, 1999). Higher organisms, such as nocardia, have been shown to develop in full-scale MBRs and produce significant foaming problems (Section 2.3.6.3).

Conditions are created in an MBR by allowing the sludge to accumulate to a maximum biomass concentration where all of the energy available is used for cell maintenance. The high sludge concentration compared to the food available creates an environment where bacteria are facing starvation conditions so the bacteria are not in a physiological state for cell growth (Muller *et al.*, 1995). Oxygen uptake rates in an MBR system compared with a conventional activated sludge system are lower, indicating that the MBR is carbon rather than oxygen limited (Witzig *et al.*, 2002). Even if the cells in an MBR system are not growing, new bacteria are constantly being introduced with the influent wastewater; since no grazing organisms exist, there must be cell decay to keep the biomass concentration constant.

The microbial community in any biological system comprises a large number of different bacterial species. In both an MBR and activated sludge system, the dominant

Table 2.4 Microbial metabolism types in wastewater biotreatment

Component	Process	Electron acceptor	Type
Organic-carbon	aerobic biodegradation	O_2	Aerobic
Ammonia	Nitrification	O_2	Aerobic
Nitrate	Denitrification	NO_3^-	Facultative
Sulphate	Sulphate reduction	SO_4^{2-}	Anaerobic
Organic-carbon	Methanogensis	CO_2	Anaerobic

group of bacteria have been shown to be β-subclass Proteobacteria (Manz *et al.*, 1994; Sofia *et al.*, 2004;); all currently characterised ammonia oxidisers belong to this group. Although these bacteria remained dominant in an MBR, a higher proportion of other bacteria (52–62%) were recorded in these studies suggesting that the long SRT shifted the microbial population away from Proteobacteria-β (Luxmy *et al.*, 2000; Sofia *et al.*, 2004). *Nitrosomonas* and *Nitrosospira* are the autotrophic ammonia-oxidising bacteria found in activated sludge, and *Nitrobacter* and *Nitrospira* the nitrite-oxidising bacteria, and it is thus between these groups that the nitrification process is carried out (Wagner *et al.*, 1996, 1998). Sofia *et al.* (2004) found the predominant nitrifiers were *Nitrosospira* and *Nitrospira*, whilst Witzig *et al.* (2002) showed no *Nitrosomonas* or *Nitrobacter* or *Nitrosospira* to be found in membrane-filtered sludge. This implies that the ammonia-oxidising bacteria are system-specific and that *Nitrospira* are responsible for the reduction of nitrite. Nitrifying autotrophs are known to be slow-growing bacteria. The long SRTs available in an MBR system are thus accepted as being highly advantageous for nitrification.

Micro-organisms can be classified according to the redox conditions in which they prevail (Table 2.4), and hence the process type, and their energy requirements. Heterotrophs use organic carbon as an energy source and for synthesis of more cellular material, and are responsible for BOD removal and denitrification. Autotrophs use inorganic reactions to derive energy, for example, oxidation of iron (II) to iron (III) or hydrogen to water, and obtain assimilable material from an inorganic source (such as carbon from carbon dioxide) to carry out such processes as nitrification, sulphate reduction and anaerobic methane formation. Autotrophs are generally less efficient at energy gathering than heterotrophs and therefore grow more slowly.

Microbial growth relies on appropriate conditions of total dissolved solids (TDS) concentration, pH and temperature. Most micro-organisms can only function in relatively dilute solutions, around neutral pH and at ambient temperature, though some can grow under extreme conditions: *Thiobacillus* growth is optimum at pH 1.5–2.0. Some MBRs are based on growth of specific cultures, such as for nitrification (Section 2.2.4.4), or recalcitrant organics biodegradation in extractive MBRs (Section 2.3.2). Classification of micro-organisms according to the temperature at which they are most active provides the terms psychrophilic, mesophilic and thermophilic for optimum growth temperatures of 15, 35 and 55°C, respectively. While most aerobic biological processes are operated at ambient temperatures, the micro-organisms usually have mesophilic temperature optima, such that pumping operations in sMBRs can provide additional benefit in raising the reactor temperature to both increase biotreatment efficacy and reduce liquid viscosity (Van Dijk and Roncken, 1997). Few examples exist of MBRs operating under thermophilic conditions, though this mode appears to offer some promise for treatment of heavy COD loads and/or recalcitrant organic matter (Section 5.4.4).

2.2.4 Process design and operation fundamentals

Monod kinetics can be used to design biological systems for a limiting substrate (S kg/m^3), usually organic carbon provided as BOD or COD. Using known biokinetic constants (Appendix B) the system kinetics and system mass balance can be used to define the rate of substrate degradation, biomass growth and sludge production. A full description of Monod kinetics for process design can be found in Metcalf and Eddy (2003).

2.2.4.1 Substrate degradation

The rate biokinetics determine the loading rate (the rate at which organic matter is introduced into the reactor, kg BOD/m^3), as determined by Monod kinetics. Accordingly, the rate of reaction is first order with respect to a limiting substrate up to a maximum specific growth rate, after which growth is unaffected by any increase in substrate concentration:

$$\mu = \frac{\mu_m S}{K_s + S} \tag{2.2}$$

where μ is the growth rate (/h), and μ_m is the maximum specific growth rate (/h), S is the limiting substrate concentration (g/m^3) and K_s is the saturation coefficient (g/m^3). It follows that there is a maximum specific substrate utilisation rate which is defined as:

$$k = \frac{\mu_m}{Y} \tag{2.3}$$

Where Y is the biomass yield (i.e. the mass of cells formed per mass of substrate consumed) (g Volatile suspended solids (VSS)/g BOD). Y can be controlled by manipulating environmental factors such as temperature and pH, but such changes are detrimental to biodegradation in the reactor (Eckenfelder and Grau, 1998). Substituting terms defined by Monod kinetics into a mass balance expression for the system and rearranging produces an expression in terms of effluent dissolved substrate S_e in g/m^3:

$$S = \frac{K_s\left(1 - k_e\theta_x\right)}{\theta_x\left(Yk - k_e\right) - 1} \tag{2.4}$$

where θ_x is the SRT or sludge age (/day) and k_e term is the death rate constant. SRT is an important design parameter used for suspended growth systems. One of the advantages of an MBR system is that all of the solids are retained by the membrane which affords the operator complete control over the SRT. The death rate constant accounts for endogenous metabolism, that is, the utilisation by cells of stored materials, and the presence of extracellular polymerics (Section 2.3.6.5) associated with the biomass. k_e also accounts for grazing of the biomass by predatory organisms, as previously discussed. k_e for conventional activated sludge and anaerobic processes is typically in the range 0.04–0.075/day (Gu, 1993; Metcalf and Eddy, 2003), and takes similar values for MBRs (Fan *et al.*, 1996; Wen *et al.*, 1999). Experiments by Huang *et al.* (2001) showed that the endogenous decay in an MBR is higher (0.05–0.32/day) than for an ASP (0.04–0.075/day).

2.2.4.2 Sludge yield

Y_{obs}, the observed yield (g/(g/day)), is always lower than Y due to the effects of cell decay (k_e). The relationship between Y_{obs} and Y is governed by the SRT, θ_x, and is defined by:

$$Y_{obs} = \frac{Y}{1 + k_e\theta_x} + \frac{f_d k_e Y \theta_x}{1 + k_e\theta_x} \qquad (2.5)$$

where f_d is the fraction of the biomass that remains as cell debris, usually 0.1–0.15 g VSS/g substrate (Metcalf and Eddy, 2003). Observed yields (Y_{obs}) are approximately 0.6/day for conventional aerobic processes and an order of magnitude lower for anaerobic ones. Y_{obs} is used to calculate the amount of heterotrophic sludge that will be produced by a biological system ($P_{x,het}$) for a given flow rate (Q m^3/day):

$$P_{x,het} = Y_{obs} Q(S - S_e) \qquad (2.6)$$

The observed yield is the increase in biomass from heterotrophic cells only. Nitrification sludge and non-biodegradable solids also impact on the total daily sludge production. P_x is the total sludge yield from substrate degradation and derives from the heterotrophic sludge yield ($P_{x,het}$) and the nitrification sludge yield ($P_{x,aut}$). The sum of all solids generated each day the non-biodegradable solids can be accounted for by:

$$X_0 = (TSS - VSS)Q \qquad (2.7)$$

The sludge production from biodegradation in an MBR can, in principle, be reduced to zero by controlling SRT (θ_x), k_e and Y. The change in θ_x has by far the greatest impact on sludge production (Xing *et al.*, 2003) and mixed liquor suspended solids (MLSS). Past experience allows designers to set a desired MLSS concentration (X g/m^3). The MLSS then affects sludge production, aeration demand (Section 2.2.5) and membrane fouling and clogging (Section 2.3.6). Using a design MLSS and SRT the aeration tank volume can be calculated by obtaining the mass of solids being aerated, and then using the MLSS to convert that mass to the volume which those solids occupy:

$$V = \frac{(P_x + X_0)\theta_x}{X} \qquad (2.8)$$

2.2.4.3 SRT and F:M ratio

The slow rate of microbial growth demands relatively long HRTs (compared with chemical processes), and hence large-volume reactors. Alternatively, retaining the biomass in the tank either by allowing them to settle out and then recycling them, as in an ASP, fixing them to porous media, such as in a TF, or selectively rejecting them, as with an MBR, permits longer SRTs without requiring the HRT to be commensurately increased. Controlling the SRT in a biological system allows the operator to control the rate of substrate degradation, biomass concentration and excess sludge production as illustrated in Equations (2.4) (2.5) and (2.8). The SRT is controlled by periodically discharging some of the solids (sludge) from the process:

$$\theta_X = \frac{VX}{Q_W X_W + Q_e X_e} \tag{2.9}$$

where V and X are the aeration tank volume (m^3) and MLSS (g/m^3), Q_W and X_W the sludge wastage rate (m^3/day) and suspended solids concentration (g/m^3), and Q_e and X_e the corresponding values for the effluent. SRT should thus in theory determine the final effluent quality, though in practice effluent quality is determined by sludge settlability. In an MBR system, no solids can pass through the membrane (i.e. $X_e = 0$), and hence the SRT is defined only by the wasted solids. If the solids wasted from the reactor are at the same concentration as those in the reactor, that is, $X_W = X$, the volume of sludge wasted Q_W becomes:

$$Q_W = \frac{V}{\theta_X} \tag{2.10}$$

An often-quoted ASP empirical design parameter is the food-to-micro-organism ratio (F:M in units of inverse time), which defines the rate at which substrate is fed into the tank (SQ, Q being the volumetric feed flow rate in m^3/day) compared to the mass of reactor solids:

$$F{:}M = \frac{SQ}{VX} \tag{2.11}$$

This relates to SRT and the process efficiency E (%) by:

$$\frac{1}{\theta_X} = Y(F{:}M)\frac{E}{100} - k_e \tag{2.12}$$

SRT values for activated sludge plants treating municipal wastewaters are typically in the range of 5–15 day with corresponding F:M values of 0.2–0.4/day. Increasing SRT increases the reactor concentration of biomass, which is often referred to as the MLSS. Conventional ASPs operating at SRTs of ~8 days have an MLSS of around 2.5 g/L, whereas one with a SRT of ~40 days might have a MLSS of 8–12 g/L. A low

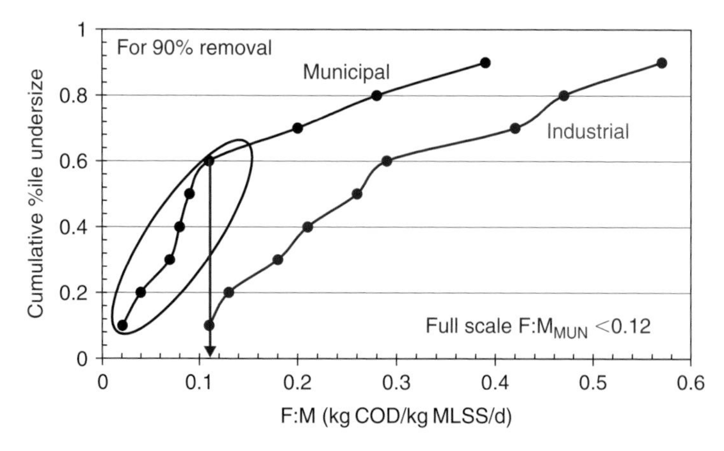

Figure 2.17 Statistical analysis of data from Stephenson et al. *(2000), municipal and industrial effluent*

F:M ratio implies a high MLSS and a low sludge yield, such that increasing SRT is advantageous with respect to waste generation. This represents one of the key advantages of MBRs, and an analysis of data from the review by Stephenson *et al.* (2000) reveals that most MBRs, where SRT can be readily extended, operate at F:M ratios of <0.12 (Fig. 2.17).

On the other hand, high MLSS values are to some extent detrimental to process performance. Firstly they would be expected to lead to an accumulation of inert compounds, reflected in a decrease in the MLVSS/MLSS ratio where MLVSS represents the volatile (organic) fraction of the MLSS, though this does not appear to be the case in practice (Huang *et al.*, 2001; Rosenburger *et al.*, 1999). Secondly, high solids levels increase the propensity for clogging or "sludging" – the accumulation of solids in the membrane channels. Lastly, and possibly most significantly, high MLSS levels reduce aeration efficiency, and this is discussed further in Section 2.2.5.

There have been a number of studies where the characteristics and performance of ASPs and MBRs have been compared when these processes operated under the same conditions of HRT and SRT. Ghyoot and Verstraete (2000), in their studies based on a skimmed milk-based analogue feed, observed sludge yields to be lower for an MBR than for an ASP (0.22 vs. 0.28 and 0.18 vs. 0.24 for operation at 12 and 24 days SRT, respectively). This trend was repeated in the work reported by Smith *et al.* (2003), who also noted the greatest impact of the membrane separation to be on K_s, which decreased from 125 ± 22 to 11 ± 1/day for the ASP compared to a corresponding increase from 2 ± 1.6 to 73 ± 22 for the MBR. Given that K_s is inversely proportional to substrate affinity, the generally lower values of K_s in the case of an MBR suggest a greater biomass substrate affinity, and also that the growth rate is less influenced by substrate concentration. Smith and co-workers proposed that this related to the difference in floc size, since the corresponding specific surface areas of the two biomasses at 30-day SRT were 0.098 m^2/g for the MBR and 0.0409 m^2/g for the ASP, revealing that the MBR biomass provides over 230% more surface area at about the same MLSS concentration.

2.2.4.4 Nitrification kinetics

Equations (2.2–2.12) are primarily concerned with the degradation of organic carbon in the feed. It is common practice to extend the SRT and HRT in the aeration basin to achieve the degradation of ammonia (NH_4-N). The effluent nitrogen concentration (N_e g/m^3) can be estimated by:

$$N_e = \frac{K_n(\mu_n + k_{e,n})}{\mu_{n,m} - k_{e,n} - \mu_n} \tag{2.13}$$

where $\mu_{n,m}$ is the maximum specific growth rate of nitrifying bacteria, K_n is the half saturation coefficient for nitrification, $k_{e,n}$ is the death rate coefficient for nitrifying bacteria and μ_n is the specific growth of nitrifying bacteria which can be found from:

$$\mu_n = \frac{1}{\theta_x} \tag{2.14}$$

Literature values for the nitrification constants along with the heterotrophic constants can be found in Appendix B. Sludge production from nitrification is given by:

$$P_{x,aut} = \frac{QY_n NO_x}{1 + k_{e,n}\theta_x} \tag{2.15}$$

where Y_n is the nitrification sludge yield (g VSS/g NH_4-N) and NO_x is the concentration of NH_4-N that is oxidised (mg/L) to form nitrate. To calculate the NO_x, a nitrogen balance can be performed on the system:

$$NO_x = N - N_e + 0.12P_x \tag{2.16}$$

where N is the influent total Kjeldahl (biochemically-oxidisable) nitrogen concentration (TKN, mg/L). NO_x is used to determine P_x. NO_x can be estimated at the first attempt and iterated to find values for NO_x and $P_{x,aut}$.

Nitrifying bacteria operate more slowly than carbon degraders such that, to achieve nitrification, a longer HRT is required; nitrifiers are slower growing and require a longer SRT. An SRT of around 10 days is required to allow full growth of the nitrifying community (Huang, 2001). Fan *et al.* (1996) reported that perfect nitrification, that is, all of influent TKN converted to NO_3^-, can be achieved in an MBR.

2.2.5 Aeration

2.2.5.1 Fundamentals

In conventional aerobic biological wastewater treatment processes, oxygen is usually supplied as atmospheric air, either via immersed air-bubble diffusers or surface aeration. Diffused air bubbles (via fine-bubble aeration) are added to the bulk liquid (as in an ASP, biological aerated filters (BAFs), fluidised bioreactors, etc.), or oxygen transfer occurs from the surrounding air to the bulk liquid via a liquid/air interface (as for a TF or rotating biological contactor (RBC)).

The oxygen requirement to maintain a community of micro-organisms and degrade BOD and ammonia and nitrite to nitrate can be found by a mass balance on the system (Metcalf and Eddy, 2003):

$$m_o = Q(S - S_e) - 1.42P_x + 4.33Q(NO_x) - 2.83Q(NO_x) \tag{2.17}$$

where m_0 is the total oxygen required (g/day). The first term in Equation (2.17) refer to substrate oxidation, the second refers to biomass respiration, the third refers to nitrification and the final term refers to denitrification. Certain terms thus disappear from the expression depending on whether or not the system is nitrifying and/or denitrifying.

2.2.5.2 Mass transfer

Mass transfer of oxygen into the liquid from air bubbles is defined by the overall liquid mass transfer coefficient (k_L m/s) and the specific surface area for mass transfer (a m^2/m^3). Because of the difficulties associated with measuring k_L and a, the two are usually combined to give the volumetric mass transfer coefficient k_La (per unit time). The standard method accepted for determining k_La in clean water is detailed in ASCE (1992). The rate of oxygen transfer into a liquid can be determined by:

$$OTR_{\text{cleanwater}} = k_La(C^* - C) \tag{2.18}$$

where C and C^* are the dissolved and saturated oxygen concentration values in kg/m^3. For pure water and equilibrium conditions C is found using Henry's Law. This can be converted to process conditions by the application of three correction factors (α, β and φ) which account for those sludge properties which impact on oxygen transfer (Section 2.2.5.3):

$$OTR_{\text{process}} = \frac{OTR_{\text{cleanwater}}}{\alpha\beta\varphi} \tag{2.19}$$

Aeration also provides agitation to ensure high mass transfer rates and complete mixing in the tank. There is thus a compromise between mixing, which demands larger bubbles, and oxygen dissolution, which demands small, indeed microscopic, bubbles (Garcia-Ochoa *et al.*, 2000). Consequently oxygen utilisation, the amount of oxygen in the supplied air which is used by the biomass, can be as low as 10%, and decreases with increasing biomass concentration (Equation (2.25)). This can be quantified by the standard aeration efficiency (kg O_2/kWh):

$$SAE = \frac{OTRxV}{W} \tag{2.20}$$

where W is the power demand. The *OTR* into the mixed liquor can be increased by using oxygen-enriched air, but this increases costs and is rarely used other than for high-strength effluents when the oxygen limitation is reached. In an iMBR, additional aeration is also required for scouring of the membrane (Section 2.3.5).

Changes in airflow have been shown to produce the largest changes in mass transfer in a coarse bubble aeration system (Ashley *et al.*, 1992), with $k_L a$ increasing with gas velocity in an airlift reactor (Lazarova *et al.*, 1997; Masoud *et al.*, 2001). Nordkvist *et al.* (2003) proposed that both the liquid and gas velocities impact on mass transfer, confirmed by experiments based on a jet loop MBR by Kouakou *et al.* (2005). However, the authors of this paper also noted a linear relationship between the mass transfer coefficient and the liquid recirculation velocity. Also, increasing horizontal velocity has been shown to increase the value of $k_L a$ in an oxygen ditch in both pilot (Gillot *et al.*, 2000) and full-scale plants (Deronzier *et al.*, 1996).

2.2.5.3 Correction for temperature and process water

φ relates to the effect of temperature on the mass transfer and is corrected by:

$$k_L a_{(T)} = k_L a_{(20°C)} \Phi^{(T-20)} \tag{2.21}$$

where T is the temperature (°C) and Φ a constant. Typical Φ values are between 1.015 and 1.040 with 1.024 being the ASCE standard (Iranpour et al., 2000) for temperature correction of viscosity:

$$\varphi = 1.024^{(T-20)} \tag{2.22}$$

Salts and particulates in wastewater both impact on the oxygen transfer rate. Comparative tests on synthetic wastewater and tap water performed by Lazarova *et al.* (1997) showed that below 2 g/L salt concentration has little effect on the oxygen transfer. Kouakou *et al.* (2005) performed comparative studies between clean water and wastewater with a salt concentration of 0.48 g/L and found the mass transfer coefficients did not significantly vary. The effect of such constituents is accounted for by the β factor which is defined as:

$$\beta = \frac{C^*_{\text{wastewater}}}{C^*_{\text{cleanwater}}} \tag{2.23}$$

and is usually around 0.95 for wastewater (EPA, 1989).

Both biomass characteristics and aeration system design impact on oxygen transfer (Mueller *et al.*, 2002). Biomass is a heterogeneous mixture of particles, microorganisms, colloids, organic polymers and cations of various sizes and surface properties which can all impact on oxygen transfer through contact area and surface energy. Bubble characteristics differ depending on the aerator type and bubble stability, the latter being influenced by the biomass characteristics and promotion of bubble coalescence. At the same time, biological and physical characteristics of the mixed liquor are affected by the shear imparted by the air flow, which can fragment flocs (Abbassi *et al.*, 1999) and cause the release of chemicals, as well as impacting on biodiversity

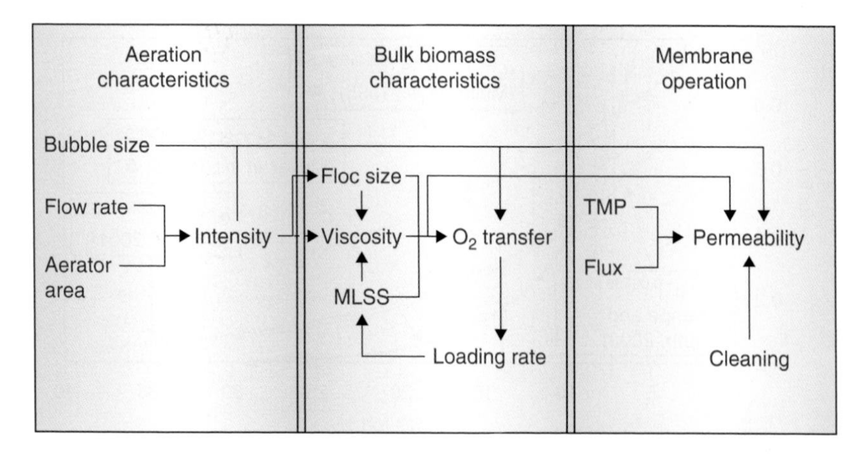

Figure 2.18 Aeration impacts in an iMBR (adapted from Germain, 2004)

(Madoni *et al.*, 1993). The inter-relationships developed between aeration and various system facets and parameters are thus complex, especially given that, for an iMBR, aeration is also used for membrane scouring (Fig. 2.18). This complex relationship is usually accounted for by the α factor. The α factor is the difference in mass transfer ($k_L a$) between clean and process water, and has the most significant impact on aeration efficiency of all three conversion factors. It is accepted that α factor is a function of SRT (mean cell retention time), air and liquid flow rate and of tank geometry for a given wastewater (Rosso and Stenstrom, 2005), and is defined as:

$$\alpha = \frac{k_L a_{\text{wastewater}}}{k_L a_{\text{cleanwater}}} \tag{2.24}$$

Wastewater composition and, in particular, the level of surfactants affect the bubble size, shape and stability. Surfactants are found in detergents of all kinds, including washing up liquid, laundry powder and soap. A high concentration of contaminants builds up on the outside of the bubble, reducing both the diffusion of oxygen into solution and the surface tension. Reduced surface tension has the beneficial effect of reducing bubble size, thereby increasing the water-air interfacial area (a). Fine bubble aeration systems are most negatively affected by surfactants, since bubbles produced are already small and cannot be further reduced in size by a reduction in surface tension (Stenstrom and Redmon, 1996). It has been shown from experiments testing oxygen transfer in waters containing different surfactants that the ratio of mass transfer from surfactant water to clean water varies between 1.03 and 0.82 (Gillot *et al.*, 2000). However, surfactants have a negative effect on ASP processes overall due to the promotion of foaming (Section 2.3.6.3).

Studies of the impact of solids concentration on oxygen transfer in biological wastewater treatment systems have all indicated a decrease in OTR with increasing solids concentration regardless of the system studied, though the relationship is system- and feedwater-dependent (Chang *et al.*, 1999; Chatellier and Audic, 2001; Fujie *et al.*, 1992; Gunder, 2001; Krampe and Krauth, 2003; Lindert *et al.*, 1992; Muller *et al.*, 1995).

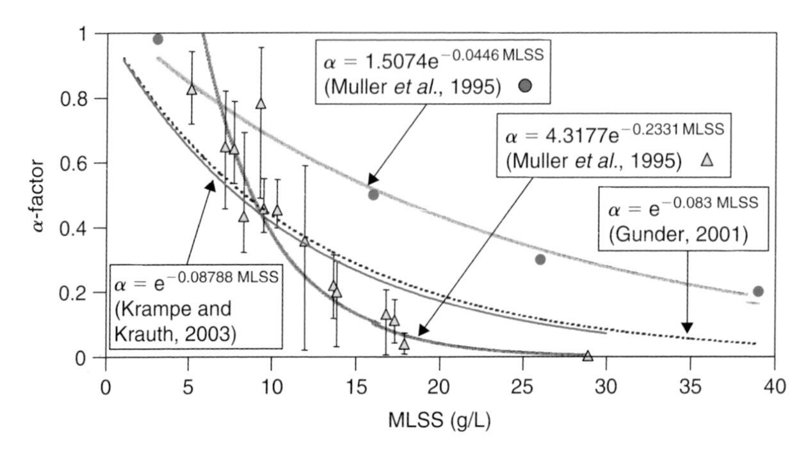

Figure 2.19 α-factor vs. MLSS concentration

In a number of studies of sewage treatment, an exponential relationship between α-factor and MLSS concentration has been observed. Muller *et al.* (1995), recorded α-factor values of 0.98, 0.5, 0.3 and 0.2 for MLSS concentrations of 3, 16, 26 and 39 g/L, respectively, yielding an exponential relationship with an exponent value of −0.045 and an R^2 of 0.99 (Fig. 2.19). Günder (2001) and Krampe and Krauth (2003) observed the same exponential trend with exponent values of −0.083 and −0.088, respectively, whereas an even higher exponent value of −0.23 was recorded by Germain (Germain, 2004). Studies on model or simplified systems by a number of authors (Freitas and Teixeira, 2001; Ozbek and Gayik, 2001; Verlaan and Tramper, 1987) appear to indicate that the principal impact of solids concentration is on the interfacial area *a*, which decreases with increasing solids level whilst leaving the mass transfer coefficient k_L largely unaffected. This has been attributed to the promotion of bubble coalescence by suspended solids (Klein *et al.*, 2002), and the effect is also aeration rate-dependent (Freitas and Teixeira, 2001). Since MBRs run at a high MLSS the aeration demand for biotreatment operation of an MBR is somewhat higher than that of an ASP.

The impact of particle size is more complex than particle concentration, since aeration, mass transfer and particle size are interrelated. For fine particles, <0.01 mm, k_La has been shown to increase with increasing solids concentration up to a certain level and remain stable, before decreasing with further increased solids concentration (Saba *et al.*, 1987; Smith and Skidmore, 1990). With larger particles, 1–3 mm, k_La appears to decrease with concentration (Hwang and Lu, 1997; Koide *et al.*, 1992; Komaromy and Sisak, 1994; Lindert *et al.*, 1992; Nakao *et al.*, 1999). Experiments examining excess sludge production, in which the DO concentration was adjusted independently of aeration intensity, indicated that higher mixing intensity and DO concentration, created by raising the airflow, had almost the same impact on floc break-up and therefore on particle size. At a sludge loading of 0.53 kg BOD_5/kg MLSS day, the excess sludge production was reduced by 22% by raising the oxygen concentration from 2 to 6 mg/L (Abbassi *et al.*, 1999). However, the principal impact of particle size in an MBR is on filter cake permeability, as indicated by the Kozeny Carman equation.

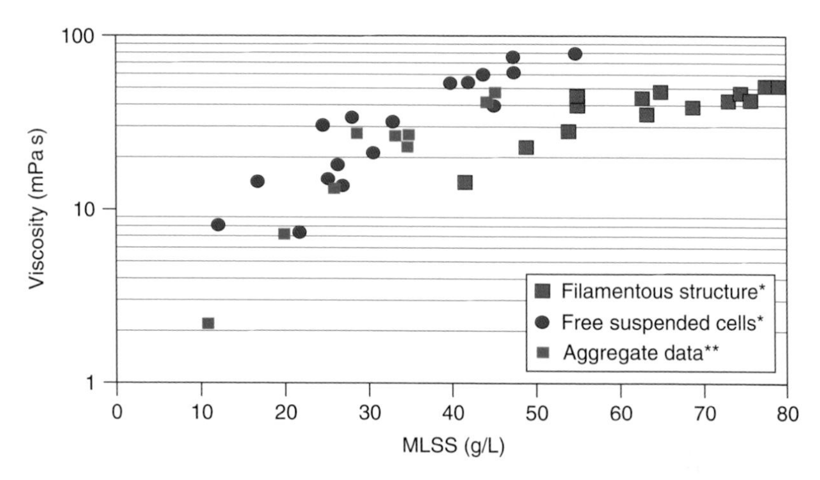

*Figure 2.20 Viscosity vs. MLSS concentration, *Rosenberger* et al. *(1999), **data from Stephenson* et al. *(2000)*

Viscosity correlations are complicated by the non-Newtonian pseudoplastic nature of the sludge, but it has nonetheless been shown to have a negative influence on the oxygen transfer coefficient (Badino *et al.*, 2001; García-Ochoa *et al.*, 2000; Jin *et al.*, 2001; Koide *et al.*, 1992; Özbek and Gayik, 2001). This has variously been attributed to bubble coalescence and solubility impacts, with larger bubbles forming (Özbek and Gayik, 2001) and greater resistance to mass transfer recorded (Badino *et al.*, 2001) at higher viscosity. Air is also less well distributed at higher viscosities, the smaller bubbles becoming trapped in the reactor (Jin *et al.*, 2001).

Correlations between the α-factor and viscosity (η kg/(m s)) have been presented, these correlations being more pronounced than those between α-factor and MLSS concentration (Wagner *et al.*, 2002). Relationships presented take the form:

$$\alpha = \eta^{-x} \tag{2.25}$$

where $x = 0.45$ (Günder, 2001) or 0.456 (Krampe and Krauth, 2003) at a shear rate of $40\ s^{-1}$ in activated sludge of high MLSS concentrations. The correlation is shear-dependent: increasing shear stress decreases viscosity (Dick and Ewing, 1967; Wagner *et al.*, 2002). An increase in aeration rate therefore offers the dual benefit to oxygen transfer in that it increases the amount of available oxygen and also decreases biomass viscosity by increasing shear stress. Viscosity increases roughly exponentially with increasing MLSS concentration (Manem and Sanderson, 1996; Rosenberger *et al.*, 1999, Fig. 2.20), impacting negatively on both oxygen transfer and membrane fouling (Section 2.3.6.3).

2.2.6 Nutrient removal

The removal of total nitrogen by biochemical means demands that oxidation of ammonia to nitrate takes place under aerobic conditions (Equations (2.27–2.29)), and that nitrate reduction to nitrogen gas takes place under anoxic conditions

(Equation (2.30)). Both these processes demand that specific micro-organisms prevail (Section 2.2.3). The exact micro-organisms responsible for denitrification (nitrate removal by biochemical reduction) are more varied – it is carried out by many different, phylogenetically-unrelated heterotrophs (Stewart, 1998; Zumft, 1992).

The biological generation of nitrate from ammoniacal nitrogen (NH_4^+) and aerobic conditions (nitrification), takes place in two distinct stages:

$$2NH_4^+ + 3O_2 \rightarrow 2NO_2^- + 2H^+ + 2H_2O \text{ (ammonia} \rightarrow \text{nitrite)} \quad (2.26)$$

$$2NO_2^- + O_2 \rightarrow 2NO_3^- \text{ (nitrite} \rightarrow \text{nitrate)} \quad (2.27)$$

Overall:

$$NH_4^+ + 2O_2 \rightarrow NO_3^- + 2H^+ + H_2O \quad (2.28)$$

Since the second step proceeds at a much faster rate than the first, nitrite does not accumulate in most bioreactors. However, since these micro-organisms are autotrophic and thus rather slow growing, they demand relatively long SRTs to accumulate and provide close to complete nitrification (i.e. above 90% ammonia removal). This presents another advantage of MBRs where long SRTs are readily attainable.

Denitrification takes place under anoxic conditions when oxidation of the organic carbon takes place using the nitrate ion (NO_3^-), generating molecular nitrogen (N_2) as the primary end product:

$$C_{10}H_{19}O_3N + 10NO_3^- \rightarrow 5N_2 + 10CO_2 + 3H_2O + NH_3 + 10OH^- \quad (2.29)$$

where in this equation "$C_{10}H_{19}O_3N$" represents wastewater.

Nitrification relies on sufficient levels of carbon dioxide, ammonia and oxygen, the carbon dioxide providing carbon for cell growth of the autotrophs. Since nitrifiers are obligate aerobes, DO concentrations need to be 1.0–1.5 mg/L in suspended growth systems for their survival. Denitrification takes place when facultative micro-organisms, which normally remove BOD under aerobic conditions, are able to convert nitrates to nitrogen gas under anoxic conditions. Denitrification requires a sufficient carbon source for the heterotrophic bacteria. This can be provided by the raw wastewater, which is why the nitrate-rich waste from the aerobic zone is recycled to mix with the raw wastewater. Complete nitrification is common in full-scale MBR municipal installations, although, since it is temperature-sensitive, ammonia removal generally decreases below 10°C. Most full-scale MBR sewage treatment plants are also designed to achieve denitrification.

Most wastewaters treated by biological processes are carbon limited, and hence phosphorus is not significantly removed. This applies as much to MBRs as to conventional plants. It appears that membrane separation offers little or no advantage regarding phosphorus removal (Yoon *et al.*, 1999). Enhanced biological phosphate removal can be achieved by the addition of an anaerobic zone at the front of an activated sludge plant and returning nitrate-free sludge from the aerobic zone (Yeoman *et al.*, 1986). This has been applied to a some full-scale MBR plant where constraints on discharged P levels have been imposed (Section 5.2.1.4, Running Springs). P removal is more commonly achieved by dosing with chemicals, such as metal coagulants or lime, that can form sparingly soluble precipitates.

2.2.7 Anaerobic treatment

Compared with aerobic processes, anaerobic biological treatment is characterised by (Stephenson *et al.*, 2000):

- a lower energy demand due to the absence of aeration
- slower microbial growth
- a lower COD removal (generally 60–90%)
- no nitrification
- greater potential for odour generation
- longer start up (months *cf.* weeks)
- higher alkalinity
- lower sludge production
- biogas (methane) generation.

Conventional anaerobic treatment process configurations are all designed to achieve both good mixing and sludge separation. A number of configurations exist:

(a) Simple contacting coupled with external sludge separation (by sedimentation, rotary vacuum filtration, etc.) and/or digestion before returning the clarified liquid to the reactor. This is a simple and relatively easily controlled process but is also made expensive by the pumping operations.
(b) Anaerobic filters, which are flooded media filters based on either packed or structured media.
(c) Upflow clarification using the upflow anaerobic sludge blanket (UASB, Fig. 2.21a) reactor (Lettinga and Vinken, 1980), in which sludge particles settle at the same rate as the water flows upwards, forming a stationary "blanket" of sludge in the reactor. This process relies on the formation of a dense granular sludge bed that is readily retained in the reactor, in much the same way as secondary clarification in the ASP relies on the growth of large settleable particles. The process is augmented in the expanded granular sludge bed (EGSB, Fig. 2.21b) which provides better influent distribution to improve contact between the sludge and wastewater and so promote more efficient use of the entire reactor volume (Seghezzo *et al.*, 1998). In this configuration, the sludge bed is expanded by operating at higher upflow rates and the reactor behaves as a completely mixed tank (Rinzema, 1988).
(d) Staged reactor systems, based on plug flow (Van Lier, 1995) and using sequentially operated reactors or compartments within a single reactor. Staged reactors include anaerobic baffled reactors (ABRs, Fig. 2.21c) in which baffles are used to direct the flow of wastewater in an upflow mode through a series of sludge blanket reactors (Metcalfe and Eddy, 2003). The process is more tolerant to non-settling particles than the UASB and EGSBs whilst still providing long solid retention times.

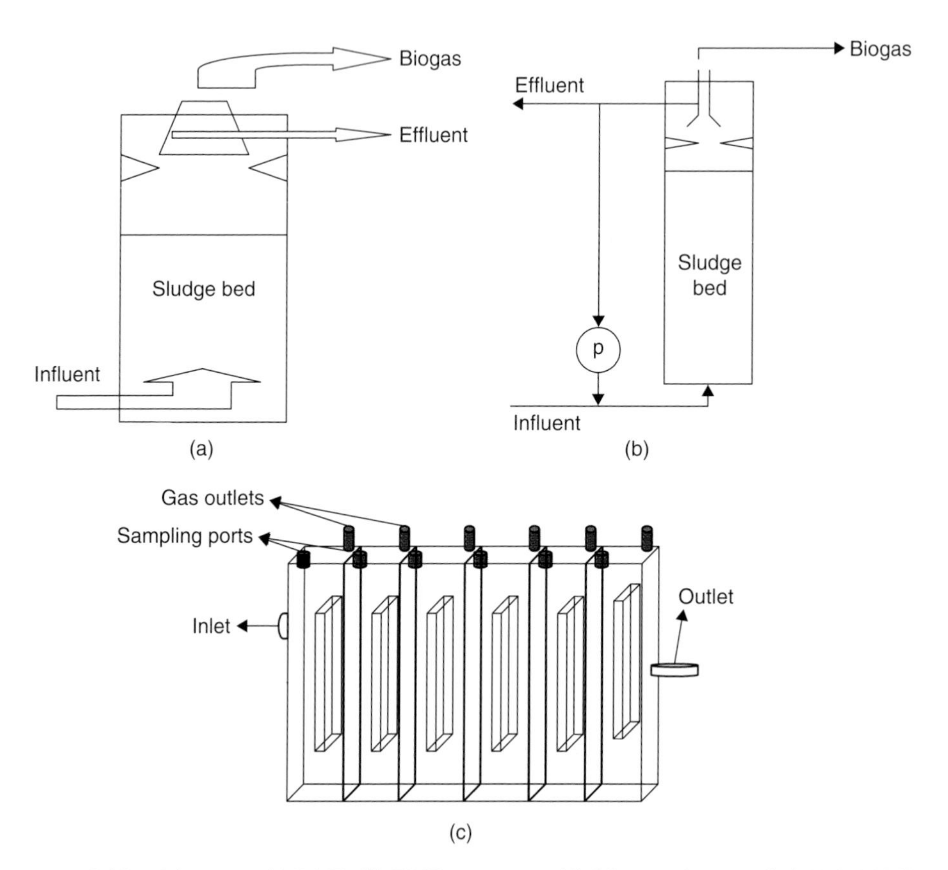

Figure 2.21 Schematics: (a) UASB, (b) EGSB reactors, modified from Seghezzo et al. *(1998, 2002) and (c) the ABR, modified from Dama* et al. *(2002)*

Anaerobic treatment is generally only considered for high-strength wastes and where low feed temperatures are less likely to be encountered. Low feed temperatures and strength imply low biomass growth yield and growth rate, such that the biomass concentration in the reactor is more difficult to sustain, particularly when substantial biomass wash-out from the reactor can occur. MBRs ameliorate this problem to a large extent, such that the range of anaerobic process operation can be extended to lower limits. This is achieved by the retention of the biomass in the reactor by the membrane independently of the HRT in the same way as for aerobic systems; significant quantities of residual organic matter are hydrolysed and biodegraded as a result. However, the fouling propensity of the bioreactor liquor is significantly higher for anaerobic treatment, such that fluxes and permeabilities are generally much lower than for the aerobic counterparts.

2.3 Membrane bioreactor technology

A classical MBR comprises a conventional ASP coupled with membrane separation to retain the biomass. Since the effective pore size can be below 0.1 μm, the MBR

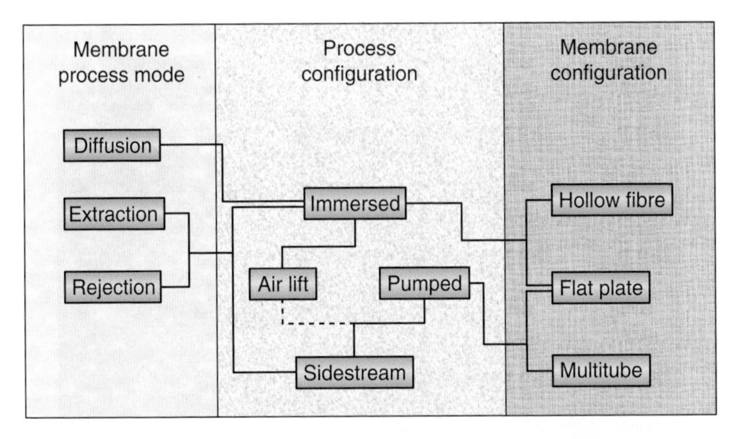

Figure 2.22 Principal configurations of MBR technologies

effectively produces a clarified and substantially disinfected effluent. In addition, it concentrates up the biomass and, in doing so, reduces the necessary tank size and also increases the efficiency of the biotreatment process. MBRs thus tend to generate treated waters of higher purity with respect to dissolved constituents such as organic matter and ammonia, both of which are removed by biotreatment. Moreover, by removing the requirement for biomass sedimentation, the flow rate through an MBR cannot affect product water quality through impeding solids settling, as is the case for an ASP. On the other hand, hydraulic and organic shocks can have other onerous impacts on the operation of an MBR.

2.3.1 MBR configurations

The word "configuration" can be used with reference to both the MBR process (and specifically how the membrane is integrated with the bioreactor) and the membrane module (Section 2.1.3). There are two main MBR process configurations (Fig. 1.6): submerged or immersed (iMBR), and sidestream (sMBR). There are also two modes of hydraulic operation: pumped and airlift. These configurations and bulk liquid transfer modes are employed commercially for what can be referred to as conventional biomass rejection MBRs, as outlined above. However, there are also two other membrane process modes, these being extractive (eMBR) and diffusive (dMBR) (Fig. 2.22), which employ a membrane for a purpose other than to separate the biomass from the treated water. Finally, whilst a number of membrane geometries and configurations exist in the membrane market place in general (Table 2.3), three predominate in existing commercial MBR technologies, these being FS, HF and MT. Examples of each type of commercial technology are detailed in Chapter 4, and a summary of the scientific and technical literature pertaining to the membrane configuration is included in Section 2.3.5.1.

iMBRs are generally less energy-intensive than sMBRs, since employing membrane modules in a pumped sidestream crossflow incurs an energy penalty due to the high pressures and volumetric flows imposed. To make the most use of this latent energy, the flow path must be as long as possible, such that as much as possible of the energy

Figure 2.23 An sMBR pilot plant. Note that the fluid path along the membrane is four times the length of one horizontal module

intrinsic in the liquid flowing at high pressure is used for permeation. To achieve a reasonable conversion of 40–50% conversion along the length of the module, a long flow path, often in excess of 20 m, is required. This then demands a large number of membrane modules in series (Fig. 2.23) incurring a significant pressure drop along the retentate flow channels.

With sMBRs, there is always a trade-off between pumping energy demand and flux. In order to maximise the flux, a high TMP is required combined with a high CFV or retentate velocity U_R. Since the energy demand is directly proportional to $Q_R \Delta P$ (retentate flow rate × pressure), it is of interest to reduce both these parameter values as much as possible. However, since Q_R determines U_R ($U_R = Q_R/A_t$, A_t being the tube cross-sectional area) and ΔP relates to TMP, reducing $Q_R \Delta P$ inevitably reduces flux. Moreover, if Q_R is reduced by decreasing the cross-sectional area A_t, this has the effect of increasing the pressure drop along the length of the module on the retentate side, since the resistance to flow is inversely proportional to A_t.

sMBRs have an inherently higher fouling propensity than iMBRs since higher flux operation always results in lower permeabilities because fouling itself increases with increasing flux, particularly above the so-called "critical flux" (Section 2.1.4.6). Moreover, it is thought that the higher shear imparted by liquid pumping of the side-stream imparts sufficient shear stress on the flocs to cause them to break-up (Tardieu *et al.*, 1999; Wisniewski and Grasmick, 1998). This both reduces particle size and promotes the release of foulant materials bound within the flocs (EPS, Section 2.3.6.3). Wisniewski and Grasmick (1998) studied the effects of the recirculation on the particle size in an sMBR. Without recirculation, floc size ranged from 20 μm to more than 500 μm. Only 15% of the particles were lower than 100 μm. With recirculation, reduction in particle size was directly proportional to the magnitude of the shear stress and the experiment time; at 5 m/s CFV 98% of the particles were smaller than 100 μm. iMBRs are therefore higher in energy efficiency, manifested as the specific energy demand in kWh/m^3 permeate product, than sMBR technologies. The immersed configuration employs no liquid pumping for permeation, instead relying on aeration to promote mass transfer of liquid across the membrane (i.e. enhancing flux) by generating significant transient shear at the membrane:solution interface (Section 2.1.4.4). Shear can also be promoted by directly moving the membrane, such as in the Huber VRM (Vacuum Rotating Membrane) system (Section 4.2.4).

Whilst sMBRs cannot provide the same low energy demand as the immersed configuration, they do offer a number of advantages:

1. Fouling has been shown to decrease linearly with increasing CFV. For example a bench-scale study revealed that CFV values of 2 and 3 m/s were sufficient to prevent the formation of reversible fouling in UF (30 kDa) and MF (0.3 μm) systems, and that fouling was suppressed for CFV values up to 4.5 m/s test (Choi *et al.*, 2005b).
2. The membranes can also be chemically cleaned "*in situ*" (CIP) easily without any chemical risk to the biomass.
3. Maintenance and plant downtime costs, particularly with reference to membrane module replacement, are generally slightly lower because of the accessibility of the modules which can be replaced in ~5 min.
4. Precipitation of sparingly soluble inorganic solids (i.e. scalants) and organic matter (gel-forming constituents) is more readily managed in sidestream MT systems by control of the hydrodynamics both during the operation and the CIP cycle.
5. It is generally possible to operate sMBRs at higher MLSS levels than HF iMBRs.
6. Aeration can be optimised for oxygen transfer and mixing, rather than demanding a compromise between membrane aeration and oxygen dissolution, as would be the case for single-tank iMBRs.

2.3.2 Extractive and diffusive MBRs

Extractive and diffusive MBR processes (eMBRs and dMBRs) are still largely at the developmental stage and are likely to be viable only for niche, high-added value applications. They have been reviewed by Stephenson *et al.* (2000). In an extractive system, specific problem contaminants are extracted from the bulk liquid across a membrane of appropriate permselectivity. The contaminant then undergoes biotreatment on the permeate side of the membrane, normally by a biofilm formed on the membrane surface. In the case of the diffusive MBRs, a gas permeable membrane is used to introduce into the bioreactor a gas in the molecular, or "bubbleless" (Ahmed and Semmens, 1992a, b; Côté *et al.*, 1988), form. This again normally feeds a biofilm at the membrane surface. Hence, both extractive and diffusive systems essentially rely on a membrane both for enhanced mass transport and as a substrate for a biofilm, and also operate by diffusive transport: the pollutant or gas for the extractive or diffusive MBR respectively travels through the membrane under a concentration gradient.

In principle, any gas can be transported across the membrane, though obviously the choices are limited if it is to be used to feed a biofilm. Diffusive systems are generally based on the transfer of oxygen across a microporous membrane and are thus commonly referred to as membrane aeration bioreactors (MABRs) (Brindle *et al.*, 1999). They present an attractive option for very high organic loading rates (OLRs) when oxygen is likely to be limiting, whilst retaining the advantages of a fixed film process (i.e. no requirement for downstream sedimentation and high OLRs). MABRs present an alternative to more classical high-gas transfer processes for oxygenation using pure oxygen such as a Venturi device. However, whereas these devices provide high levels of oxygenation (i.e. high OTRs, Equation (2.18)), it is not necessarily the case that they also provide high levels of utilisation by the biomass (oxygen utilisation efficiency (OUE)). MABRs, on the other hand, have been shown to provide 100% OUEs (Ahmed and Semmens, 1992b; Pankhania *et al.*, 1994, 1999) and organic removal rates of 0.002–0.005 $kg\,m^{-2}\,d^{-1}$ from analogue effluents (Brindle *et al.*, 1998; Suzuki *et al.*, 1993; Yamagiwa *et al.*, 1994) and OLRs of almost 10 $kg\,m^{-3}\,d^{-1}$ (Pankania *et al.*, 1994) – around five times that of conventional MBRs. This means much less membrane area is required to achieve organic removal, but removal efficiencies also tend to be lower.

Extractive MBRs allow the biodegradable contaminant to be treated *ex situ*. This becomes advantageous when the wastewater requiring biotreatment is particularly onerous to micro-organisms which might otherwise be capable of degrading the organic materials of concern. Examples include certain industrial effluents having high concentrations of inorganic material, high acidity or alkalinity, or high levels of toxic materials. Extraction of priority pollutants specifically using a permselective membrane, such as a silicone rubber membrane used to extract selectively chlorinated aromatic compounds from effluents of low pH or of high ionic strength (Livingston, 1993b; Livingston, 1994), allows them to be treated under more benign conditions than those prevailing *in situ*.

Whilst the diffusive and extractive configurations offer specific advantages over biomass separation MBRs, they are also subject to one major disadvantage. Neither process presents a barrier between the treated and untreated stream. This means that little or no rejection of micro-organisms takes place and, in the case of diffusive systems, there is a risk of sloughing off of biomass into the product stream in the same way as in the case of a TF. On the other hand, both configurations offer promise for the particular case of nitrate removal.

2.3.3 Denitrification

The three alternative membrane process modes can all be employed for the removal of nitrate from potable water supplies. Denitrification is the biochemical reduction of nitrate (Equation (2.30)). This process is conventionally configured as a packed bed in which denitrification is achieved by the biofilm formed on the packing material. Full-scale schemes for potable duty based on this technology can nonetheless encounter problems of (a) sloughed biomass and (b) residual organic carbon (OC) arising in the treated product.

Biological anoxic denitrification is extensively employed in wastewater treatment but various configurations have been trialled for drinking water denitrification (Matějů *et al.*, 1992; Soares, 2000). Its full-scale application has been limited, however, because of poor retention of both the microbial biomass and the electron donor – an

Table 2.5 System facets of denitrification MBR configurations

Configuration	Advantages	Disadvantages
Extractive microporous	Separation of biomass and carbon source from product water	Requires further downstream processing Carbon source breakthrough Pumping costs
Extractive ion-exchange	Dense membrane significantly reduces risk of carbon source breakthrough	Requires further downstream processing Potentially complex operation Unknown impact of fouling Comparatively high membrane cost Pumping costs
Diffusive	Non-toxic and low cost electron donor Good nitrate removal Low biomass yield	Requires further downstream processing Biomass breakthrough Potential for fouling to limit mass transfer Health and safety risk with respect to hydrogen gas dissolution Autotrophs, slow to adapt
Biomass rejection	Retention of biomass/ active denitrifiers Limited further downstream processing High rate nitrate removal Proven at full scale Appropriate dose control to limit breakthrough Comparatively low cost Comparatively simple to operate	Potential for carbon source breakthrough Limited knowledge of fouling potential

organic chemical dosed to substitute for the $C_{10}H_{19}O_3N$ in Equation (2.21). Electron donors trialled have included methanol (Mansell and Schroeder, 1999), ethanol (Fuchs *et al.*, 1997), acetic acid (Barrieros *et al.*, 1998), hydrogen (Haugen *et al.*, 2002) and sulphur (Kimura *et al.*, 2002), all designed to promote the appropriate heterotrophic (organic carbon-based electron donor) or autotrophic conditions necessary for denitrification, and each having its own limitations. As already stated, MBRs can be employed in three different configurations to augment denitrification (Table 2.5):

- selective extraction of nitrate with porous (Fuchs *et al.*, 1997; Mansell and Schroeder, 1999) or dense (ion-exchange) membranes (Velizarov *et al.*, 2003);
- supply of gas in molecular form (Ho *et al.*, 2001; Lee and Rittmann, 2002), or
- rejection of biomass (Nuhoglu *et al.*, 2002; Urbain *et al.*, 1996).

2.3.3.1 Extractive microporous MBR

In this configuration (Fig. 2.24a), also known as a "confined cell" or "fixed membrane biofilm reactor", nitrate is extracted from the pumped raw water by molecular diffusion through a physical barrier to a recirculating solution containing the denitrifying biomass. Pressure should ideally be equalised to reduce the influence of diffusion (Mansell and Schroeder, 2002). Various materials have been researched to separate effectively the solutions, including calcium alginate gel, polyacrylamide/alginate copolymer, an agar/microporous membrane composite structure and various microporous membranes

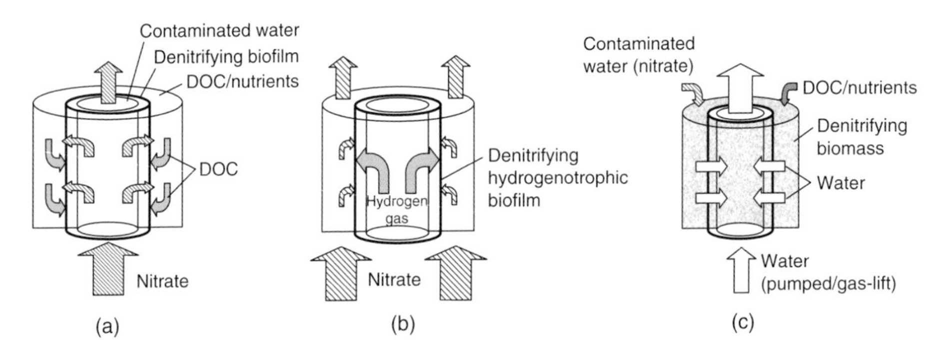

Figure 2.24 System configurations, denitrifying MBR: (a) nitrate extraction (eMBR), (b) biomass rejection (rMBR) and (c) membrane hydrogenation (MHBR)

(Mansell and Schroeder, 2002). Membrane configurations have typically consisted of either an FS (Reising and Schroeder, 1996) or tubular (Ergas and Rheinheimer, 2004) type. The advantage of this process is that both the electron donor and the heterotrophic denitrifying biomass are separated from the product water. Whilst the membrane can permit electron donor transport, biofilm formation should theoretically aid donor retention (Fuchs *et al.*, 1997).

Between 90% and 99% removal of nitrate has been reported at nitrate levels as high as 200 mg NO_3^--N/L (Ergas and Rheinheimer, 2004; Fuchs *et al.*, 1997; Mansell and Schroeder, 1999). The main limitation of this system appears to be permeation of the electron donor (such as methanol) into the product water, with 8% transfer and 4 mg total organic carbon (TOC)/L product water concentration being respectively reported by Ergas and Rheinheimer (Ergas and Rheinheimer, 2004) and Mansell and Schroeder (Mansell and Schroeder, 1999) in controlled addition experiments. It has been suggested that this problem can be ameliorated by continuous, rather than batch, operation and appropriate control of biofilm growth (Reising and Schroeder, 1996), a postulate corroborated to some extent by studies by Fuchs *et al.* (1997). Problems of organic carbon breakthrough into the product water can be obviated by using hydrogen as the electron donor, coupled with a bicarbonate carbon source (Mansell and Schroeder, 2002). However, the process then becomes limited by dissolution of hydrogen.

2.3.3.2 Extractive ion-exchange MBR

This configuration is identical to the extractive process except that the microporous membrane is replaced by an ion-exchange membrane, which is then, in principle, more selective for nitrate, which is removed under a concentration gradient. By appropriate membrane selection, the electron donor concentration in the product water can be reduced to below 1 mg/L (Fonseca *et al.*, 2000; Velizarov *et al.*, 2000/2001) coupled with 85% nitrate removal at a feed concentration of 135–350 mgNO_3^-/L (Fonseca *et al.*, 2000). Thus, the ion-exchange membrane provides lower nitrate removal rates than a microporous membrane but greater rejection of low-molecular-weight organic carbon. However, as with the extractive technology, the use of the membrane to simply extract nitrate implies that further processing of the product water is required.

Moreover, the potential for membrane fouling from both organic materials and hardness (Oldani *et al.*, 1992) has not yet been explored. The relative expense of the ion-exchange membrane (Crespo *et al.*, 2004) is likely further to constrain development of this process for the duty of drinking water denitrification.

2.3.3.3 Diffusive MBRs

As already stated, the use of hydrogen (H_2) as the electron donor combined with either carbon dioxide or bicarbonate as the carbon source (Mansell and Schroeder, 2002) obviates the organic carbon contamination issue, since hydrogen gas does remain dissolved in the water. Such autotrophic (or "hydrogenotrophic") denitrification can be considered both inexpensive and non-toxic (Haugen *et al.*, 2002), as well as producing a relatively low biomass yield (Lee and Rittmann, 2002). As with MABRs, diffusive H_2 MBRs typically employ microporous HF membranes or silicon tubes (Ho *et al.*, 2001) to deliver gas directly to biomass (in this case a denitrifying biofilm) attached to the shell-side of the membrane (Fig. 2.24b) providing up to 100% gas transfer (Mo *et al.*, 2005). Although around 40% slower than heterotrophic denitrification according to some batch measurements (Ergas and Reuss, 2001), high nitrate removal rates have none-the-less been reported in hydrogenotrophic MBRs (Haugen *et al.*, 2002; Ho *et al.*, 2001). These are apparently attained through high H_2 gas mass transfer rates sustained by limiting fouling. Fouling is mainly manifested as thick and dense biofilms (Roggy *et al.*, 2002) and, possibly, scalants (Ergas and Reuss, 2001; Lee and Rittmann, 2002) at the membrane surface, though precipitation of mineral deposits appears not to adversely affect H_2 transfer in all studies (Lee and Rittmann, 2003; Roggy *et al.*, 2002).

The disadvantage of gas transfer MBRs is, as with the extractive processes, that the membrane is not used for filtration, but it also does not retain the biomass. As such, product water is prone to "sloughed" biomass and other organic matter in the same way as any fixed film biological process. Also, regulation of the gas flux through the membrane due to the partial pressure drop along the membrane fibre length (Ahmed and Semmens, 1992) can produce uneven biofilm growth. In addition, doubts about the safety of dosing water with hydrogen remain, notwithstanding the apparent near quantitative retention of hydrogen by the biomass. Finally, the poor adaptability of the autotrophic bacteria under drinking water denitrification conditions demonstrated in several studies by long acclimatisation periods of 40 and 70 days (Ergas and Reuss, 2001; Ho *et al.*, 2001) casts doubt on the suitability of this configuration for full-scale operation.

2.3.3.4 Biomass rejection MBR

In the conventional configuration the membrane is actually used to filter the water, and both nitrate and the electron donor enter the developed biofilm in the same direction (Fig. 2.24c). Both heterotrophic and autotrophic systems have been investigated using acetate (Barrieros *et al.*, 1998), ethanol (Chang *et al.*, 1993; Delanghe *et al.*, 1994; Urbain *et al.*, 1996) and elemental sulphur (Kimura *et al.*, 2002) as electron donors, though studies based on an iMBR specifically for drinking water duties appear to be limited to Kimura *et al.* (2002). These authors used rotating disc modules to "control" fouling, thus avoiding aeration and maintaining an anoxic environment, and sustained a flux of 21 LMH for 100 days with limited fouling. Fouling in denitrification MBRs has not been characterised, though reported data suggest the biomass has a higher fouling propensity than that generated from sewage treatment (Delanghe *et al.*, 1994; Urbain *et al.*, 1996). This can apparently to some extent be controlled by reducing the flux to below 10 LMH and operating at crossflows of 2 m/s in sMBRs (Urbain *et al.*, 1996). Nitrate removal efficiencies of up to 98.5% have generally been reported in these studies but, as with previous configurations reported, several investigators reported organic carbon (Delanghe *et al.*, 1994) and elevated assimilable organic carbon (AOC) concentrations (Kimura *et al.*, 2002) in the product water.

A full-scale 400 m^3/day (0.4 megalitres per day (MLD)) nitrate removal MBR process was constructed in Douchy, France, incorporating powdered activated carbon (PAC) dosing for pesticide removal. Stabilised fluxes between 60 and 70 $l\,m^{-2}h^{-1}$ were obtained at full scale and, contrary to previous investigations, having optimised C:N dosing, treated water of low organic carbon concentration as well as tri-halo methane formation potential (THMFP) was reported. The author hypothesised that the low effluent organic content was a consequence of effective membrane rejection of biomass byproducts of high-molecular-weight organic matter (Urbain *et al.*, 1996).

2.3.3.5 *Hybrid MBR systems*

A system ingeniously using electrolysis to generate hydrogen and feed a biofilm on a granular activated carbon (GAC) support, coupled with a downstream membrane to filter the water, has been trialled (Prosnansky *et al.*, 2002). This system provided treated water nitrate levels of 5–10 mg NO_3^--N/L once optimised by employing high-specific area GAC. The low nitrate removal rates were a consequence of influent DO concentration affecting hydrogen dissolution, difficulties with pH control, hydrodynamic limitations and the influence of the anode on nitrate migration on increasing electric field intensity. Furthermore, the process is somewhat limited in application due to an intensive energy requirement and, once again, formation of hydrogen bubbles which impose a safety risk.

More recent research (Mo *et al.*, 2005) has focused on incorporating both gas transfer and immersed pressure-driven membranes into the same reactor. The authors focused treatment on suspended biomass rather than biofilms to minimise mass transfer problems previously reported with biofilm development (Crespo *et al.*, 2004; Ergas and Reuss, 2001). Nitrate loading rates between 24 and 192 mg NO_3^--N/(L/day) were trialled, with all but the higher loadings resulting in 100% removal performance. However, average effluent dissolved organic carbon (DOC) concentrations of approximately 8 mg/L were also recorded, possibly due to the regular mechanical removal of biofilm from the membrane surface which would otherwise act to reject organic matter.

2.3.3.6 *Synopsis*

The use of MBRs for drinking water denitrification is very much at the research and development stage. Three different MBR configurations have been studied for this application and, as yet, only one full-scale plant has been installed. Challenges for all three configurations remain, specifically contamination of the treated water by organic carbon arising either from the electron donor in a heterotrophic system or from the biomass. There are also issues with nitrite (NO_2^-) arising in the effluent due to incomplete nitrification, and this becomes a serious limitation, particularly in European Union (EU) countries which are subject to more rigorous limits on nitrite levels than the US (0.03 vs. 1 mg/L as NO_2^--N). If the membrane is not used for direct filtration, as is the case for diffusive and extractive denitrification MBRs, further downstream processing is required for colour, taste and turbidity improvements and for disinfection. There are additionally health and safety concerns to address with the hydrogenotrophic systems. Notwithstanding the elegance of the diffusive and extractive systems, it is possible that the conventional biomass rejection configuration may hold the most promise since, as a barrier system, it also achieves disinfection. However, with only one full-scale plant in existence, there is clearly further development required in this area.

2.3.4 Elements of an immersed biomass-rejection MBR

MBRs employing immersed membranes to reject biomass represent the most widely employed of all MBR configurations, since they incur the lowest specific energy demand and therefore become the most economically viable for large-scale applications. There are essentially five key elements of the iMBR process which are key to its design and operation (Fig. 2.25). These are:

1. The membrane, its design and the sustaining of permeability,
2. Feedwater, its characteristics and its pretreatment,
3. Aeration of both membrane and the bulk biomass,
4. Sludge withdrawal and residence time,
5. Bioactivity and nature of the biomass.

These elements are obviously largely inter-related (Fig. 2.26), in particular the latter three which obviously relate to operation. The rate at which sludge is withdrawn controls the residence time (i.e. the SRT) which then determines the concentration of the biomass (or, strictly speaking, the mixed liquor). The MLSS concentration then impacts both on the biological properties, that is, the bioactivity and microbial speciation (Section 2.2.3), and also on the physical properties such as the viscosity and oxygen transfer (Section 2.2.5). The feedwater chemistry provides the biggest impact on MBR operation, in that the membrane fouling propensity of the mixed

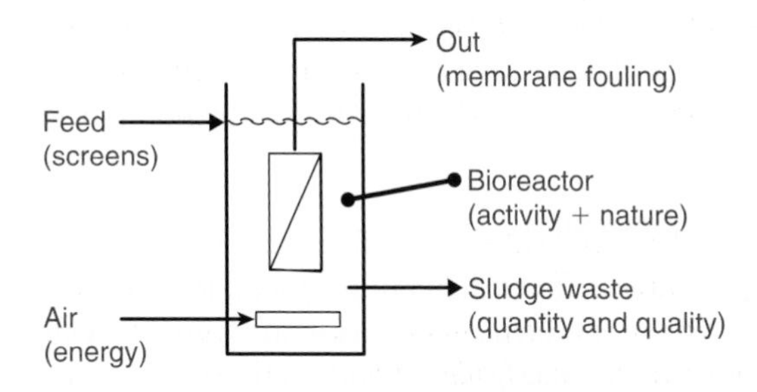

Figure 2.25 Elements of an MBR

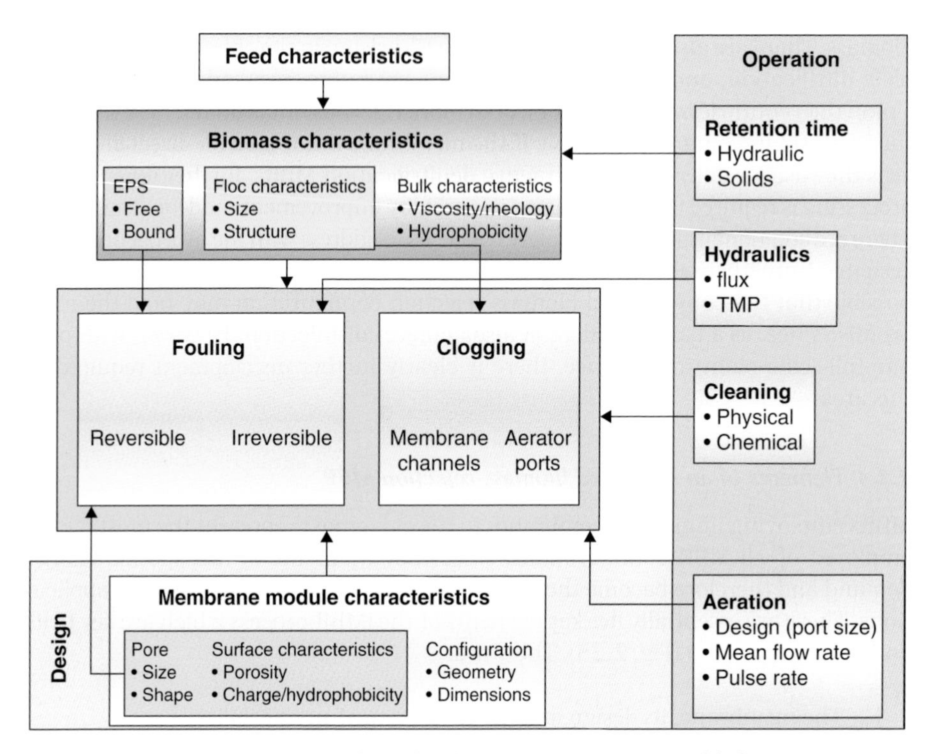

Figure 2.26 Inter-relationships between MBR parameters and fouling

liquor is generally mainly dictated by the nature of the feedwater from which it is generated. Similarly, the rigour of the pretreatment of the feedwater by screening has a significant impact on the clogging propensity.

Whilst governing principles and the nature of inter-relationships can be appreciated (Fig. 2.26), actual operating conditions and the associated absolute operating parameter values can generally only be arrived at heuristically. Having said this, an understanding of the fundamentals of MBR design, operation and maintenance can proceed through a comprehensive examination of the biological, chemical and physical phenomena occurring in MBRs, since these interact to generate fouling through a number of mechanisms. In the following sections, the elements of the iMBR are considered in turn, namely the membrane itself (Section 2.3.5), the feedwater and biomass characteristics (Section 2.3.6) and the operation and maintenance aspects (Section 2.3.7), with a view to appraising mechanisms of fouling (Section 2.3.8) and, ultimately, developing methods for its control (Section 2.3.9).

2.3.5 Membrane characteristics

Key membrane design parameters are configuration, that is, the geometry and flow direction (Section 2.1.3), the surface characteristics (normally denoted by the pore size and material but also including such things as the surface charge, hydrophobicity and porosity, pore tortuosity and shape, and crystallinity), and the inter-membrane

Table 2.6 Effect of pore size on MBR hydraulic performances

Membranes tested	Optimum	Test duration	Other	Reference
0.1, 0.22, 0.45 μm	0.22 μm	20 h	–	Zhang *et al.* (2006)
20, 30, 50, 70 kDa	70 kDa	110 min	Concentrated	He *et al.* (2005)
	50 kDa	110 days	feed, anaerobic	
70 kDa, 0.3 μm	70 kDa	8 h	–	Choi *et al.* (2005a)
30 kDa, 0.3 μm	30 kDa	2 h	CFV = 0.1 m/s	Choi *et al.* (2005b)
	0.3 μm		CFV = 3.5 m/s	
0.1, 0.2, 0.4, 0.8 μm	0.8 μm	n/a	–	Lee *et al.* (2005)
200 kDa, 0.1, 1 μm	1 μm	3 h	Flux-step test	Le-Clech *et al.* (2003c)
0.3, 1.5, 3, 5 μm	5 μm	25 min	–	Chang *et al.* (2001b)
	0.3 μm	45 days		
0.4, 5 μm	0.4 μm	1 day	–	Gander *et al.* (2000)
	No effect	From 50 days		
0.01, 0.2, 1 μm	No effect	A few hours	Flux-step test	Madaeni *et al.* (1999)
200 kDa, 0.1, 1 μm	0.1 μm	n/a	Anaerobic	Choo and Lee (1996a)
0.05, 0.4 μm	0.05 μm	n/a	–	Chang *et al.* (1994)

separation. The pore size of commercial MBR materials tends to be in the coarse UF to fine MF region (Section 4.7), since experience indicates that this pore size range offers sufficient rejection and reasonable fouling control under the conditions employed. The range of organic membrane materials employed is also, in practice, limited to those polymers which are:

(a) sufficiently mechanically and chemically robust to withstand the stresses imposed during the filtration and cleaning cycles,
(b) readily modified to provide a hydrophilic surface, which then makes them more resistant to fouling, particularly by EPS (Section 2.3.6.5),
(c) readily attached to a substrate to provide the mechanical integrity required, and
(d) manufactured at a relatively low cost.

Point (d) is especially important in the case of iMBRs, since these operate at relatively low fluxes and so demand much larger membrane areas than sMBRs.

2.3.5.1 Physical parameters

Pore size The effects of pore size on membrane fouling are strongly related to the feed solution characteristics and, in particular, the particle size distribution (Section 2.3.6.1). This has led to conflicting trends reported in the literature (Table 2.6), with no consistent general trend noted between pore size and hydraulic performance. This can, in part, be attributed to the complex and changing nature of the biological suspension in MBR systems and the comparatively large pore size distribution the membranes used (Chang *et al.*, 2002a; Le-Clech *et al.*, 2003c), along with operational facets such as the system hydrodynamics and the duration of the test. A direct comparison of MF and UF membranes at a CFV of 0.1 m/s has shown an MF membrane to provide a hydraulic resistance of around twice that of a UF membrane (Choi *et al.*, 2005b). Interestingly, the DOC rejection of both membranes was similar following 2 h of operation, indicating the dynamic membrane layer formed on the membranes to have provided the perm-selectivity rather than the membrane substrate itself.

Conventional wisdom considers smaller pores to afford greater protection of the membrane by rejecting a wider range of materials, with reference to their size, thus increasing cake (or fouling layer) resistance. Compared to that formed on membranes having larger pores, the layer is more readily removed and less likely to leave residual pore plugging or surface adsorption. It is the latter and related phenomena which cause irreversible and irrecoverable fouling. However, when testing membranes with pores ranging from 0.4 to 5 μm, Gander *et al.* (2000) conversely observed greater initial fouling for the larger pore-size membranes and significant flux decline when smaller pore-size membrane were used over an extended period of time, though these authors used isotropic membranes without surface hydrophilicisation.

Characterisation of the distribution of MW compounds present in the supernatant of MBRs operated with membranes of four pore sizes (ranging from 0.1 to 0.8 μm) has also been presented (Lee *et al.*, 2005). Although providing a lower fouling rate, the 0.8 μm pore-size MBR nonetheless had a slightly higher supernatant concentration of most of the macromolecules. According to these results, it seems unlikely for the small differences in MW distribution to cause significant variation in fouling rates observed between the four MBR systems. In another study based on short-term experiments, sub-critical fouling resistance and fouling rate increased linearly with membrane resistance ranging from 0.4 to $3.5 \times 10^{9}\,m^{-1}$, corresponding to membrane pore size from 1 down to 0.01 μm (Le-Clech *et al.*, 2003c). These results suggest that a dynamic layer is created of greater overall resistance for the more selective membranes operating under sub-critical conditions, and supports the notion that larger pores decrease deposition onto the membrane at the expense of internal adsorption. Long-term trials have revealed that progressive internal deposition eventually leads to catastrophic increase in resistance (Cho and Fane, 2002; Le-Clech *et al.*, 2003b; Ognier *et al.*, 2002a), as discussed in Section 2.1.4.6. Tests conducted using a very porous support for the formation of a dynamic membrane have yielded reasonable removal efficiencies and permeabilities (Wu *et al.*, 2004), and full-scale installations now exist based on this approach (Section 5.2.5).

Porosity/pore size distribution/roughness Membrane roughness and porosity were identified as possible causes of differing fouling behaviour observed when four MF membranes with nominal pore sizes between 0.20 and 0.22 μm were tested in parallel (Fang and Shi, 2005). The track-etched membrane, with its dense structure and small but uniform cylindrical pores, provided the lowest resistance due to its high surface isoporosity whereas the other three membranes were more prone to pore fouling due to their highly porous network. Although all membranes were of similar nominal pore size, the PVDF, mixed cellulose esters (MCE) and PES membranes resulted in relative pore resistance of 2, 11 and 86% of the total hydraulic resistance respectively. It was suggested that membrane microstructure, material and pore openings all affected MBR fouling significantly (Fang and Shi, 2005). Comparison between two microporous membranes prepared by stretching demonstrated fouling to be influenced by pore aspect ratio (i.e. pore surface length/pore surface width). Whilst both membranes had the same average pore size and pure water flux, reduced fouling was observed with the membrane having higher pore aspect ratio (i.e. more elliptical and less circular) (Kim *et al.*, 2004). Surface roughness has been implicated as promoting fouling from studies on an anaerobic MBR system (He *et al.*, 2005), although in this study both pore aspect ratio, i.e. more elliptical size and membrane morphology were changed simultaneously.

Membrane configuration As already discussed (Section 2.3.1) the immersed process configuration is generally favoured over the pumped sidestream configuration for medium to large-scale domestic wastewater treatment (Fane *et al.*, 2002; Gunder and Krauth, 1999; Judd, 2005; Le-Clech *et al.*, 2005b). This relates mainly to the impact of aeration, which suppresses fouling through generating shear (Section 2.3.6.4).

iMBR membranes are largely configured either as HF or FS whereas sMBRs are either FS or MT (Fig. 2.5). Whilst HF modules are generally less expensive to manufacture, allow high membrane density and tolerate vigorous backflushing, they are also less readily controlled hydrodynamically than FS or MT membranes where the membrane channel width is well defined. A discussion of the relative merits of FS and HF membranes was initiated by Gunter and Krauth (1998), who demonstrated the superior hydraulic performance (i.e. higher permeability) attainable from the FS membrane. On the other hand, in a comparative study of FS and HF membranes of the same pore size (0.4 μm) used for anaerobic treatment (Hai *et al.*, 2005), the authors found the FS membrane to foul slightly more than the HF membrane and the permeability was not recovered following cleaning with water.

An important parameter for HF systems is the packing density, discussed in Section 4.8 with reference to commercial membrane products. The separation distance between adjacent membranes has a direct impact on clogging, shear and aeration energy demand. For a given liquid upflow rate, as provided by airlift, increasing the separation reduces the risk of clogging by gross solids. Reducing the separation will also, for a given bubble volume, retard the rising bubble as a consequence of the gas:membrane contact area increasing, thereby increasing the downward drag force. This might then be expected to decrease the flux because shear forces are reduced as a result. On the other hand, for a given liquid upflow velocity and thus the same shear, increasing channel width also increases the aeration energy demand since a larger volume of liquid (dictated by the channel width) is being passed over the same membrane area; it is the volumetric flow rate that determines energy demand.

Experiments conducted on a bundle of nine fibres revealed the module performance to be significantly worse than that of module based on an individual fibre (Yeo and Fane, 2004), with the surrounding fibres being less productive. At high feed concentrations and low CFVs, the surrounding fibres became completely blocked and eventually produced negligible flux. A lowering of the packing density by 30% was advised to allow the bundle to perform similarly to individual fibres, since this reduced the impact of cake layers forming on adjacent fibres and allowed greater shear to be generated (Yeo and Fane, 2004). A mathematical model based on substrate and biomass mass balance also revealed the significant role played by packing density in overall MBR performance and the maintenance of MLSS in particular (Vigneswaran *et al.*, 2004).

The effects of other membrane characteristics, including HF orientation, size and flexibility, have been reviewed (Cui *et al.*, 2003). For HF membranes used for yeast filtration, higher critical fluxes were measured for membranes of smaller diameter (0.65 mm) and greater length (80 cm) (Wicaksana *et al.*, 2006), though contradictory results showing slightly higher permeabilities for shorter membranes (0.3 *cf.* 1.0 m) have also been reported (Kim and DiGiano, 2004). A significant impact of membrane length is the pressure drop due to lumen-side permeate flow. Ideally, the resistance to permeate flow from the membrane to the outlet of the element should be small compared with that offered by the membrane itself. If this is not the case, then the hydraulic losses across the permeate side of the membrane element may be sufficient to produce a significant flux distribution across the membrane surface. Significant pressure losses (up to 530 mbar) have been measured for fibres longer than 15 cm; below this critical length, pressure loss was reported as being at less than 110 mbar (Kim *et al.*, 2004). Further discussion of fouling distribution in HFs can be found elsewhere (Chang and Fane, 2002; Chang *et al.*, 2002b; Lipnizki and Field, 2001; Ognier *et al.*, 2004; Zheng *et al.*, 2003; Zhongwei *et al.*, 2003).

Anaerobic MBRs Research into membrane fouling of anaerobic membrane bioreactors (anMBRs) has been limited. Whilst fouling mechanisms may be similar to aerobic MBRs, the nature of the foulants can be expected to be different and, as with the aerobic systems, to change with feedwater characteristics, membrane surface and membrane module properties and process operating conditions. A recent review by Bérubé *et al.* (2006) has been conducted, and aspects of this review are summarised below and in Table 2.7.

Cake layers on membranes in anMBRs have been reported to contain both organic matter and precipitated struvite (Choo and Lee, 1996a), though fouling of organic membranes appears to be governed by biological/organic interactions with the membrane rather than by struvite formation. Choo *et al.* (2000) observed no difference in fouling rate when ammonia, a component of struvite, was removed from the mixed liquor prior to filtration using an organic membrane. It also appears that internal

Table 2.7 Anaerobic MBR studies: optimal membrane pore size

Mixed liqour[1]/ feed[2]	Mode	Mean particle dia. (μm)	Reactor volume (L)	Material type	Area (m^2)	Pore size (μm) Tested	Optimal	Reference
Anaerobic digestion[a,1]	Batch	3.28–23.3	0.18	PVDF	0.09	0.015–1	0.1	Choo and Lee (1996b)
Sludge digester[b,1], Synthetic[c,2]	Continuous sMBR	13	10	ZrO_2–TiO_2	N/a	0.05–0.2	0.14	Elmaleh and Abdelmoumni (1998)
Synthetic[2]	Continuous iMBR.	N/a	N/a	N/a	N/a	0.45 & 0.6	0.6	Chung *et al.* (1998)

[a]Broth was fractionated into constituent parts (dissolved, fine colloids, cell suspension).
[b]Supernatant.
[c]Acetic acid/nutrients in tap water.

fouling (i.e. membrane pore plugging) by soluble and colloidal material is less significant than cake layer fouling (Choo and Lee, 1996a; Kang *et al.*, 2002; Lee *et al.*, 2001c) on organic membranes. However, for ceramic membranes, which provide higher fluxes and for which cake layers are much thinner (especially at high crossflows), the bulk of the fouling has been attributed to internal fouling by struvite. This has been concluded from SEM studies coupled with a magnesium mass balance (Yoon *et al.*, 1999), and from observed impacts of ammonia level (Choo *et al.*, 2000). As with aerobic processes, hydrophobicity suppresses fouling on anMBR polymeric membranes to some extent, and membrane charge may also be important in determining fouling (Kang *et al.*, 2002) unless the ionic strength is high enough to compress the double layer and thus nullify charge repulsion (Fane *et al.*, 1983).

Membrane pore size effects also follow similar patterns to those of aerobic systems in that large pores provide greater initial fluxes but more rapid subsequent flux decay (Choo and Lee, 1996b; He *et al.*, 1999, 2005; Imasaka *et al.*, 1989; Saw *et al.*, 1986), which has been attributed to either internal or surface pore plugging. However, the optimum pore size appears to depend on the liquor characteristics. Elmaleh and Abdelmoumni (1997), investigating the impact of pore size on the anMBR steady-state permeate flux, recorded highest steady-state fluxes at a pore size of ~0.45 μm for an anaerobic mixed liquor, compared with 0.15 μm for a mixed microbial population of methanogens.

The long-term impact of UF membrane pore size on hydraulic performance has been assessed by He and co-workers for an anaerobic MBR (He *et al.*, 2005). The lowest MWCO-rated membrane tested (20 kDa) yielded the largest permeability loss within the first 15 min of filtration when compared to 30, 50 and 70 kDa membranes. However, when operated for an extended time (over 100 days) with regular hydraulic and chemical cleaning, the largest MWCO membrane (70 kDa) experienced the greatest fouling rate, as 94% of its original permeability was lost, compared to only a 70% performance decrease for the other three membranes. The 30 and 50 kDa membranes thus provided the best overall hydraulic performance, possibly indicating an optimum membrane pore size for a given application. These results also reveal the impact of test duration. Similar temporal trends where long-term fouling was exacerbated by larger pores has been demonstrated for microfiltration membranes ranging from 1.5 to 5 μm pore size for aerobic MBRs (Chang *et al.*, 2001b).

Although both sidestream and immersed process configurations have been studied for anMBR applications, as with aerobic systems sidestream anMBRs have the longest history. They generally operate at CFVs and TMPs of 1–5 m/s and 2–7 bar, respectively to provide reasonable fluxes. Much lower pressures (0.2–1 bar) arise in immersed systems, though these are still higher than in corresponding aerobic iMBRs. CFVs in immersed systems have been reported as being less than 0.6 m/s (Bérubé and Lei, 2004). Stuckey and Hu (2003) reported that slightly higher permeate fluxes could be maintained for an HF compared with an FS membrane element in an immersed anMBR.

2.3.5.2 Chemical parameters

Since hydrophobic interactions take place between solutes, microbial cells of the EPS and the membrane material, membrane fouling is expected to be more severe with hydrophobic rather than hydrophilic membranes (Chang *et al.*, 1999; Madaeni *et al.*, 1999; Yu *et al.*, 2005a; Yu *et al.*, 2005b). In the literature, changes in membrane hydrophobicity are often linked with other membrane modifications such as pore size and morphology, which make the correlation between membrane hydrophobicity and fouling more difficult to assess. In a recent anMBR study, for example, the contact angle measurement demonstrated that the apparent hydrophobicity of PES membranes decreased (from 55 to 47°) with increasing MWCO (from 20 to 70 kDa membranes, respectively) (He *et al.*, 2005). The effect of membrane hydrophobicity in an aerobic MBR, from a comparison of two UF membranes of otherwise similar characteristics, revealed greater solute rejection and fouling and higher cake resistance for the hydrophobic membrane (Chang *et al.*, 2001a). It was concluded that the solute rejection was mainly due to the adsorption onto or sieving by the cake deposited on the membrane, and, to a lesser extent, direct adsorption into membrane pores and at the membrane surface. It has also been suggested (Fang and Shi, 2005) that membranes of greater hydrophilicity are more vulnerable to deposition of foulants of hydrophilic nature, though in this study the most hydrophilic membrane was also the most porous and this can also enhance fouling (Section 2.3.5.1).

Although providing superior chemical, thermal and hydraulic resistance, the use of ceramic membranes in MBR technologies is limited by their high cost to niche applications such as treatment of high-strength industrial waste (Luonsi *et al.*, 2002; Scott *et al.*, 1998) and anaerobic biodegradation (Fan *et al.*, 1996) in sMBRs. A direct comparison of a 0.1 μm ceramic and 0.03 μm polymeric multi-channel membrane modules operated in sidestream air-lift mode showed the former to operate without fouling up to at least 60 LMH, the highest flux tested, whereas for the latter criticality was indicated at ~36 LMH (Judd *et al.*, 2004). Novel stainless steel membrane modules have recently been shown to provide good hydraulic performance and fouling recovery when used in an anaerobic MBR (Zhang *et al.*, 2005).

Since fouling is expected to be more severe at higher hydrophobicities, efforts have naturally been focused on increasing membrane hydrophilicity by chemical surface modification. Recent examples of MBR membrane modification include NH_3 and CO_2 plasma treatment of PP HFs (Yu *et al.*, 2005a; Yu *et al.*, 2005b) to functionalise the surface with polar groups. In both cases, membrane hydrophilicity significantly increased and the new membranes yielded better filtration performance and flux recovery than those of unmodified membranes. In another study, addition of TiO_2 nanoparticles to the casting solution and direct pre-filtration of TiO_2 allowed the preparation of two types of TiO_2-immobilised UF membrane, respectively comprising entrapped and deposited particles, which were used in MBR systems (Bae and Tak, 2005). A lower flux decline was reported for the TiO_2-containing membranes compared to the unmodified materials, the surface-coated material providing the greatest fouling mitigation. When MBR membranes were precoated with ferric hydroxide flocs and compared to an unmodified MBR, both effluent quality and productivity were found to increase (Zhang *et al.*, 2004).

Whilst many of the scientific studies of MBR membrane surface characterisation and/or modification relate to fouling by EPS, it appears that in practice both the choice of membrane material and the nominal membrane pore size are limited. Commercially-available membranes and MBR systems are reviewed in Chapter 4 and their characteristics summarised in Annex 3 and Table 4.5.

2.3.6 *Feed and biomass characteristics*

2.3.6.1 Feed nature and concentration

Whilst membrane fouling in physical wastewater filtration depends directly on the water quality (Fuchs *et al.*, 2005; Judd and Jefferson, 2003; Schrader *et al.*, 2005), MBR membrane fouling is mostly affected by the interactions between the membrane and biological suspension rather than feed water (Choi *et al.*, 2005a). More recalcitrant feedwaters, such as landfill leachate (Section 5.4.3), may undergo more limited biochemical transformation such that the membrane is challenged in part by the raw, unmodified feed. Biological transformations which take place which are influenced both by the operating conditions and the feedwater quality (Jefferson *et al.*, 2004; Le-Clech *et al.*, 2003b).

2.3.6.2 Biomass foulants

Two types of foulant study dominate the MBR scientific literature: characterisation and identification. Characterisation refers to properties (usually relating to membrane permeability) the foulant demonstrates either *in situ*, that is, within the MBR, or *ex situ* in some bespoke or standard measurement, such as capillary suction time (CST) or specific resistance to filtration (SRF). Identification refers to physical and/or chemical classification of the foulant, invariably through extraction and isolation prior to chemical analysis. Of course, foulant isolates may also be characterised in the same way as the MBR biomass.

In general, foulants can be defined in three different ways (Table 2.8):

1. practically, based on permeability recovery,
2. mechanistically, based on fouling mechanism,
3. by material type, based on chemical or physical nature or on origin.

Table 2.8 Foulant definitions

Practical	Mechanism	Foulant material type
Reversible/temporary: • Removed by physical cleaning Irreversible/permanent: • Removed by chemical cleaning Irrecoverable/absolute[a]: • Not removed by any cleaning regime	Pore blocking/filtration models (Fig. 2.9): • Complete blocking • Standard blocking • Intermediate blocking • Cake filtration	Size: • Molecular, macro-molecular, colloidal or particulate Surface charge/chemistry: • Positive or negative (cationic or anionic) Chemical type: • Inorganic (e.g. scalants) or organic (e.g. humic materials, EPS) • Carbohydrate or protein (fractions of EPS) Origin: • Microbial (autochthonous), terrestrial (allochthonous) or man-made (anthropogenic) • (Extracted) EPS ((e)EPS) or soluble microbial product (SMP)[b]

[a]Irrecoverable fouling is long-term and insidious.
[b]eEPS refers to microbial products directly associated with the cell wall; SMP refers to microbial products unassociated with the cell (Fig. 2.29).

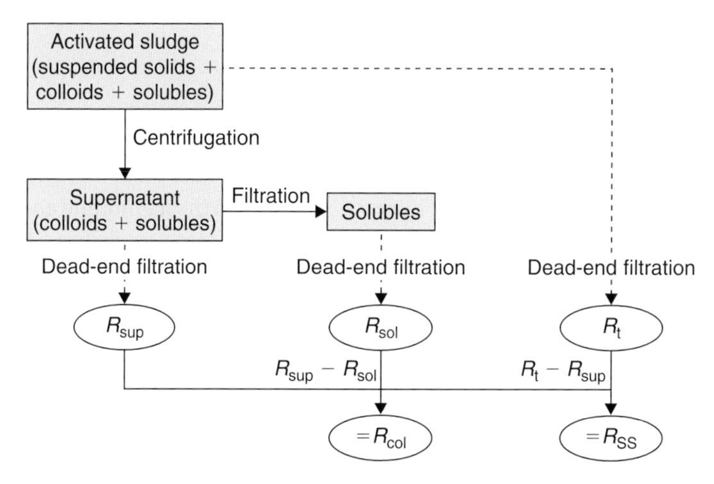

Figure 2.27 Experimental method for the determination of the relative fouling propensity for the three physical biomass fractions

Of these, evidence suggests that it is the physical nature, and specifically the size, of the foulant that has the greatest impact on its fouling propensity. Hence, activated sludge biomass can be fractionated into three categories: suspended solids, colloids and solutes. The fractionation methodology critically affects the measurements made. Typically, the biomass sample is centrifuged. The resulting supernatant is then filtered with a dead-end membrane cell, with the calculated hydraulic resistance being attributed to colloidal and soluble matter combined (R_{col} and R_{sol}, respectively, Fig. 2.27). Another portion of the biomass suspension is then microfiltered at a nominal pore size of ~0.5 μm and the fouling properties of this supernatant (R_{sol}) attributed solely to the soluble matter. The relative fouling contributions of the suspended and colloidal matter can then be calculated (Bae and Tak, 2005). The resistance provided by colloidal matter has also been attributed to the difference between the levels of TOC present in the filtrate passing through 1.5 μm filtration paper and in the permeate collected from the MBR membrane (0.04 μm) (Fan *et al.*, 2006).

Fractionation methods may vary slightly for different studies, but results are often reported in terms of hydraulic resistances for suspended solids, colloids and soluble matter, the sum of which yields the resistance of the activated sludge. Although an interesting approach to the study of MBR fouling, fractionation neglects coupling or synergistic effects which may occur among different biomass components, and also with operating determinants. Such interactions are numerous and include feedwater quality (Li *et al.*, 2005b), membrane permeability, particle size and hydrodynamics conditions (Bae and Tak, 2005) (Fig. 2.26). An attempt to compare results obtained from different studies is depicted in Fig. 2.28, where relative fouling resistance contributions have been calculated.

The relative contribution of the biomass supernatant to overall fouling ranges from 17% (Bae and Tak, 2005) to 81% (Itonaga *et al.*, 2004). Such variation is probably attributable to the different operating conditions and biological state of the suspended biomass. It appears from these data that fouling by suspended solids is rather

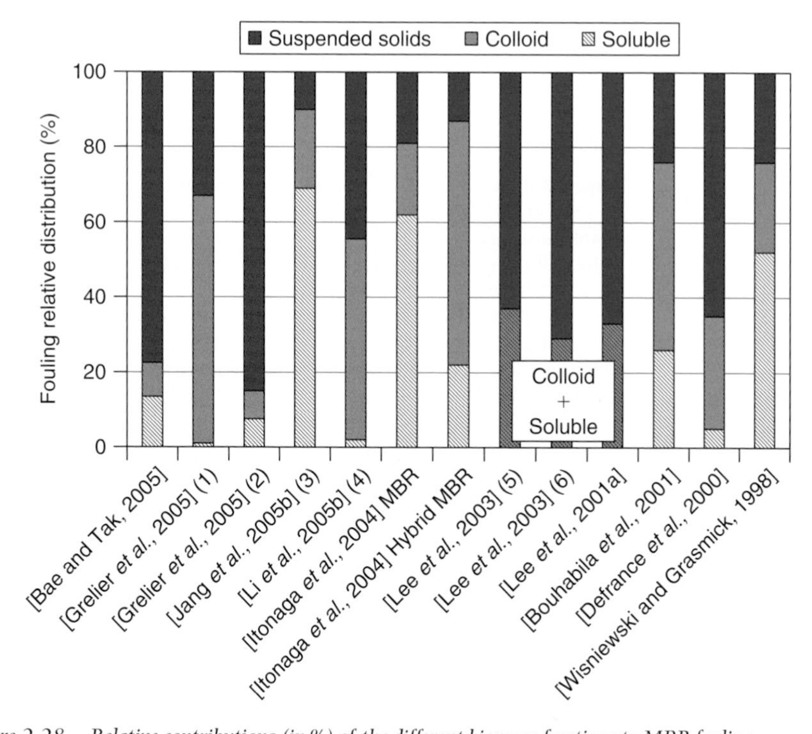

Figure 2.28 *Relative contributions (in %) of the different biomass fractions to MBR fouling*

(1–2): For SRT increase from 8(1) to 40 days (2)

(3): F:M ratio of 0.5, results based on modified fouling index

(4): Based on flux reduction after 600 min of each fraction filtration

(5–6): For SRT increase from 20 (5) to 60 days (6).

less than that of the supernatant. The latter is generally regarded as comprising soluble microbial product (SMP), which is soluble and colloidal matter that derives from the biomass (Section 2.3.6.5). With respect to fouling mechanisms, soluble and colloidal materials are assumed to be responsible for membrane pore blockage, whilst suspended solids account mainly for the cake layer resistance (Itonaga *et al.*, 2004). However, since iMBRs are typically operated at a modest flux, the cake tends not to form and deposition of physically smaller species is more likely to take place.

2.3.6.3 Biomass (bulk) parameters

MLSS concentration Whilst suspended solids concentration may seem intuitively to provide a reasonable indication of fouling propensity, the relationship between MLSS level and fouling propensity is rather complex. If other biomass characteristics are ignored, the impact of increasing MLSS on membrane permeability can be either negative (Chang and Kim, 2005; Cicek *et al.*, 1999), positive (Defrance and Jaffrin, 1999; Le-Clech *et al.*, 2003c), or insignificant (Hong *et al.*, 2002; Lesjean *et al.*, 2005), as indicated in Table 2.9. The existence of a threshold above which the MLSS

Table 2.9 Influence of shift in MLSS concentration (g/L) on MBR fouling

MLSS shift (g/L)	Details	Reference
Fouling increase		
0.09–3.7	Cake resistance: 21–54 10^{11}/m and α: 18.5–0.7 10^8 m/kg	Chang and Kim (2005)
2.4–9.6	Total resistance: 9–22 10^{11}/m	Fang and Shi (2005)
7–18	Critical flux: 47–36 LMH (for SRT of 30–100 days)	Han *et al.* (2005)
2.1–9.6	Critical flux: 13–8 LMH	Bin *et al.* (2004)
1–10	Critical flux: 75–35 LMH	Madaeni *et al.* (1999)*
2–15	"Limiting flux": 105–50 LMH	Cicek *et al.* (1998)*
1.6–22	"Stabilized flux": 65–25 LMH	Beaubien *et al.* (1996)*
Fouling decrease		
3.5–10	Critical flux: >80, <60 LMH	Defrance and Jaffrin (1999)*
No (or little) effect		
4.4–11.6	No impact between 4 and 8 g/L, Slightly less fouling for 12 g/L	Le-Clech *et al.* (2003c)
4–15.1	Critical flux decreased from 25 to 22 LMH	Bouhabila *et al.* (1998)
3.6–8.4		Hong *et al.* (2002)

*sMBR

concentration has a negative influence has been reported (30 g/L, according to Lubbecke *et al.*, 1995). A more detailed fouling trend has been described (Rosenberger *et al.*, 2005), in which an increase in MLSS reduced fouling at low MLSS levels (<6 g/L) whilst exacerbating fouling at MLSS concentrations above 15 g/L. The level of MLSS did not appear to have a significant effect on membrane fouling between 8 and 12 g/L. In another study of MLSS concentration impacts (WERF, 2005), it was concluded that hydrodynamics (more than MLSS concentration) control the critical flux at MLSS levels above 5 g/L.

Contradictory trends from data obtained in the same study are apparent. For example, the cake resistance (R_c) has been observed to increase and the specific cake resistance (ψ, the resistance per unit cake depth) to decrease with increasing MLSS, indicating that the bulk cake becomes more permeable. Bin *et al.* (2004) observed the permeate flux to decrease (albeit at a reduced fouling rate) with increasing MLSS. This was attributed to the rapid formation of a fouling cake layer (potentially protecting the membrane) at high concentration, while progressive pore blocking created by colloids and particles was thought to take place at lower MLSS concentrations when the membrane was less well protected. This may well explain the subcritical fouling behaviour depicted in Fig. 2.14.

Empirical relationships predicting flux from MLSS level have been proposed in a number of papers (Fang and Shi, 2005; Krauth and Staab, 1993; Sato and Ishii, 1991; Shimizu *et al.*, 1996). However, these equations have limited use as they are generally obtained under very specific conditions and are based on a limited number of operating parameters, whilst other parameters are disregarded. A mathematical expression linking MLSS concentration, EPS and TMP with specific cake resistance has been proposed by Cho and co-workers (Cho *et al.*, 2005a). In this study, specific resistance changed little at MLSS levels between 4 and 10 g/L at constant EPS and TMP. The experimental method used for adjusting MLSS concentration can significantly impact on biomass characteristics, since biomass solids levels can be raised by sedimentation coupled with decantation without acclimatization, yielding different biological characteristics (Cicek *et al.*, 1999). MLSS concentration also impacts on removal; an optimum concentration of 6 g/L has been identified based on the removal of COD (Ren *et al.*, 2005) and phage (Wong *et al.*, 2004).

The lack of a clear correlation between MLSS concentration and any specific foulant characteristic(s) indicates that MLSS concentration alone is a poor indicator of biomass fouling propensity. Some authors (Brookes *et al.*, 2003b; Jefferson *et al.*, 2004) have recommended using fundamental operating parameters such as HRT and SRT as foulant level indicators, supported by relatively stable foulant levels and characteristics under steady-state conditions. Current studies tend to point to non-settlable/colloidal organic substances, rather than MLSS concentration, as being primary indicators of fouling propensity in MBRs (Section 2.3.6.5).

Viscosity As with a conventional ASP, biomass viscosity closely relates to its concentration (Section 2.2.5.3) and contributes to fouling (Yeom *et al.*, 2004). Whilst viscosity has been reported to increase roughly exponentially with MLSS concentration (Manem and Sanderson, 1996; Rosenberger *et al.*, 1999, Fig. 2.19), a critical MLSS concentration exists below which viscosity remains low and rises only slowly with the concentration and above which the exponential relationship is observed (Itonaga *et al.*, 2004). This critical value, generally between 10 and 17 g/L, depending on feedwater quality and process operating conditions, also exists for CST: viscosity correlations (Brookes *et al.*, 2003a). MLSS viscosity impacts both on flux and air bubble size and can dampen lateral movement of HFs in immersed bundles (Wicaksana *et al.*, 2006).

Temperature Temperature impacts on membrane filtration through permeate viscosity (Mulder, 2000). A similar temperature correction can be applied as that used for $k_{L}a$ (Equation (2.15), Section 2.2.5.3) (Rautenbach and Albrecht, 1989):

$$J = J_{20} 1.025^{(T-20)} \tag{2.30}$$

where J is the flux at the process temperature T in °C. This correction is not comprehensive, however, as has been demonstrated through normalisation of flux data obtained at different temperatures (Jiang *et al.*, 2005). The greater than expected normalized resistance at lower temperatures has been explained by a number of contributing phenomena:

1. impacts of the viscosity of the sludge, rather than that of the permeate, which increases more significantly than permeate viscosity and reduces the shear stress generated by coarse bubbles as a result;

2. intensified deflocculation at low temperatures, reducing floc size (Section 2.3.6.4) and releasing EPS to the solution;
3. particle back transport velocity, which decreases linearly with temperature according to Brownian diffusion;
4. biodegradation of COD, which decreases with temperature and results in a higher concentration of unbiodegraded solute and particulate COD (Fawehinmi *et al.*, 2004; Jiang *et al.*, 2005).

All of these factors directly impact on membrane fouling and, as such, more extensive deposition of foulant materials on the membrane surface is to be expected at lower temperatures.

Higher operating temperatures can be maintained in anMBRs than in aerobic ones (Baek and Pagilla, 2003), and this both decreases viscosity and apparently lowers levels of SMP (Fawehinmi *et al.*, 2004; Schiener *et al.*, 1998). Higher temperatures, at least within the mesophilic temperature range, also result in increased COD removal, as demonstrated with a conventional UASB (Singh and Virarghavan, 2003).

Dissolved oxygen The bioreactor DO concentration is controlled by the aeration rate (Section 2.2.5), which provides oxygen to the biomass, and is also used for membrane fouling control. DO impacts on MBR fouling through system biology, e.g. biofilm structure, SMP levels and floc size distribution (Lee *et al.*, 2005). Higher DO levels generally provide better filterability, as manifested in filter cakes of lower specific resistance due to larger particles (Kang *et al.*, 2003). Against this, some researchers have noted a decrease with DO concentration in mixed liquor dissolved COD levels (Ji and Zhou, 2006). Moreover, the contribution of SMP concentration to membrane filterability was found to be less than that of particle size and porosity (Kang *et al.*, 2003). Ji and Zhou claimed aeration rate to directly control the quantity and composition of SMP, EPS and total polymeric substances in the biological flocs and ultimately the ratio of protein/carbohydrate deposited on the membrane surface. In a study of anoxic and aerobic sludge filterability, distinct carbohydrate structures were observed and used to explain the different fouling rates obtained for the two systems. The effect of oxygen limitation causing a lowering of the cell surface hydrophobicity has been reported as another potential cause of MBR fouling (Jang *et al.*, 2005a). Suspended air, on the other hand, does not appear to contribute significantly to fouling (Jang *et al.*, 2004).

As the thickness of the biological fouling layer increases with extended MBR filtration, some biofilm regions have been observed to become anaerobic (Zhang *et al.*, 2006), therefore impacting on membrane fouling differently to a wholly aerobic film. Endogenous decay, similar to that expected within the fouling layer, was simulated and revealed the foulant levels and, specifically, the carbonate fraction of the EPS (Section 2.3.6.5) to significantly increase. Since the transition between aerobic to anaerobic conditions appears to produce a large amount of EPS, this phenomenon could also contribute to MBR fouling.

Foaming Foaming in activated sludge plants is caused by high SRTs, warm temperatures, low F:M ratios and high MLSS levels, as well as oil and grease and/or surfactants in the influent. Abundance of actinomycetes such as *Nocardia* or *Microthrix* are commonly related to foaming in activated sludge plants, and have been identified in a full-scale MBR plant subject to variable OLRs (Smith, 2006). However, foam in MBR plants has been observed in the absence of actinomycetes. The degree of foaming is reported as being related to the protein EPS concentrations (Nakajima and Mishima, 2005). Foaming sludges also appear to yield lower membrane permeabilities (Chang and Lee, 1998), attributed to the higher hydrophobicity of foaming activated sludge (Section 2.3.6.4). Foaming thus provides an indication of sludge fouling propensity.

2.3.6.4 Floc characteristics

Floc size Comparison of the aggregate size distribution of ASP and MBR sludges (Cabassud *et al.*, 2004) revealed a distinct difference in terms of mean particle sizes (160 and 240 μm, respectively). A bimodal distribution was observed for the MBR sludge (5–20 and 240 μm), the high concentration of small colloids, particles and free bacteria being caused by their complete retention by the membrane. In another study where partial characterisation of the MBR flocs up to 100 μm was carried out, floc sizes ranging from 10 to 40 μm were reported with a mean size of 25 μm (Bae and Tak, 2005). Floc size distributions reported for three MBRs operated at different SRTs were similar, although the mean floc size increased slightly from 5.2 to 6.6 μm for SRTs increasing from 20 to 60 days (Lee *et al.*, 2003).

Given the large size of the flocculant solids compared to the membrane pore size, pore plugging by the flocs themselves is not possible. Flocs are also to some extent impeded, by drag forces and shear-induced diffusion, from depositing on the membrane surface. They nonetheless contribute to fouling through production of EPS and also directly affect clogging of the membrane channels. The interaction between EPS levels and floc size is discussed in Section 2.3.6.5, and the use of ancillary materials to suppress fouling through floc size and structure discussed in Section 2.3.9.5.

Hydrophobicity and surface charge A number of reports can be found in the literature providing evidence of membrane fouling by highly hydrophobic flocs. Relative floc hydrophobicity can be directly measured by bacterial adhesion/partition using hydrocarbons such as hexane (Jang *et al.*, 2005b), or estimated by contact angle determination (Yu *et al.*, 2005b). Although the direct effect of floc hydrophobicity on MBR fouling is difficult to assess, hydrophobicity measurement of sludge and EPS solutions has been carried out and briefly reported by Jang and co-workers (Jang *et al.*, 2005a; Jang *et al.*, 2005b). EPS level and the filamentous index (a parameter related to the relative presence of filamentous bacteria in sludge) directly influence biomass floc hydrophobicity and zeta potential. Excess growth of filamentous bacteria has been reported to yield higher EPS levels, lower zeta potentials, more irregular floc shape and higher hydrophobicity (Meng *et al.*, 2006). Sludge of higher foaming propensity, attributed to its hydrophobic nature, has been shown to produce a flux decline 100 times greater than that of non-foaming sludge (Chang and Lee, 1998). The anionic nature of the functional groups of natural organic materials means that charge and zeta potential of activated sludge flocs (and EPS) tend to be in the −0.2 to −0.7 meq/g VSS and from −20 to −30 mV regions, respectively (Lee *et al.*, 2003;

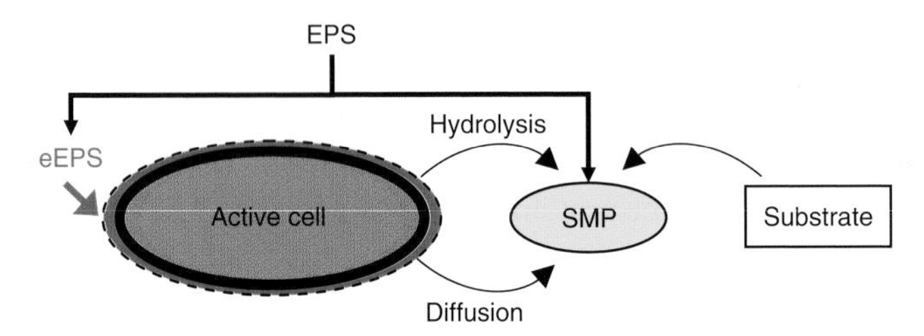

Figure 2.29 Simplified representation of EPS and SMP

Liu and Fang, 2003). Increasing SRT has been shown to produce an increase in both contact angle and surface charge, with an apparently strong correlation with fouling propensity (Lee *et al.*, 2003).

2.3.6.5 Extracellular polymeric substances

Extracted EPS (eEPS) and SMP As already stated, membrane fouling in MBRs has been largely attributed to EPS (Chang and Lee, 1998; Cho and Fane, 2002; Nagaoka *et al.*, 1996; Nagaoka *et al.*, 1998; Rosenberger and Kraume, 2002), the construction materials for microbial aggregates such as biofilms, flocs and activated sludge liquors. The term would be bound EPS, if in situ, or extracted EPS ("eEPS") if removed (Fig. 2.29–30) "EPS" is used as a general term which encompasses all classes of autochthonous macromolecules such as carbohydrates, proteins, nucleic acids, (phosphor)lipids and other polymeric compounds found at or outside the cell surface and in the intercellular space of microbial aggregates (Flemming and Wingender, 2001). They consist of insoluble materials (sheaths, capsular polymers, condensed gel, loosely bound polymers and attached organic material) secreted by the cell, shed from the cell surface or generated by cell lysis (Jang *et al.*, 2005a). Functions of the EPS matrix include aggregation of bacterial cells in flocs and biofilms, formation of a protective barrier around the bacteria, retention of water and adhesion to surfaces (Laspidou and Rittmann, 2002). With its heterogeneous and changing nature, EPS can form a highly hydrated gel matrix in which microbial cells are embedded (Nielson and Jahn, 1999) and can thus help create a significant barrier to permeate flow in membrane processes. Finally, bioflocs attached to the membrane can provide a major nutrient source during biofilm formation on the membrane surface (Flemming *et al.*, 1997). Their effects on MBR filtration have been reported for more than a decade (Ishiguro *et al.*, 1994) and have received considerable attention in recent years (Chang *et al.*, 2002a).

Analysis of EPS relies on its extraction from the sludge flocs (Fig. 2.29). So far, no standard method of extraction exists, making comparison of reported data generated from different extraction methods difficult. The latter include cation exchange resin (Frolund *et al.*, 1996; Gorner *et al.*, 2003; Jang *et al.*, 2005a), heating (Morgan *et al.*, 1990) and organic solvent (Zhang *et al.*, 1999). The relative efficacies of these techniques, along with a number of others, have been compared (Liu and Fang, 2003); results suggest formaldehyde extraction to be the most effective in extracting EPS. However, because of its simplicity, the heating method is sometimes preferred

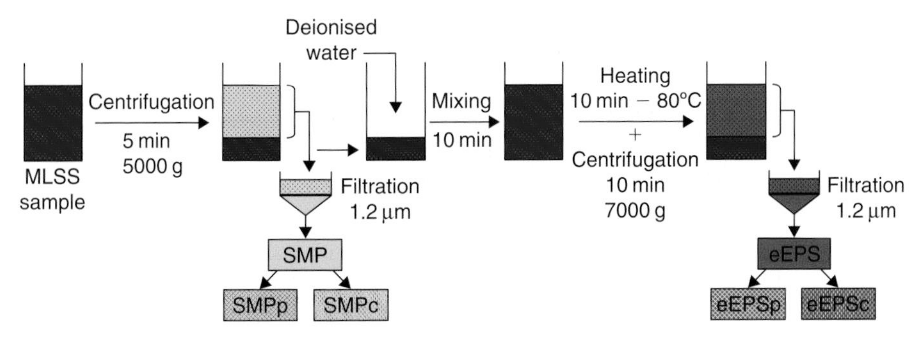

Figure 2.30 Candidate method for EPS and SMP extractions and measurements

(Fig. 2.30). Regardless of the extraction method used, a distinction can be made between EPS which derives directly from the active cell wall and that which is not associated with the cell but is solublised in the mixed liquor. The former is usually referred to as "EPS" in the literature, although a less ambiguous term would be "eEPS" (*extracted* EPS, Fig. 2.29). The latter is normally termed SMP and invariably refers to clarified biomass, although for some more recalcitrant feedwaters, clarified biomass will inevitably contain feedwater constituents which remain untransformed by the biotreatment process. SMP concerns soluble cellular components released during cell lysis, which then diffuse through the cell membrane and are lost during synthesis or are excreted for some purpose (Laspidou and Rittmann, 2002; Li *et al.*, 2005a). In MBR systems, they can also be provided from the feed substrate. It is now widely accepted that the concepts of soluble EPS and SMP are identical (Jang *et al.*, 2005a; Laspidou and Rittmann, 2002; Rosenberger *et al.*, 2005).

Protein and carbohydrate EPS fractions Typically, the EPS solution is characterised according to its relative content of protein (EPSp) and carbohydrate (EPSc), measured by the respective photometric methods of Lowry (Lowry *et al.*, 1951) and Dubois (Dubois *et al.*, 1956). Reported data are summarised in Table 2.10. While EPSp generally has hydrophobic tendencies, EPSc is more hydrophilic (Liu and Fang, 2003) and may therefore interact more strongly with the membrane. The EPS solution can also be characterized in terms of its TOC level (Cho *et al.*, 2005b; Nagaoka and Nemoto, 2005) and, less frequently, its hydrophobicity by measurement of the ultraviolet absorbance per unit TOC concentration, the specific UV absorbance (SUVA), (Ahn *et al.*, 2005). In many reported cases, EPSp (with a maximum concentration of 120 mg/gSS) is greater than EPSc (maximum concentration of 40 mg/gSS) and the total concentration range reported is surprisingly narrow: 11–120 mg/L for EPSp and 7–40 mg/L for EPSc. Sludge flocs have also been characterised in terms of protein and carbohydrate levels, through colorimetric analysis carried out directly on the washed biomass (Ji and Zhou, 2006), without any correlation evident between these indicators and MBR fouling propensity. Finally, the measurement of humic substances, generally overlooked for protein and carbohydrate, have revealed that they arise in significant concentrations in activated liquors (Liu and Fang, 2003) and may demand more attention in future research on MBR fouling.

Table 2.10 Concentration of EPS components in different MBR systems (mg/gSS unless otherwise stated)

EPSp	EPSc	Other	Details	Reference
25–30	7–8	Humic: 12–13	R, (10)	Cabassud *et al.* (2004)
29	36	SUVA: 2.8–3.1 L/m mg	S	Ahn *et al.* (2005)*
120	40		S, (∞)	Gao *et al.* (2004b)
31–116	6–15	TOC: 37–65	Four pilot-scale plants, Municipal	Brookes *et al.* (2003b)
20	14		Pilot-scale plant, Industrial	
11–46	12–40	TOC: 44–47	Three full-scale plants, Municipal	
25	9	TOC: 42	Full-scale plants, Industrial	
		EPSp + EPSc = 8		Jang *et al.* (2005a)
30–36	33–28		(20–60)	Lee *et al.* (2003)
73	30		S, (∞)	Le-Clech *et al.* (2003b)
60	17		R, (∞)	Le-Clech *et al.* (2003b)
		TOC: 250 mg/L	S, MLSS: 14 g/L	Nagaoka and Nemoto (2005)
		TOC: 26–83 mg/gVSS	(8–80)	Cho *et al.* (2005b)
116–101	22–24		S, (20)	Ji and Zhou (2006)

*Anaerobic UASB and aerobic MBR (Ahn *et al.*, 2005).
S: Synthetic wastewater; R: Real wastewater; (SRT in days in parenthesis; ∞ = infinite SRT – no wastage).

A functional relationship between specific resistance, MLVSS, TMP, permeate viscosity and EPS has recently been obtained by dimensional analysis (Cho *et al.*, 2005b). EPS was found to have no effect on the specific resistance below 20 and above 80 mgEPS/gMLVSS, but played a significant role on MBR fouling between these two limits.

Analysis of EPS isolates is normally by UV absorbance, though more extensive analysis has been conducted by a number of authors. In a recent study based on an intermittently aerated MBR, the EPS fraction was found to feature three main peaks at 100, 500 and 2000 kDa following gel chromatographic analysis. EPS larger than 1000 kDa in MW were assumed to be mainly responsible for MBR fouling (Nagaoka and Nemoto, 2005). High performance size exclusion chromatography (HPSEC), a technique more widely used for potable raw water analysis to analyse allochthonous natural organic matters (NOM) (Nissinen *et al.*, 2001), has been applied to EPS. Analysis of EPS fractions obtained from MBRs at different locations revealed their EPS profiles to be similar (Brookes *et al.*, 2003a; Brookes *et al.*, 2003b; Jefferson *et al.*, 2004). This would seem to corroborate previous findings from ASP sludge based on size exclusion chromatography combined with infrared micro-spectroscopy techniques (Gorner *et al.*, 2003), where EPS chromatographs exhibited seven distinct peaks. Analysis revealed 45–670 kDa MW proteins and 0.5–1 kDa MW carbohydrates to be present. The existence of low-MW proteins associated with carbohydrates was proposed as being pivotal in floc formation and may therefore be expected to play a significant part in MBR membrane fouling. Reported studies of EPS

Table 2.11 Concentration of SMP components (in mg/L and *mg/gSS)

SMPp	SMPc	Other	Operating conditions	Reference
8	25	Humic substance: 36	R, (10)	Cabassud *et al.* (2004)
		TOC: Up to 8 mg/L	S, (∞)	Gao *et al.* (2004b)
0.5–9*	n.d. –10*	4–37* (TOC)	Four pilot-scale plants, Municipal	Brookes *et al.* (2003b)
0.5–1*	n.d.	11*	Three full-scale plants, Municipal	
0.5*	n.d.	1.5*	Full-scale plant, Industrial	
		TOC: 30–70 mg/L	S, (∞), MLSS 15 g/L	Liu *et al.* (2005)
23	7		R, (not available)	Evenblij and van der Graaf (2004)
		DOC: 5 mg/L	S, (20)	Shin and Kang, 2003
		TOC: 8–10 mg/L	R, (21)	Tao *et al.* (2005)
10–34	5–33		R, (from 40 to 8)	Grelier *et al.*, (2005)
4.5–6	4.5–3.7		S, (20)	Ji and Zhou (2006)

n.d.: Non detected; S: Synthetic wastewater; R: Real wastewater; SRT are given in days in bracket, ∞: infinite SRT (i.e. no wastage).

characterisation pertaining to flocculation, settling and dewatering in conventional ASP technologies (Liu and Fang, 2003; Yin *et al.*, 2004) may therefore be germane to MBR technologies.

Since the EPS matrix features in floc formation (Liu and Fang, 2003) and specifically the hydrophobic interactions between microbial cells, a decrease in EPS levels may be expected to cause floc deterioration, as indicated by the results from a comparative study of nitrification/denitrification in an MBR (Jang *et al.*, 2005a). This would seem to imply that too low an EPS level is detrimental to MBR performance, though there is no firm experimental evidence to prove this.

Many operating parameters including gas sparging, substrate composition (Fawehinmi *et al.*, 2004) and OLR (Cha *et al.*, 2004; Ng *et al.*, 2005) appear to affect EPS characteristics in the MBR, but SRT is probably the most significant (Hernandez Rojas *et al.*, 2005). A decrease in EPS levels has been observed for extended SRTs, with this reduction becoming negligible at SRTs greater than 30 days (Brookes *et al.*, 2003b). Lee and co-workers (Lee *et al.*, 2003) observed an increase in protein concentration (along with stable carbohydrate levels) when SRT was increased.

Soluble microbial products Whilst the impact of dissolved matter on fouling has been studied for over a decade, the concept of SMP fouling in the MBR is a relatively new one (Chang *et al.*, 2002a), with available data being reported within the last few years (Table 2.11). Experiments recently conducted with a dual compartment MBR, where the membrane was challenged ostensibly with the mixed liquor supernatant (i.e. the SMP) rather than the whole biomass (Ng *et al.*, 2005), have revealed greater filtration resistance from the SMP than from the biomass at 4 g/L MLSS concentration. This implies that SMP characteristics have a significant impact on membrane permeability. During filtration, SMP materials are thought to adsorb onto the membrane surface, block membrane pores and/or form a gel structure on the membrane surface where they provide a possible nutrient source for biofilm formation and a hydraulic resistance to permeate flow (Rosenberger *et al.*, 2005). SMP materials appear to be retained at or near the membrane. Biomass fractionation studies conducted by Lesjean and co-workers (Lesjean *et al.*, 2005) revealed levels of carbohydrates, proteins and organic colloids to be higher in the SMP than in the permeate, a finding similar to those previously reported (Brookes *et al.*, 2003a; Evenblij and van der Graaf, 2004).

Three methods of separating the water phase from the biomass, so as to isolate the SMP, have been investigated. Simple filtration through filter paper (12 μm) was shown to be a more effective technique than either centrifugation or sedimentation (Evenblij and van der Graaf, 2004). It is likely that removal of colloidal material would demand more selective pre-filtration, e.g. 1.2 μm pore size (Figure 2.30). As with EPS, the SMP solution can be characterised with respect to its relative protein and carbohydrate content (Evenblij and van der Graaf, 2004), TOC level (Gao *et al.*, 2004b) or with SUVA measurement (Shin and Kang, 2003), as well as MW distribution. HPSEC analysis conducted on SMP solutions has revealed the SMP MW distribution to be different significantly across a range of full-scale reactors operated under different conditions, unlike the MW distribution for the eEPS fraction (Brookes *et al.*, 2003b). However, the SMP solution fingerprint was largely unchanged in weekly analysis conducted on a single reactor, indicating no significant change in SMP characteristics for biomass acclimatised to specific operating conditions. When compared to eEPS MW distribution, the SMP solution featured generally larger macromolecules.

Comparison between acclimatised sludges obtained from MBR and ASP pilot plants revealed similar levels of EPSp, EPSc and EPS humic matter (Cabassud *et al.*, 2004). The membrane did not seem to affect the floc EPS content. However, corresponding levels of the SMP fractions were significantly higher for the MBR sludge. Critical flux tests carried out under the same conditions for both MBR and ASP sludge revealed a higher fouling propensity of the MBR sludge over that of the ASP; critical flux values were around 10–15 and 32–43 LMH, respectively. Since the measured levels of EPS were unchanged, it was surmised that the higher fouling propensity related to the SMP level. During this study, Cabassud and co-workers observed significant biological activity in the MBR supernatant, indicating the presence of free bacteria which may have contributed to fouling.

A number of different studies have indicated a direct relationship between the carbohydrate level in SMP fraction and MBR membrane fouling directly (Lesjean *et al.*, 2005), or fouling surrogates such as filtration index and CST (Evenblij *et al.*, 2005a; Grelier *et al.*, 2005; Reid *et al.*, 2004; Tarnacki *et al.*, 2005), critical flux (Le-Clech *et al.*, 2005b) and permeability (Rosenberger *et al.*, 2005). The hydrophilic nature of carbohydrate may explain the apparently higher fouling propensity of SMPc over that of SMPp, given that proteins are more generally hydrophobic than carbohydrates. Strong interaction between the hydrophilic membrane generally used in MBRs and hydrophilic organic compounds may be the cause of the initial fouling observed in MBR systems. However, the nature and fouling propensity of SMPc has been observed to change during unsteady MBR operation (Drews *et al.*, 2005) and, in this specific study, it was not possible to correlate SMPc to fouling. Thus far, correlation of MBR membrane fouling with SMP protein has not been widely reported although, since a significant amount of protein is retained by the membrane – from 15%, according to Evenblij and van der Graaf (2004), to 90% (Drews *et al.*, 2005) – it must be presumed that such materials have a role in fouling. It was recently reported that the specific resistance increased by a factor of 10 when the SMPp increased from 30 to 100 mg/L (Hernandez Rojas *et al.*, 2005). Against this, analysis of the fouling layer has revealed higher levels of carbohydrate and lower protein concentrations compared to those in the mixed liquor (Chu and Li, 2005; Zhang *et al.*, 2006), tending to reinforce the notion that SMPc is more significant than SMPp in MBR membrane fouling. Humic matter, on the other hand, may not significantly contribute to fouling due to the generally lower MW of these materials (Drews *et al.*, 2005).

Many research studies have been based on synthetic/analogue wastewaters. Those analogues comprising the most basic constituents, such as glucose, are very biodegradable and, as such, would be expected to yield rather lower SMP levels than those arising in real systems. Since it may be assumed that there are almost no substrate residuals from glucose in the supernatant, the less biodegradable SMP induced by cell lysis or cell release would account for most of the supernatant EPS measured in such analogue-based studies and may explain the reduced influence of SMP compared with that of EPS reported in some of these studies (Cho *et al.*, 2005b). SUVA measurements carried out on MBR mixed liquor supernatant has confirmed the presence of organic matter originating from the decayed biomass and of larger MW and greater aromaticity and hydrophobicity than that of the analogue wastewater feed (Shin and Kang, 2003). This would seem to confirm that fouling materials are generated by biological action and arise as SMP, though once again the chemical nature of these products is obviously affected by the chemical nature of the feed.

In another important study based on synthetic wastewater, Lee *et al.* (2001a) revealed that levels of soluble organic matter in isolation cannot be used to predict MBR fouling. By comparing filterabilities of attached and suspended growth microorganisms, Lee and co-workers observed the rate of membrane fouling of the attached growth system (0.1 g/L MLSS and 2 g/L attached biomass) to be about 7 times higher than that of a conventional suspended growth MBR at 3 g/L MLSS. With similar soluble fraction characteristics in both reactors, it was concluded that the discrepancy arose from the formation of a protective dynamic membrane created by suspended solids in the suspended growth system, a conclusion subsequently corroborated by the work of Ng *et al.* (2005).

As expected, many operating parameters affect SMP levels in MBRs. As for EPS, SMP levels decrease with increasing SRT (Brookes *et al.*, 2003b). For SRTs ranging from 4 to 22 days, SMPp and SMPc levels have been reported to decrease by factors of 3 and 6, respectively (Grelier *et al.*, 2005).

2.3.6.6 Anaerobic systems

Studies have generally shown the membrane permeability to decline with MLSS for a number of different anaerobic matrices, including synthetic sewage (Stuckey and Hu, 2003), digested sludge (Saw *et al.*, 1986) and distillery wastewater (Kitamura *et al.*, 1996). Specific resistance has been shown to increase linearly with normalised EPS levels between 20 to 130 mg/SSg (Fawehinmi *et al.*, 2004). It is recognised that it is the colloidal material which is mainly responsible for fouling in an anMBR (Choo and Lee, 1996b, 1998), as has been shown in aerobic MBRs, and the colloid concentration is higher in anaerobic than aerobic systems. As with all membrane systems, colloidal matter is transported more slowly back into the bulk solution than coarser particulate materials due to the lower diffusion rates (Choo and Lee, 1998), which means they tend to collect at the membrane surface and form a low-permeability fouling layer. They are also of a size which can plug the membrane pores, particularly for the larger pores of microfiltration membranes, if able to migrate into the membrane.

Whilst far less characterisation of foulants has been conducted for anaerobic MBRs than for aerobic systems, reported trends tend to suggest that it is the colloidal component of the SMP fraction which is the dominant component in membrane fouling. The nature and concentration of the SMP in the anaerobic mixed liquor is dependent on the feedwater, the concentration being a function of the feedwater COD (Barker and Stuckey, 2001). However, it is generally higher in organic concentration than the SMP from aerobic systems and less anaerobically than aerobically biodegradable (Barker *et al.*, 2000). Also, SMP composition changes as a result of permeation (Stuckey, 2003), implying that some SMP components are adsorbed onto the membrane. SMP levels also appear to increase with increasing HRTs in conventional systems (Barker *et al.*, 2000), attributable to more extensive biomass decay to colloidal and soluble products. Levels may also increase with decreasing loading rate, since anMBR membrane permeability appears to increase with this parameter (Kayawake *et al.*, 1991) at low-intermediate loading rate values – 1.5–10 *cf.* <1.5 kg/(m^3/day) (Hernández *et al.*, 2002). The negative impact of lower temperatures (Section 2.2.7.2) may also be a reflection of slower biodegradation of fouling constituents of the SMP fraction (Barker *et al.*, 2000).

The use of supplementary dosing with PAC to ameliorate fouling has been extensively studied in membrane filtration of potable water and in aerobic MBRs, and such studies have also been conducted on anMBRs (Park *et al.*, 1999). It has been suggested (Choo and Lee, 1996b) that the addition of an adsorbent or a coagulant can enhance the permeate flux by agglomerating colloids to form larger particles of lower fouling propensity. The coarser and more rigid particles additionally improve scouring of the membrane surface. Dosing of anMBRs with ion-exchange resin has also been studied (Imasaka *et al.*, 1989), with beneficial effects noted only at very high concentrations of 5 wt%.

2.3.7 Operation

2.3.7.1 Membrane aeration or gas scouring

Aerobic systems Aeration is arguably the most important parameter in the design and operation of an MBR. As already stated, aeration is required for biotreatment (Section 2.2.5), floc agitation and membrane scouring (Dufresne *et al.*, 1997) and it is not necessarily essential or desirable to employ the same aerator for both duties. Ostensibly, air is used to lift the mixed liquor through the membrane module channels. However, the gas bubbles additionally enhance membrane permeation

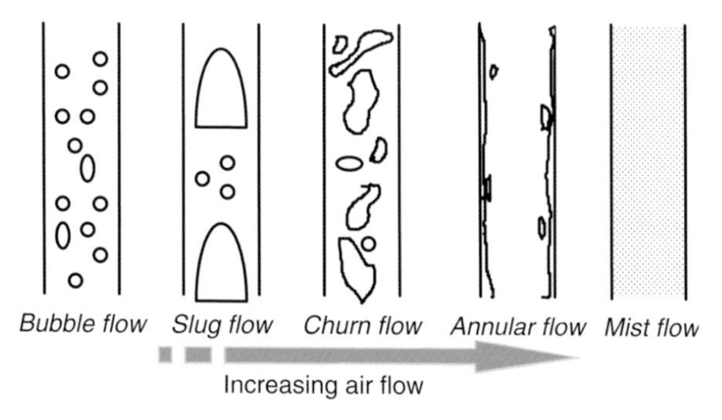

Figure 2.31 *Air–liquid flow regimes in a cylindrical channel (Judd* et al., *2001)*

(Cui *et al.*, 2003) by inducing liquid flow fluctuations and local tangential shear transients, the shear rate γ (/s) being given by:

$$\gamma = \frac{\kappa U_{\mathrm{L}}}{\delta} \qquad (2.31)$$

where U_{L} is the liquid CFV (m/s), δ is the separation (m) and κ is a constant depending on the membrane geometry. The effect is to increase back transport and promote mass transfer of liquid through the membrane (Section 2.1.4.4). Tangential shear at the membrane surface prevents large particle deposition on the membrane surface. However, since it is proportional to the cube of particle diameter, lateral migration velocity for smaller particles is much less, leading to more severe membrane fouling by fine materials (Choo and Lee, 1998).

It has long been recognised through studies of model systems (Cabassud *et al.*, 1997; Ghosh and Cui, 1999; Cui and Wright, 1994; Mercier *et al.*, 1997) as well as MBRs themselves(Le-Clech *et al.*, 2003a, b) that gas bubbles (or "slugs") passing up through a tubular membrane are able to enhance the flux over that attainable from liquid crossflow at the same velocity. This type of two-phase air–liquid flow is termed "slug flow" (Fig. 2.31) and represents the most effective type of air–liquid flow for promoting flux. Much work has been conducted, principally by Cui and his various co-workers, to model membrane aeration in channel flow. Thus far, models have been produced which describe the spatial variation of shear with time for rising bubbles as a function of bubble (or slug) size, channel dimension and geometry for Newtonian fluids. It is also possible, within certain boundary conditions, to relate γ to the flux, J, from first principles, provided assumptions can be made about the particle size and concentration, the system hydrodynamics and the fluid and membrane homogeneity. Such assumptions, however, are not pertinent to an iMBR where three-phase flow prevails in a highly heterogeneous non-Newtonian fluid containing solutes, colloids and particulates. Moreover, the system becomes yet more complicated when the geometry deviates from well-defined channels, as provided by FS or tubular configurations, to HF modules.

Aeration also affects HF iMBR performance by causing fibre lateral movement (or sway) (Côté *et al.*, 1998; Wicaksana *et al.*, 2006), which imparts shear at the membrane surface through the relative motion of the membrane and the surrounding liquid.

In the case of HFs, effective distribution of air over the whole element cross-section and length becomes particularly challenging. For MT membrane modules in particular, provided an air bubble of diameter greater than that of the tube diameter is introduced into the tube, then air scouring of the entire membrane surface is assured. This is not necessarily the case for the FS and HF configurations, and HF systems additionally provide no fixed channel for the air bubble to travel up; this appears to impact on membrane permeability. On the other hand, experimental studies and heuristic investigations reveal FS systems to generally demand higher aeration rates than HF systems to sustain higher membrane permeabilities, and this is reflected in aeration demand data from pilot-scale studies and full-scale operating plant (Section 3.3.1.1). Some HF systems are operated with intermittent aeration, lowering the aeration demand further, and aeration demand may also be lowered by stacking the membrane modules such that the same volume of air is passed over twice the membrane area.

A number of authors (Le-Clech *et al.*, 2003c; Liu *et al.*, 2003; Psoch and Schiewer, 2005b; Ueda *et al.*, 1997) have demonstrated that flux increases roughly linearly with aeration rate up to a threshold value beyond which no further increase in permeability takes place. It follows that operation is sub-optimal if the aeration rate, and specifically the approach velocity, exceeds this threshold value. Intense aeration may also damage the floc structure, reducing floc size and releasing EPS in the bioreactor (Ji and Zhou, 2006; Park *et al.*, 2005b) in the same way as has been reported for CFV in sMBRs (Section 2.3.1). Given that aeration lifts the sludge through the module, a relationship must exist between gas and liquid velocity (U_G and U_L), respectively. Determination of U_L induced by aeration can be difficult; techniques such as electromagnetic flow velocimetry (Sofia *et al.*, 2004), particle image velocimetry (Yeo and Fane, 2004), and constant temperature anemometry (Le-Clech *et al.*, 2006), have all been used for liquid velocity estimation in iMBRs. Based on short-term critical flux tests, a direct comparison between immersed and sMBRs showed that similar fouling behaviour was obtained when the two configurations were respectively operated at a superficial gas velocity (U_G) of 0.07–0.11 m/s and CFV of 0.25–0.55 m/s (Le-Clech *et al.*, 2005b). An increase of U_G in the iMBR was also found to have more effect in fouling removal than a similar rise of CFV in the sidestream configuration.

In practice, much development of commercial systems has been focused on reducing aeration whilst maintaining membrane permeability, since membrane aeration contributes significantly to energy demand (though not generally as much as biochemical aeration demand). A key parameter is thus the specific aeration demand (SAD), either with respect to membrane area (SAD_m in Nm^3 air/(h m^2)) or permeate volume (SAD_p Nm^3 air/m^3 permeate). The latter is a useful unitless indicator of aeration efficiency, and values for this parameter, which can range between 10 and 100, are now often quoted by the membrane suppliers. Further discussion of specific aeration demand is provided in Chapter 3 and values from case studies included in Chapter 5.

Anaerobic and anoxic systems Gas sparging to maintain a high membrane permeability, as used in immersed aerobic systems, is more problematic in anMBRs since air cannot be used routinely. Air sparging has been shown to be effective in an MBRs (Lee *et al.*, 2001c) but the duration of air sparging was necessarily brief in this study (5 s every 10 min), providing little permeability promotion overall. Sparging with head space gas has been shown to be effective for immersed polymeric (Fawehinmi *et al.*, 2004; Stuckey and Hu, 2003) and sidestream ceramic (Kayawake *et al.*, 1991) membranes. As with the aerobic MBRs described above, a maximum permeability was reached at a certain gas flow rate (Imasaka *et al.*, 1989; Stuckey and Hu, 2003).

In common with all membrane processes (Section 2.1.4.5), increasing crossflow increases flux in sidestream anMBRs by suppressing the fouling layer concentration polarisation (Grethlein, 1978; Imasaka *et al.*, 1989; Saw *et al.*, 1986). However, a plateau has been reported at Reynolds numbers beyond 2000-or-so where no further increase in permeability takes place (Choo and Lee, 1998; Choo *et al.*, 2000; Elmaleh and Abdelmoumni, 1997, 1998). For ceramic membranes, where the fouling layer is minimal, high crossflows have been reported as having a detrimental effect because the thinning cake layer offers less protection against internal fouling (Choo and Lee, 1998; Choo *et al.*, 2000; Kang, 1996). Elmaleh and Abdelmoumni (1997) have reported close to zero fouling for crossflows above 3 m/s in an MT organic membrane module sidestream anMBR, with flux increasing linearly with shear stress up to this point. Baffles were shown by these authors to increase flux by promoting shear, the effect being greatest in the transition region between laminar and turbulent flow. However, the increase in flux attained by these measures is normally at the expense of a punitive increase in energy demand (Bourgeous *et al.*, 2001) and non-uniform, and thus sub-optimal, TMP distribution (Lee *et al.*, 1999). High-shear operation might also be expected to impact negatively on floc size and biomass bioactivity (Brockmann and Seyfried, 1996; Choo and Lee, 1998; Ghyoot and Verstraete, 1997) with, at the highest shears, cell lysis taking place, though it has been concluded by Elmaleh and Abdelmoumni (1997) that such effects are less severe for anaerobic than aerobic biomass.

Reports of operation at different TMPs indicate that, as with most membrane separation processes, membrane resistance determines flux at low TMPs, with no impact of crossflow or MLSS concentration above 2.5 g/L (Beaubien *et al.*, 1996). At higher TMPs, crossflow (and thus surface shear) becomes important (Beaubien *et al.*, 1996; Zhang *et al.*, 2004), the flux increasing linearly with CFV (Beaubien *et al.*, 1996), the slope decreasing with increasing MLSS partly due to viscosity effects. At very high TMPs, permeate flux has been shown to decrease with increasing TMP due to compaction of the fouling layer (Elmaleh and Abdelmoumni, 1997). However, this effect appears to depend on the membrane filter; Saw *et al.* (1986), filtering anerobic sludge, observed that at very high TMPs the permeate flux decreased with TMP for an MF membrane but was constant for an 8–20 kDa MWCO UF membrane. The authors suggested that this was due to the impact of the membrane substrate on the fouling layer structure, but a more likely explanation is migration of fines through the cake at higher TMPs into the more porous MF membrane, causing pore plugging (Beaubien *et al.*, 1996).

The use of extended intermittent aeration has been reported for nitrification–denitrification MBR systems (Nagaoka and Nemoto, 2005; Yeom *et al.*, 1999). In this less common scenario, a single tank was used for both anoxic and aerobic biological degradation. Filtration was carried out in only the aerobic phase to take advantage of the anti-fouling properties of the air scouring, since severe fouling has been reported when aeration ceases (Jiang *et al.*, 2005; Psoch and Schiewer, 2005b).

2.3.7.2 *Solid retention time (SRT)*

SRT impacts on fouling propensity through MLSS concentration, which increases with increasing SRT, and in doing so reduces the F:M ratio (Equation (2.11)) and so alters the biomass characteristics. Extremely low SRTs of $\sim$2 days have been shown to increase the fouling rate almost 10 times over that measured at 10 days, with the F:M ratio correspondingly increasing from 0.5 to 2.4 g COD/(g VSS/day) and the MLSS increasing only slightly from 1.5 to 1.2 g/L (Jang *et al.*, 2005b). In practice, the F:M ratio is generally maintained at below 0.2/day.

Operation at long SRTs minimises excess sludge production but the increase in MLSS level which inevitably takes place presents problems of clogging of membrane channels, particularly by inert matter such as hair, lint and cellulosic matter (Le-Clech *et al.*, 2005a), membrane fouling and reduced aeration efficiency, as manifested in the α-factor (Fig. 2.19). Even after increasing membrane aeration by 67%, fouling of an HF sMBR has been reported to almost double on increasing the SRT from 30 to 100 days, producing a corresponding increase in MLSS levels from 7 to 18 g/L and a decrease in F:M ratio from 0.15 to 0.05 kg COD/kg MLSS/day (Han *et al.*, 2005). At infinite SRT, most of the substrate is consumed to ensure the maintenance needs and the synthesis of storage products. The very low apparent net biomass generation observed can also explain the low fouling propensity observed for high SRT operation (Orantes *et al.*, 2004). In such cases sludge production is close to zero.

Scientific studies indicate that SRT is a key parameter in determining fouling propensity through MLSS and EPS fraction concentrations. On this basis, an optimum SRT can be envisaged where foulant concentrations, in particular in the SMP fraction, are minimised whilst oxygen transfer efficiency remains sufficiently high and membrane clogging at a controllable level. In practice, SRT tends not to be rigorously controlled. Moreover, SRT probably has less of an impact on fouling than feedwater quality and fluctuations therein.

2.3.7.3 *Unsteady-state operation*

Unsteady-state operation arising from such things as variations in feedwater quality (and so organic load), feedwater and/or permeate flow rate (and hence hydraulic load) and aeration rate are all known to impact on MBR membrane fouling propensity, along with other dynamic effects (Table 2.12). In an experiment carried out with a large pilot-scale MBR in which the effects of unstable flow and sludge wastage were assessed (Drews *et al.*, 2005), it was established that the level of carbohydrate in the supernatant before and after each sludge withdrawal increased. Whilst the increase following wastage was thought to be due to the sudden stress experienced by cells due to biomass dilution (which in extreme cases is known to lead to foaming in full-scale plant), increase before sludge withdrawal was attributed to the high MLSS concentration and the resulting low DO level in the bioreactor. It was concluded that unsteady-state operation changed the nature and/or structure (and fouling propensity) of the carbohydrate rather than the overall EPS formation. These findings corroborated results previously reported on effects of transient conditions

Table 2.12 Dynamic effects

Determinants	Variables
Flow rate	Ultimate flux and rate of change
Feedwater quality	Ultimate composition and rate of change
MLSS dilution	Dilution factor and rate of concentration change
(Partial) aeration loss	Percentage and period of reduction
Backflush/cleaning loss	Period of loss
Hydraulic shock	Rate and level of flow increase
Saline intrusion	Ultimate concentration factor and rate of concentration change

in feeding patterns: the addition of a pulse of acetate in the feedwater has been shown to significantly decrease the MBR biomass filterability due to the increase in SMP levels produced (Evenblij *et al.*, 2005b).

The effects of starvation conditions on the biological suspension have been assessed by incorporating different substrate impulses in batch tests (Lobos *et al.*, 2005). Exogenous phases were followed by starvation periods, both characterized by the S:X (substrate to biomass concentration ratio) where high ratios led to multiplication of bacteria cells whilst at low ratios MLVSS decreased, SMPp production was absent and bacteria lysis ceased. S:X closely relates to F:M ratio (Equation (2.11)), and the low F:M values generally used in MBRs are thus theoretically close to starvation conditions which are in turn likely to be beneficial to MBR operation on the basis of the reduced SMPp production and correspondingly reduced fouling.

The principal period of unsteady-state operation is during start-up when the system is acclimatising. Cho *et al.* (2005b) reported temporal changes of the bound EPS levels when the MBR was acclimatised at three different SRTs (8, 20, 80 days). As expected from general trends described in Section 2.3.6.5, the EPS concentration was lower at the longer SRT (83 vs. 26 mgTOC/gSS for SRTs of 8 and 80 days, respectively). An initial latent phase was observed in which EPS concentration did not vary significantly. However, EPS levels increased exponentially after 40 days of operation at an SRT of 8 days, and after 70 days when the MBR was operated at 20-day SRT. No change in EPS levels was observed during the 80 days of operation at 80-day SRT. For another MBR operated at infinite SRT, no significant changes in SMP concentration during 100 days of operation were observed, over which time period the MLSS increased from 1.8 to 4.5 g/L (Jinhua *et al.*, 2004). In a further study, following a latent phase of 30 days, MLSS and SMP levels started to significantly increase and stabilised after 140 days of operation at infinite SRT, whereas EPS levels increased continuously from the start but also stabilized after 140 days (Gao *et al.*, 2004a). Nagaoka and Nemoto (2005) observed an increase in MLSS concentration from 4 to 14 g/L over 100 days along with a steady increase in EPS (from 50 to 250 mgTOC/L). There therefore appears to be no distinct pattern regarding foulant species generation and start-up, other than a general trend of more stable foulant levels at longer SRTs.

The generation of foulants arising from changes in salinity have been studied by Reid (2006). According to established literature on the ASP extending back to the 1960s (Ludzack and Noran, 1965; Tokuz and Eckenfelder, 1979), changes in salinity have a greater impact on biotreatment efficacy, as manifested in the outlet

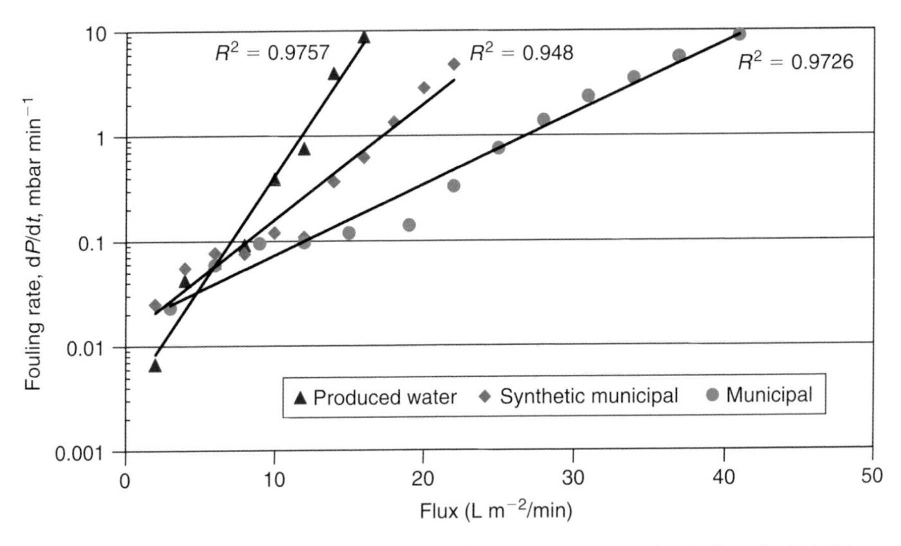

Figure 2.32 Fouling rate as a function of flux, flux step experiments (Le-Clech et al., *2003b)*

organic carbon concentration, than high salinity levels *per se*. According to Reid, SMP and EPS turbidity, EPSp and SMPc all increased when a shock load of sodium chloride was administered to an MBR in a way designed to mimic saline intrusion in coastal MBRs. As with other studies (Section 2.3.6.5), permeability decline correlated with SMPc.

2.3.8 Fouling mechanisms in MBRs

MBRs are routinely operated under notionally constant flux conditions with convection of foulant towards the membrane surface therefore maintained at a constant rate determined by the flux. Since fouling rate increases roughly exponentially with flux (Fig. 2.32), sustainable operation dictates that MBRs should be operated at modest fluxes and preferably below the so-called critical flux (Section 2.1.4.6). As noted previously, even sub-critical flux operation can lead to fouling according to a two-stage pattern (Brookes *et al.*, 2004; Ognier *et al.*, 2001; Wen *et al.*, 2004): a low TMP increase over an initial period followed by a rapid increase after some critical time period. Pollice *et al.* (2005) reviewed the sub-critical fouling phenomenon, introducing the parameters t_{crit} and dTMP/dt to represent the critical time over which low-fouling operation at a rate of dTMP/dt is maintained (Table 2.13). Prior to these two filtration stages, a conditioning period is generally observed (Zhang *et al.*, 2006). The three-stage process, wherein various mechanisms prevail, is summarised in Fig. 2.33.

2.3.8.1 Stage 1: Conditioning fouling

The initial conditioning stage arises when strong interactions take place between the membrane surface and the EPS/SMP present in the mixed liquor. Ognier *et al.* (2002a)

Table 2.13 Sub-critical long-term parameters, from Pollice *et al*. (2005)

Flux (LMH)	dTMP/dt (kPa/h)	t_{crit} (hours)	Reference
17	0.005	>600	(Wen *et al*., 2004)
22	0.011	1200	(Wen *et al*., 2004)
25	0.024	300	(Wen *et al*., 2004)
30	0.072	250	(Wen *et al*., 2004)
n.a.	0.023	350	(Frederickson and Cicek, 2004)
20	–	600	(Lee and Choi, 2004)
8	–	350	(Li *et al*., 2005c)
30	0.036	360	(Cho and Fane, 2002)
10	0.036	550	(Ognier *et al*., 2002b)
8	0.03	72	(Brookes *et al*., 2003a)
7	0.006	96	(Le-Clech *et al*., 2003b)
9	0.004	240	(Le-Clech *et al*., 2003b)
18	0.104	48	(Le-Clech *et al*., 2003b)
12	0.0002	300	(Rosenberger *et al*., 2002)

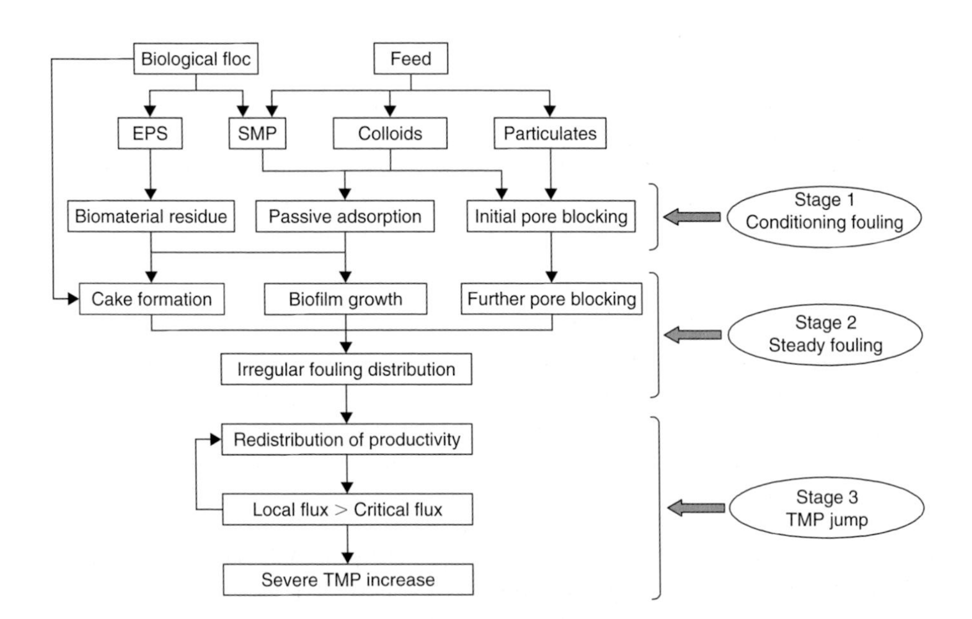

Figure 2.33 Fouling mechanisms for MBR operated at constant flux (adapted from Zhang et al., *2006)*

described rapid irreversible fouling in this initial stage, and passive adsorption of colloids and organics have been observed even for zero-flux operation and prior to particle deposition (Zhang *et al*., 2006). Another detailed study based on passive adsorption revealed the hydraulic resistance due to this process to be almost independent of tangential shear, and the initial adsorption to account for 20–2000% of the clean membrane resistance depending on the membrane pore size (Ognier *et al*., 2002b).

In a more recent study, the contribution of conditioning fouling to overall resistance was found to become negligible once filtration takes place (Choi *et al.*, 2005a). By applying a vacuum (rather than suction) pump coupled with air backflushing, Ma *et al.* (2005) were able to reduce colloidal adsorption onto the membrane (Ma *et al.*, 2005). These studies suggest that colloid adsorption onto new or cleaned membranes coupled with initial pore blocking may be expected in MBRs (Jiang *et al.*, 2005). The intensity of this effect depends on membrane pore size distribution, surface chemistry and especially hydrophobicity (Ognier *et al.*, 2002a). In a test cell equipped with direct observation through a membrane operating with crossflow and zero flux, flocculant material was visually observed to temporarily land on the membrane (Zhang *et al.*, 2006). This was defined as a random interaction process rather than a conventional cake formation phenomenon. While some flocs were seen to roll and slide across the membrane, biological aggregates typically detached and left a residual footprint of smaller flocs or EPS material. Biomass approaching the membrane surface was then able to attach more easily to the membrane surface to colonise it and contribute to Stage 2.

2.3.8.2 Stage 2: Slow fouling

Even when operated below the critical flux for the biomass, temporary attachment of the floc can contribute to the second fouling stage. After Stage 1, the membrane surface is expected to be mostly covered by SMP, promoting attachment of biomass particulate and colloidal material. Because of the low critical flux measured for SMP solutions, further adsorption and deposition of organics on the membrane surface may also occur during Stage 2. Since adsorption can take place across the whole surface and not just on the membrane pore, biological flocs may initiate cake formation without directly affecting the flux in this initial stage. Over time, however, complete or partial pore blocking takes place. The rate of EPS deposition, and resulting TMP rise, would then be expected to increase with flux leading to a shorter Stage 2. Such fouling would prevail even under favourable hydrodynamic conditions providing adequate surface shear over the membrane surface. However, since uneven distribution of air and liquid flow is to be expected in iMBRs, correspondingly inhomogeneous fouling must take place.

2.3.8.3 Stage 3: TMP jump

With regions of the membrane more fouled than others, permeability is significantly less in those specific locations. As a result, permeation is promoted in less fouled areas of the membrane, exceeding a critical flux in these localities. Under such conditions, the fouling rate rapidly increases, roughly exponentially with flux according to data of Le-Clech *et al.* (2003b) (Fig. 2.33). The sudden rise in TMP or "jump" is a consequence of constant flux operation, and several mechanisms can be postulated for the rapid increase in TMP under a given condition. As with classical filtration mechanisms (Fig. 2.9), it is likely that more than one mechanism will apply when an MBR reaches the TMP jump condition, and a number of models can be considered:

(i) *Inhomogeneous fouling (area loss) model*: This model was proposed to explain the observed TMP profiles in nominally sub-critical filtration of upflow anaerobic sludge (Cho and Fane, 2002). The TMP jump appeared to coincide with a measured loss of local permeability at different positions along the membrane, due to slow fouling by EPS. It was argued that the flux redistribution (to maintain the constant average flux) resulted in regions of sub-critical flux and consequently in rapid fouling and TMP rise.

(ii) *Inhomogeneous fouling (pore loss) model*: Similar TMP transients have been observed for the crossflow MF of a model biopolymer (alginate) (Ye *et al.*, 2005). These trends revealed the TMP transient to occur with relatively simple feeds. The data obtained have been explained by a model that involves flux redistribution among open pores. Local pore velocities eventually exceed the critical flux of alginate aggregates that rapidly block the pores. This idea was also the base of the model proposed by Ognier *et al.* (2004). While the "area loss" model considers macroscopic redistribution of flux, the "pore loss" model focuses on microscopic scale. In MBR systems, it is expected that both mechanisms occur simultaneously.

(iii) *Critical suction pressure model*: The two-stage pattern of a gradual TMP rise followed by a more rapid increase has been observed from studies conducted based on dead-end filtration of a fine colloid by an immersed HF. A critical suction pressure it is suggested coagulation or collapse occurs at the base of the cake, based on membrane autopsy evaluations supplemented with modelling (Chang *et al.*, 2005). A very thin dense layer close to the membrane surface, as observed in the study, would account for the rapid increase in resistance leading to the TMP jump. Although this work was based on dead-end rather than crossflow operation, the mechanism could apply to any membrane system where fouling continues until the critical suction pressure is reached, whereupon the depositing compound(s) coalesce or collapse to produce a more impermeable fouling layer.

(iv) *Percolation theory*: According to percolation theory, the porosity of the fouling layer gradually reduces due to the continuous filtration and material deposition within the deposit layer. At a critical condition, the fouling cake loses connectivity and resistance, resulting in a rapid increase in TMP. This model has been proposed for MBRs (Hermanowicz, 2004), but indicates a very rapid change (within minutes), which is not always observed in practice. However, the combination of percolation theory with the inhomogeneous fouling (area loss) model could satisfy the more typically gradual inclines observed for TMP transients. Similarly, fractal theory was successfully applied to describe cake microstructure and properties and to explain the cake compression observed during MBR operation.

(v) *Inhomogeneous fibre bundle model*: Another manifestation of the TMP transient has been observed for a model fibre bundle where the flow from individual fibres was monitored (Yeo and Fane, 2004). The bundle was operated under suction at constant permeate flow, giving constant average flux, and the flow was initially evenly distributed amongst the fibres. However, over time the flows became less evenly distributed so that the standard deviation of the fluxes of individual fibres started to increase from the initial range of 0.1–0.15 up to 0.4. Consequently, the TMP rose to maintain the average flux across the fibre bundle, mirroring the increase in the standard deviation of the fluxes. At some point, both TMP and standard deviation rose rapidly. This is believed to be due to flow mal-distribution within the bundle leading to local pore and flow channel occlusion. It was possible to obtain more steady TMP and standard deviation profiles when the flow regime around the fibres was more rigorously controlled by applying higher liquid and/or air flows.

2.3.9 Fouling control and amelioration in MBRs

Whilst an understanding of fouling phenomena and mechanisms may be enlightening, control of fouling and clogging in practice is generally limited to five main strategies:

1. applying appropriate pretreatment to the feedwater,
2. employing appropriate physical or chemical cleaning protocols,
3. reducing the flux,
4. increasing the aeration,
5. chemically or biochemically modifying the mixed liquor.

All of the above strategies are viable for full-scale operating MBRs, and each are considered in turn below.

2.3.9.1 Feed pretreatment

It is generally recognised that the successful retrofitting of an ASP or SBR with an MBR is contingent on upgrading the pretreatment and, specifically, the screening. Whilst an MBR can effectively displace primary sedimentation, biotreatment and secondary solid–liquid separation, as well as tertiary effluent polishing, classical screens of around 6 mm rating are normally insufficient for an MBR. Such relatively coarse screens increase the risk of clogging of the membrane module retentate flow channels, especially by hairs in municipal wastewaters, which aggregate and clog both the membrane interstices and aeration ports. HF membranes have a tendency for aggregates of hair and other debris to collect at the top of the membrane element. Hairs may then become entwined with the filaments and are not significantly removed by backflushing. FS membrane clogging occurs when debris agglomerate at the channel edges and entrance. If the aeration fails to remove these aggregates, sludge accumulates above the blockage, increasing the affected excluded area. Fibres collecting in the aeration system can change the flow pattern and volume of air to the membranes, reducing the degree of scouring. As a result of the decreased scouring, membrane fouling is increased. Aerators are thus normally designed to resist clogging and/or allow periodic flushing with water.

Since HF modules are more susceptible to clogging and the impact is rather more severe, for such modules screens normally rated at between 0.8 and 1.5 mm are usually employed. FS modules are slightly more tolerant of clogging, despite being non-backflushable, and screens of 2–3-mm rating are normally adequate for MBRs of this membrane configuration.

2.3.9.2 Employing appropriate physical or chemical cleaning protocols

Cleaning strategies have been outlined in Section 2.1.4.3, and protocols applied in practice are detailed in Chapters 4 (for comparative pilot plant studies) and 5 (full-scale reference sites). A summary of these data is presented in Section 3.3.1.2, along with the implications of the physical and chemical cleaning protocols on process design and operation. Results from selected bench-scale studies are given below.

Physical cleaning Key general cleaning parameters are duration and frequency, since these determine process downtime. For backflushing, a further key parameter is the backflush flux, generally of 1–3 times the operational flux and determined by the backflush TMP. Less frequent, longer backflushing (600 s filtration/45 s back-flushing) has been found to be more efficient than more frequent but shorter back-flushing (200 s filtration/15 s backflush) (Jiang *et al.*, 2005). In another study based on factorial design, backflush frequency (between 8 and 16 min) was found to have more effect on fouling removal than either aeration intensity (0.3 to 0.9 m^3/h per m^2 membrane area) or backflush duration (25–45 s) for an HF iMBR (Schoeberl *et al.*, 2005). Hence, although more effective cleaning would generally be expected for more frequent and longer backflushing, the possible permutations need exploring to minimise energy demand. This has been achieved through the design of a generic control system which automatically optimised backflush duration according to the monitored TMP value (Smith *et al.*, 2005). However, increasing backflush flux leads to more loss product and reduces the net flux.

Air can also be used to affect backflushing (Sun *et al.*, 2004) or to enhance back-flush with water. Up to 400% in the net flux over that attained from continuous operation has been recorded using an air backflush, although in this case 15 min of air backflush were required every 15 min of filtration (Visvanathan *et al.*, 1997). Whilst air backflushing is undoubtedly effective, anecdotal evidence suggests that it can lead to partial drying out of some membranes, which can then produce embrittlement and so problems of membrane integrity.

Membrane relaxation encourages diffusive back transport of foulants away from the membrane surface under a concentration gradient, which is further enhanced by the shear created by air scouring (Chua *et al.*, 2002; Hong *et al.*, 2002). Detailed study of the TMP behaviour during this type of operation revealed that, although the fouling rate is generally higher than for continuous filtration, membrane relaxation allows filtration to be maintained for longer periods of time before the need for cleaning arises (Ng *et al.*, 2005). Although some authors have reported that this type of operation may not be economically feasible for operation of large-scale MBRs (Hong *et al.*, 2002), relaxation is almost ubiquitous in modern full-scale iMBRs (Chapter 5). Recent studies assessing maintenance protocols have tended to combine relaxation with backflushing for optimum results (Vallero *et al.*, 2005; Zhang *et al.*, 2005).

In practice, physical cleaning protocols tend to follow those recommended by the suppliers. Relaxation is typically applied for 1–2 min every 8–15 min of operation, both for FS and HF systems. For HF systems, backflushing, if employed, is usually applied at fluxes of around 2–3 times the operating flux and usually supplements rather than displaces relaxation. It is likely that operation without backflushing, whilst notionally increasing the risk of slow accumulation of foulants on or within

Table 2.14 Examples of intensive chemical cleaning protocols, four MBR suppliers

Technology	Type	Chemical	Concentration (%)	Protocols
Mitsubishi	CIP	NaOCl	0.3	Backflow through membrane (2 h) + soaking (2 h)
		Citric acid	0.2	
Zenon	CIA	NaOCl	0.2	Backpulse and recirculate
		Citric acid	0.2–0.3	
Memcor	CIA	NaOCl	0.01	Recirculate through lumens, mixed liquors and in-tank air manifolds
		Citric acid	0.2	
Kubota	CIP	NaOCl	0.5	Backflow and soaking (2 h)
		Oxalic acid	1	

Exact protocol for chemical cleaning can vary from one plant to another (Section 3.3.1.2).
CIP: Cleaning in place, without membrane tank draining; chemical solutions generally backflushed under gravity in-to-out.
CIA: Cleaning in air, where membrane tank is isolated and drained; module rinsed before soaking in cleaning solution and rinsed after soaking to remove excess reagent.

the membrane, conversely largely preserves the biofilm on the membrane which affords a measure of protection. This fouling layer is substantially less permeable and more selective than the membrane itself, and thus can be beneficial to the process provided the total resistance it offers does not become too great.

Chemical cleaning Physical cleaning is supplemented with chemical cleaning to remove "irreversible" fouling (Fig. 2.10), this type of cleaning tending to comprise some combination of:

- CEB (on a daily basis),
- Maintenance cleaning with higher chemical concentration (weekly),
- Intensive (or recovery) chemical cleaning (once or twice a year).

Maintenance cleaning is conducted *in situ* and is used to maintain membrane permeability and helps reduce the frequency of intensive cleaning. It is performed either with the membrane *in situ*, a normal CIP, or with the membrane tank drained, sometimes referred to as "cleaning in air" (CIA). Intensive, or recovery, cleaning is either conducted *ex situ* or in the drained membrane tank to allow the membranes to be soaked in cleaning reagent. Intensive cleaning is generally carried out when further filtration is no longer sustainable because of an elevated TMP. Recovery chemical cleaning methods recommended by suppliers (Table 2.14) are all based on a combination of hypochlorite for removing organic matter, and organic acid (either citric or oxalic) for removing inorganic scalants. Whilst some scientific studies of the impacts of chemical cleaning on the MBR system, such as the microbial community (Lim *et al.*, 2004), have been conducted, there has been no systematic study comparing the efficacy of a range of cleaning reagents or cleaning conditions on permeability recovery. Some experiments with augmented cleaning, such as sonically-enhanced processes (Fang and Shi, 2005; Lim and Bai, 2003), have been conducted, however.

Whilst ultrasonic cleaning can undoubtedly enhance flux recovery, tests conducted in potable water suggest that it can result in adverse impacts on membrane integrity (Masselin *et al.*, 2001).

Maintenance cleaning, usually taking 30–60 minutes for a complete cycle, is normally carried out every 3–7 days at moderate reagent concentrations of 200–500 mg/L NaOCl for classical aerobic MBRs. Recovery cleaning employs rather higher reagent concentrations of 0.2–0.3 wt% NaOCl, coupled with 0.2–0.3 wt% citric acid or 0.5–1 wt% oxalic acid (Section 3.3.1.2). Membrane cleaning studies on anaerobic systems have generally indicated that a combination of caustic and acid washes are required to remove organic and inorganic (namely struvite) foulants from organic anMBR membranes (Choo *et al.*, 2000; Kang *et al.*, 2002; Lee *et al.*, 2001c). For inorganic membranes, acid washing has been found to be less effective, and this has been attributed to surface charge effects (Kang *et al.*, 2002).

2.3.9.3 Reducing the flux

Reducing the flux always reduces fouling but obviously then impacts directly on capital cost through membrane area demand. A distinction must be made, however, between operating (i.e. gross) flux and net flux (the flux based on throughput over a complete cleaning cycle), as well as peak and average flux.

There are essentially two modes of operation of an MBR regarding operating flux, which then determine the cleaning requirements (Sections 2.1.4.3 and 2.3.9.2) and thus net flux:

- *Sustainable permeability operation*: In this instance, the conditions are chosen so as to maintain stable operation (little or negligible increase in TMP at constant flux) over an extended period of time (i.e. several weeks or months) with only moderate remedial measures (namely relaxation), if any. All immersed FS and all sidestream systems operate under these conditions, with sMBRs operating continuously (i.e. without relaxation) between chemical cleans.
- *Intermittent operation*: In this mode of operation, the operational flux is above that which can be sustained by the filtration cycle operating conditions and, as a result, intermittent remedial measures are employed. These comprise relaxation supplemented with backflushing and, usually, some kind of maintenance chemical cleaning procedure. All immersed HF systems operate in this manner.

Modern practice appears to favour operation at net fluxes of around 25 LMH for municipal wastewater, incorporating physical cleaning every 10–12 minutes, regardless of membrane configuration. Maintenance cleaning, if employed, adds insignificantly to downtime. The greatest impact on operating vs. net flux is therefore peak loading, normally from storm waters. It is these increased hydraulic loads, coupled with feedwater quality fluctuations, which represent one of the major challenges to MBR design and operation.

2.3.9.4 Increasing aeration

Whilst increasing aeration rate invariably increases the critical flux up to some threshold value, increasing membrane aeration intensity is normally prohibitively expensive. As already stated (Section 2.3.7.1), much attention has been focused on commercial development of efficient and effective aeration systems to reduce the specific aeration demand, with possibly the most important publications arising in the patent literature (Côté, 2002; Miyashita *et al.*, 2000) and including cyclic aeration (Rabie *et al.*, 2003) and jet aeration (Fufang and Jordan 2001, 2002). The use of uniformly distributed fine air bubbles from 0.5 mm ports has been shown to provide greater uplift and lower resistance compared to a coarse aerator having 2 mm ports at similar aeration rates (Sofia *et al.*, 2004). In the same study, a bi-chamber (a riser and down-comer) in an FS MBR has been shown to play a significant role in inducing high CFVs. The use of a variable aeration rate to increase the flux during peak loads has been reported for short-term tests (Howell *et al.*, 2004). However, a recent study from Choi and co-workers (Choi *et al.*, 2005a) carried out with an sMBR indicated tangential shear (imparted by liquid crossflow) to have no effect on flux decline when pseudo-steady-state is reached (i.e. once the fouling layer governs permeability).

2.3.9.5 Chemically or biochemically modifying the mixed liquor

The biomass quality can be controlled biochemically, through adjustment of the SRT (Section 2.3.6.3) or chemically. In practice, SRT is rarely chosen on the basis of foulant concentration control, instead a target value is almost invariably based on membrane module clogging propensity and biomass aeration efficiency. However, studies have shown that a modicum of fouling control can be attained through the addition of chemicals.

Coagulant/flocculant Ferric chloride and aluminium sulphate (alum) have both been studied in relation to membrane fouling amelioration, most extensively for potable systems but also for MBRs. In MBR-based trials, addition of alum to the reactor led to a significant decrease in SMPc concentration, along with an improvement in membrane hydraulic performances (Holbrook *et al.*, 2004). Small biological colloids (from 0.1 to 2 μm) have been observed to coagulate and formed larger aggregate when alum was added to MBR activated sludge (Lee *et al.*, 2001b). Although more costly, dosing with ferric chloride was found to be more effective than alum. Ferric dosing of MBRs has been used for enhancing the production of iron-oxidizing bacteria responsible for the degradation of gaseous H_2S (Park *et al.*, 2005a). In this study, specific ferric precipitates like ferric phosphate and K-jarosite ($K\text{-}Fe_3(SO_4)_2(OH)_6$) have been observed to foul the membrane. Pretreatment of the effluent by pre-coagulation/sedimentation has revealed some fouling reduction (Adham *et al.*, 2004), and pre-clarification is employed at some sewage treatment works. In a recent example, the ferric dosing was shown to control both irreversible fouling and suspension viscosity (Itonaga *et al.*, 2004). Precoating of MBR membranes with ferric hydroxide has also been studied as a means of increasing permeability and improving permeate quality (Zhang *et al.*, 2004). In this study, additional ferric chloride was added to remove non-biodegradable organics which accumulated in the bioreactor.

Adsorbent agents Addition of adsorbents into biological treatment systems decreases the level of organic compounds. Dosing with PAC produces biologically activated carbon (BAC) which adsorbs and degrades soluble organics and has been shown to be effective in reducing SMP and EPS levels in a comparative study of a sidestream and immersed hybrid PAC–MBR (Kim and Lee, 2003). Decreased membrane fouling has also been demonstrated in studies of the effects of dosing MBR supernatant at up to 1 g/L PAC (Lesage *et al.*, 2005) and dosing activated sludge itself (Li *et al.*, 2005c), for which an optimum PAC concentration of 1.2 g/L was recorded. In the latter study, floc size distribution and apparent biomass viscosity were identified as being the main parameters influenced, resulting in a reduced cake resistance, when PAC was dosed into the bioreactor. Conversely, no significant improvement in performance was recorded when a concentration of 5 g/L of PAC was maintained in the bioreactor without sludge wastage (Ng *et al.*, 2005). It was postulated that, under these conditions, the PAC was rapidly saturated with organic pollutants and that fouling suppression by PAC relies on its regular addition brought about by lower SRTs.

Experiments conducted with different system configurations based on immersed HF membranes allowed direct comparison of hydraulic performances for pre-flocculation and PAC addition. Under the operating conditions employed, pre-flocculation provided higher fouling mitigation than that of PAC addition (Guo *et al.*, 2004). However, the use of both strategies simultaneously provided the greatest permeability enhancement (Cao *et al.*, 2005; Guo *et al.*, 2004).

A detailed mathematical model has been proposed for predicting performances for hybrid PAC–MBR systems (Tsai *et al.*, 2004). The model encompasses sub-processes such as biological reaction in bulk liquid solution, film transfer from bulk liquid phase to the biofilm, diffusion with biological reaction inside the biofilm, adsorption equilibria at the biofilm–adsorbent interface and diffusion within the PAC particles. Numerous other studies in which the use of PAC has been reported for fouling amelioration have generally been limited in scope and have not addressed the cost implications of reagent usage and sludge disposal. Tests have been performed using zeolite (Lee *et al.*, 2001b) and aerobic granular sludge, with an average size around 1 mm (Li *et al.*, 2005b) to create granular flocs of lower specific resistance. Granular sludge was found to increase membrane permeability by 50% but also lower the permeability recovery from cleaning by 12%, which would be likely to lead to unsustainable operation.

A novel "membrane performance enhancer" MPE50, a cationic polymer-based compound, has recently been developed by the company Nalco for use in MBRs. The addition of 1 g/L of of the reagent directly to the bioreactor led to the reduction of SMPc from 41 to 21 mg/L (Yoon *et al.*, 2005). The interaction between the polymer and the soluble organics in general, and SMPc in particular, was identified as being the main mechanism responsible for the performance enhancement. In another example, an MBR operated at an MLSS level as high as 45 g/L yielded a lower fouling propensity when 2.2 g/L of polymer was dosed into the bioreactor. The product has been tested at full scale (Section 5.2.1.4).

2.4 Summary

Membrane separation processes applied to MBRs have conventionally been limited to MF and UF for separation of the permeate product from the bioreactor MLSS. Other processes, in which the membrane is used to support a biomass and facilitate gas transfer into the biofilm (Section 2.3.2–2.3.3), have not reached the commercial stage of development. Membrane module configurations employed for biomass separation MBRs are limited to FS and HF for immersed processes (where the membrane is placed in the tank), and mainly MTs where it is placed outside the tank. The latter provide shear through pumping, as with most other membrane processes, whereas immersed processes employ aeration to provide shear. Shear enhancement is critical in promoting permeate flux through the membrane and suppressing membrane fouling, but generating shear also demands energy.

A considerable amount of research has been devoted to the study of membrane fouling phenomena in MBRs, and there is a general consensus that fouling constituents originate from the clarified biomass. Many authors that have employed standard chemical analysis on this fraction have identified the carbohydrate fraction of the SMPs (SMPc) arising from the bacterial cells as being mainly responsible for fouling, rather than suspended solid materials. However, recent attempts to predict fouling rates by EPS/SMP levels have not translated well across different plants or studies since biomass characteristics vary significantly from one plant to another. Moreover, achieving a consensus on the relative contributions of candidate foulants to membrane fouling is constrained by the different analytical methodologies and instruments employed.

There are also cross-disciplinary issues in the area of membrane fouling. There appears to be little interconnection between foulant analysis in the wastewater and potable applications, and membrane cleaning between the industrial process and municipal water and wastewater sectors. Studies in the potable area tend to point to colloidal materials and Ca-organic carboxylate complexation as being the two key foulant types, and this may apply as much to wastewater as potable water membrane applications. Within the municipal sector, the number of studies devoted to characterisation of foulants vastly exceeds that for optimising chemical cleaning, notwithstanding the fact that it is the latter which controls irrecoverable fouling and so, ultimately, membrane life. Membrane cleaning in industrial process water applications, however, is rather more advanced – dating back to the 1980s – with protocols arguably developed on a more scientific basis than those in the municipal sector.

Dynamic effects exert the greatest influence on consistency in MBR performance, ultimately leading to equipment and/or consent (i.e. target product water quality) failures. Specifications for full-scale MBR installations are generally based on conservative estimates of hydraulic and organic (and/or ammoniacal) loading. However, in reality, these parameters fluctuate significantly. Moreover, even more significant and potentially catastrophic deterioration in performance can arise through equipment malfunction and operator error. Such events can be expected to produce, over short periods of time:

- decreases in the MLSS concentration (either through loss of solids by foaming or by dilution with feedwater),
- foaming problems, sometimes associated with the above,
- loss of aeration (through control equipment malfunction or aerator port clogging),
- loss of permeability (through misapplication of backflush and cleaning protocol, hydraulic shocks or contamination of the feed with some unexpected component).

Lastly, the vast majority of all studies in this area have focused on fouling and, specifically, the characterisation of polymeric substances and mostly ignored the potentially more serious problem of clogging. Clogging can arise both in the membrane channels and in the aerator ports, in both cases impacting deleteriously on flux distribution and thus fouling rates. In this respect, developing methods of ensuring homogeneity of air distribution will advance both fouling and clogging control. Once again, though, the number of MBR papers published which have been devoted specifically to aerator design are limited.

The nature of irrecoverable fouling and clogging is that they can only be studied over extended time periods which are simply not conducive to academic research. Research into fouling characterisation is likely to continue for some time and new membranes and systems are being developed from research programmes globally. However, much practical information can be obtained from the examination of pilot- and full-scale plant data (Chapters 3 and 5), and it is also instructive and expedient to consider the attributes of existing individual commercial technologies (Chapter 4).

References

Abbassi, B., Dullstein, S. and Rabiger, N. (1999) Minimization of excess sludge production by increase of oxygen concentration in activated sludge flocs; experimental and theoretical approach. *Water Res.*, **34**, 139–146.

Adham, S., DeCarolis, J.F. and Pearce, W. (2004) Optimization of various mbr systems for water reclamation – phase iii, Desalination and Water Purification Research and Development Program, Final Report No. 103, Agreement No. 01-FC-81–0736.

Ahmed, T. and Semmens, M.J. (1992a) The use of independently sealed microporous membranes for oxygenation of water: model development. *J. Membrane Sci.*, **69**, 11–20.

Ahmed, T. and Semmens, M.J. (1992b) Use of sealed end hollow fibres for bubbleless membrane aeration: experimental studies. *J. Membrane Sci.*, **69**, 1–10.

Ahn, Y.T., Kang, S.T., Chae, S.R., Lim, J.L., Lee, S.H. and Shin, H.S. (2005) Effect of internal recycle rate on the high-strength nitrogen wastewater treatment in the combined ubf/mbr system. *Water Sci. Technol.*, **51**, 241–247.

ASCE (1992) ASCE Standard – Measurement of Oxygen Transfer in Clean Water, ANSI/ASCE 2–91, (2nd edn). Reston VA.

Ashley, K.I., Mavinic, D.S. and Hall, K.J. (1992) Bench scale study of oxygen transfer in coarse bubble diffused aeration. *Water Res.*, **26**, 1289–1295.

Badino, J.A.C., Facciotti, M.C.R. and Schmidell, W. (2001) Volumetric oxygen transfer coefficients (kLa) in batch cultivations involving non-Newtonian broths. *Biochem. Engng. J.*, **8**, 111–119.

Bae, T.-H. and Tak, T.-M. (2005) Interpretation of fouling characteristics of ultrafiltration membranes during the filtration of membrane bioreactor mixed liquor. *J. Membrane Sci.*, **264**, 151–160.

Baek, S.H. and Pagilla, K. (2003) Comparison of aerobic and anaerobic membrane bioreactors for municipal wastewater treatment, *Proceedings 76th Water Environment Federation Technical Exposition and Conference* [CD-ROM]; Los Angeles, Ca.

Barker, D.J. and Stuckey, D.C. (2001) Modeling of soluble microbial products in anaerobic digestion: the effect of feed strength and composition. *Water Environ. Res.*, **73**, 173.

Barker, D.J., Salvi, S.M., Langenhoff, A.A.M. and Stuckey, D.C. (2000) Soluble microbial products in ABR treating low-strength wastewater. *J. Environ. Eng.*, **26**, 239.

Barrieros, A.M., Rodrigues, C.M., Crespo, J.P.S.G. and Reis, M.A.M. (1998) Membrane bioreactor for drinking water denitrification. *Bioprocess Eng.*, **18**, 297–302.

Beaubien, A., Baty, M., Jeannot, F., Francoeur, E. and Manem, J. (1996) Design and operation of anaerobic membrane bioreactors: Development of a filtration testing strategy. *J. Membrane Sci.*, **109**, 173–184.

Bérubé, P., Hall, E. and Sutton, P. (2006) Mechanisms governing the permeate flux in an anaerobic membrane bioreactor treating low-strength/municipal wastewaters: a literature review. *Water Env. Res.*, in press.

Bérubé, P.R. and Lei, E. (2004) Impact of membrane configuration and hydrodynamic conditions on the permeate flux in submerged membrane systems for drinking water treatment, *Proceedings AWWA Water Quality Technology Conference* [CD-Rom]; San Antonio, TX.

Bin, C., Xiaochang, W. and Enrang, W. (2004) Effects of tmp, mlss concentration and intermittent membrane permeation on a hybrid submerged mbr fouling, *Proceedings of Water Environment-Membrane Technology Conference*, Seoul, Korea.

Bouchez, T., Patreau, D., Dabert, P., Juretschko, S., Wagner, M., Godon, J., Delgenès, J. and Moletta, R. (1998) Bioaugmentation of a nitrifying sequencing batch reactor with the aerobic denitrifying bacteria Microvirgula aerodenitrificans: effects on nitrogen removal and consequences in terms of microbial community, *Proceedings of the European Conference on New Advances in Biological Nitrogen and Phosphorus Removal for Municipal or Industrial Wastewaters*, October, Narbonne, France, 273–277.

Bouhabila, E.H., Ben Aim, R. and Buisson, H. (1998) Microfiltration of activated sludge using submerged membrane with air bubbling (application to wastewater treatment). *Desalination*, **118**, 315–322.

Bouhabila, E.H., Ben Aïm, R. and Buisson, H. (2001) Fouling characterisation in membrane bioreactors. *Sep. Purif. Tech.*, 22–23, 123–132.

Bourgeous, K.N., Darby, J.L. and Tchobanoglous, G. (2001) Ultrafiltration of wastewater: effects of particles, mode of operation, and backwash effectiveness. *Water Res.*, **35**, 77.

Bowen, W. R., Calvo, J. I. and Hernandez, A. (1995) Steps of membrane blocking in flux decline during protein microfiltration. *J. Membrane Sci.*, **101**, 153–165.

Brindle, K., Stephenson, T. and Semmens, M.J. (1998) Nitrification and oxygen utilisation in a membrane aeration bioreactor. *J. Membrane Sci.*, **144**, 197–209.

Brindle, K., Stephenson, T. and Semmens, M.J. (1999) Pilot plant treatment of a high strength brewery wastewater using a membrane aeration bioreactor. *Water Environ. Res.*, **71**, 1197–1204.

Brockman, M. and Seyfried, C.F. (1996) Sludge activity and cross-flow microfiltration – a non-beneficial relationship. *Water Sci. Technol.*, **34**(9), 205.

Brookes, A., Jefferson, B., Le-Clech, P. and Judd, S. (2003a) Fouling of membrane bioreactors during treatment of produced water, *Proceedings of International Membrane Science and Technology Conference (IMSTEC)*, Sydney, Australia.

Brookes, A., Judd, S., Reid, E., Germain, E., Smith, S., Alvarez-Vazquez, H., Le-Clech, P., Stephenson, T., Turra, E. and Jefferson, B. (2003b) Biomass characterisation in membrane bioreactors, *Proceedings of International Membrane Science and Technology Conference (IMSTEC)*, Sydney, Australia.

Brookes, A., Jefferson, B. and Judd, S.J. (2004) Sub-critical fouling in a membrane bioreactor: impact of flux and MLSS, *IWA 4th Word Water Congress*, Marrakech, September.

Cabassud, C., Laborie, S. and Lainé, J.M. (1997) How slug flow can enhance the ultrafiltration flux in organic hollow fibres. *J. Membrane Sci.*, **128**, 93.

Cabassud, C., Massé, A., Espinosa-Bouchot, M. and Spérandio, M. (2004) Submerged membrane bioreactors: Interactions between membrane filtration and biological activity, *Proceedings of Water Environment – Membrane Technology Conference*, Seoul, Korea.

Cao, J.-H., Zhu, B.-K., Lu, H. and Xu, Y.-Y. (2005) Study on polypropylene hollow fiber based recirculated membrane bioreactor for treatment of municipal wastewater. *Desalination* **183**, 431–438.

Cha, G.-C., Jeong, T.-Y., Yoo, I.-K. and Kim, D.-J. (2004) Kinetics characteristics of smp and ecp in relation to loading rate in an mbr process, *Proceedings of Water Environment – Membrane Technology Conference*, Seoul, Korea.

Chang, I.-S. and Kim, S.-N. (2005) Wastewater treatment using membrane filtration–effect of biosolids concentration on cake resistance. *Proc. Biochem.*, **40**, 1307–1314.

Chang, I.S. and Lee, C.H. (1998) Membrane filtration characteristics in membrane-coupled activated sludge system – the effect of physiological states of activated sludge on membrane fouling. *Desalination*, **120**, 221–233.

Chang, I.-S., Choo, K.-H., Lee, C.-H., Pek, U.-H., Koh, U.-C., Kim, S.-W. and Koh, J.-H. (1994) Application of ceramic membrane as a pretreatment in anaerobic digestion of alcohol-distillery wastes. *J. Membrane Sci.* **90**, 131–139.

Chang, I.S., Lee, C.H. and Ahn, K.H. (1999) Membrane filtration characteristics in membrane-coupled activated sludge system: the effect of floc structure on membrane fouling. *Sep. Sci. Technol.*, **34**, 1743–1758.

Chang, I.-S., Bag, S.-O. and Lee, C.-H. (2001a) Effects of membrane fouling on solute rejection during membrane filtration of activated sludge. *Proc. Biochem.*, **36**, 855–860.

Chang, I.-S., Gander, M., Jefferson, B. and Judd, S.J. (2001b) Low-cost membranes for use in a submerged mbr. *Proc. Safety Env. Protect.*, **79**, 183–188.

Chang, I.-S., Le Clech, P., Jefferson, B. and Judd, S. (2002a) Membrane fouling in membrane bioreactors for wastewater treatment. *J. Env. Eng. ASCE.*, **128**, 1018–1029.

Chang, S. and Fane, A.G. (2002) Filtration of biomass with laboratory-scale submerged hollow fibre modules – effect of operating conditions and module configuration. *J. Chem. Technol. Biotechnol.*, **77**, 1030–1038.

Chang, S., Fane, A.G. and Vigneswaran, S. (2002b) Modeling and optimizing submerged hollow fiber membrane modules. *AICHE J.*, **48**, 2203–2212.

Chang, S., Fane, A.G. and Waite, T.D. (2005) Effect of coagulation within the cakelayer on fouling transitions with dead-end hollow fiber membranes, *Proceedings of International Congress on Membranes and Membrane Processes (ICOM)*, Seoul, Korea.

Chang, J., Manem, J. and Beaubien, A. (1993) Membrane bioprocesses for the denitrification of drinking water supplies. *J. Membrane Sci.*, **80**, 233–239.

Chatellier, P. and Audic, J.M. (2001) Mass balance for on-line alpha k(L)a estimation in activated sludge oxidation ditch. *Water Sci. Technol.*, **44**, 197–202.

Cho, B.D. and Fane, A.G. (2002) Fouling transients in nominally sub-critical flux operation of a membrane bioreactor. *J. Membrane Sci.*, **209**, 391–403.

Cho, J., Song, K.-G. and Ahn, K.-H. (2005a) The activated sludge and microbial substances influences on membrane fouling in submerged membrane bioreactor: Unstirred batch cell test. *Desalination*, **183**, 425–429.

Cho, J.W., Song, K.G., Lee, S.H. and Ahn, K.H. (2005b) Sequencing anoxic/anaerobic membrane bioreactor (sam) pilot plant for advanced wastewater treatment. *Desalination*, **178**, 219–225.

Choi, H., Zhang, K., Dionysiou, D.D., Oerther, D.B. and Sorial, G.A. (2005a) Effect of permeate flux and tangential flow on membrane fouling for wastewater treatment. *Sep. Purif. Technol.*, **45**, 68–78.

Choi, H., Zhang, K., Dionysiou, D.D., Oerther, D.B. and Sorial, G.A. (2005b) Influence of cross-flow velocity on membrane performance during filtration of biological suspension. *J. Membrane Sci.*, **248**, 189–199.

Choo, K.H. and Lee, C.H. (1996a) Membrane Fouling Mechanisms in the Membrane-Coupled Anaerobic Bioreactor. *Water Res.*, **30**, 1771.

Choo, K.H. and Lee, C.H. (1996b) Effect of anaerobic digestion broth composition on membrane permeability. *Water Sci. Technol.*, **34**(9), 173–179.

Choo, K.-H. and Lee, C.-H. (1998) Hydrodynamic behavior of anaerobic biosolids during crossflow filtration in the membrane anaerobic bioreactor. *Water Res.*, **32**, 3387–3397.

Choo, K.H., Kang, I.J., Yoon, S.H., Park, H., Kim, J.H., Adlya, S. and Lee, C.H. (2000) Approaches to membrane fouling control in anaerobic membrane bioreactors. *Water Sci. Technol.*, **41**(10–11), 363.

Chu, H.P. and Li, X.Y. (2005) Membrane fouling in a membrane bioreactor (mbr): sludge cake formation and fouling characteristics. *Biotech. Bioeng.*, **90**, 323–331.

Chua, H.C., Arnot, T.C. and Howell, J.A. (2002) Controlling fouling in membrane bioreactors operated with a variable throughput. *Desalination*, **149**, 225–229.

Chung, Y.C., Jung, J.Y., Ahn, D.H. and Kim, D.H. (1998) Development of two phase anaerobic reactor with membrane separation system. *J. Environ. Sci. Health, A*, 33, 249.

Cicek, N., Franco, J.P., Suidan, M.T. and Urbain, V. (1998) Using a membrane bioreactor to reclaim wastewater. *J. Am. Water Works Assoc.*, **90**, 105–113.

Cicek, N., Franco, J.P., Suidan, M.T., Urbain, V. and Manem, J. (1999) Characterization and comparison of a membrane bioreactor and a conventional activated-sludge system in the treatment of wastewater containing high-molecular-weight compounds. *Water Env. Res.*, **71**, 64–70.

Côté, P. (2002) Inverted air box aerator and aeration method for immersed membrane, US Patent 6,863,823.

Côté, P., Bersillon, J.L. and Faup, G. (1988) Bubble free aeration using membranes: process analysis. *J. Water Pollut. Control Fed.*, **60**, 1986–1992.

Côté, P., Buisson, H. and Praderie, M. (1998) Immersed membranes activated sludge process applied to the treatment of municipal wastewater. *Water Sci. Technol.*, **38**, 437.

Crespo, J.G., Velizarov, S. and Reis, M.A. (2004) Membrane bioreactors for the removal of anionic micropollutants from drinking water. *Biochem. Eng.*, **15**, 463–468.

Cui, Z.F. and Wright, K.I.T. (1994) Gas–liquid 2-phase cross-flow ultrafiltration of Bsa and Dextran Solutions. *J. Membrane Sci.*, **90**, 183–189.

Cui, Z.F., Chang, S. and Fane, A.G. (2003) The use of gas bubbling to enhance membrane processes. *J. Membrane Sci.*, **221**, 1–35.

Defrance, L. and Jaffrin, M.Y. (1999) Reversibility of fouling formed in activated sludge filtration. *J. Membrane Sci.*, **157**, 73–84.

Defrance, L., Jaffrin, M.Y., Gupta, B., Paullier, P. and Geaugey, V. (2000) Contribution of various species present in activated sludge to membrane bioreactor foulong. *Bioresource Technol.*, **73**, 105–112.

Delanghe, B., Nakamura, F., Myoga, H., Magara, Y. and Guibal, E. (1994) Drinking water denitrification in a membrane bioreactor. *Water Sci. Tech.*, **30**(6), 157–160.

Deronzier, G., Gillot, S., Duchène, Ph. and Héduit, A. (1996) Influence de la vitesse horizontale du fluide sur le transfert d'oxygène en fines bulles dans les basins d'aération. *Tribune de l'eau*, 5–6, 91–98.

Dick, R.J. and Ewing, B.B. (1967) The rheology of activated sludge. *J. Water Poll. Cont. Fed.*, **39**, 543–560.

Drews, A., Vocks, M., Iversen, V., Lesjean, B. and Kraume, M. (2005) Influence of unsteady membrane bioreactor operation on eps formation and filtration resistance, *Proceedings of International Congress on Membranes and Membrane Processes* (ICOM), Seoul, Korea.

Dubois, M., Gilles, K.A., Hamilton, J.K., Rebers, P.A. and Smith, P. (1956) Colorimetric method for determination of sugars and related substances. *Anal. Chem.*, **28**, 350–356.

Dufresne, R., Lebrun, R.E. and Lavallee, H.C. (1997) Comparative study on fluxes and performances during papermill wastewater treatment with membrane bioreactor. *Can. J. Chem. Engng.*, **75**, 95–103.

Eckenfelder, W. and Grau, P. (1998) Activated Sludge Process Design and Control: Theory and Practice (2nd edn). Techonomic, Lancaster Pa.

Elmaleh, S. and Abdelmoumni, L. (1997) Cross-Flow Filtration of an Anaerobic Methanogenic Suspension. *J. Mem. Sci.*, **131**, 261.

Elmaleh, S. and Abdelmoumni, L. (1998) Experimental test to evaluate performance of an anaerobic reactor provided with an external membrane unit. *Water Sci. Technol.*, **38**, 8–9, 385.

EPA (1989) EPA/ASCE Design manual on fine pore aeration. Cincinnati, Ohio.

Ergas, S.J. and Reuss, A.F. (2001) Hydrogenotrophic denitrification of drinking water using a hollow fibre membrane bioreactor. *J. Water Supply Res. Technol.*, **50.3**, 161–171.

Ergas, S.J. and Rheinheimer, D.E. (2004) Drinking water denitrification using a membrane bioreactor. *Water Res.*, **38**, 3225–3232.

Evenblij, H. and van der Graaf, J. (2004) Occurrence of eps in activated sludge from a membrane bioreactor treating municipal wastewater. *Water Sci. Technol.*, **50**, 293–300.

Evenblij, H., Geilvoet, S., van der Graaf, J. and van der Roest, H.F. (2005a) Filtration characterisation for assessing mbr performance: three cases compared. *Desalination*, **178**, 115–124.

Evenblij, H., Verrecht, B., van der Graaf, J. and Van der Bruggen, B. (2005b) Manipulating filterability of mbr activated sludge by pulsed substrate addition. *Desalination*, **178**, 193–201.

Fan, X.-J., Urbain, V., Qian, Y. and Manem, J. (1996) Nitrification and mass balance with a membrane bioreactor for municipal wastewater treatment. *Water Sci. Technol.*, **34**, 129–136.

Fan, F., Zhou, H. and Husain, H. (2006) Identification of wastewater sludge characteristics to predict critical flux for membrane bioreactor processes. *Water Res.*, (submitted).

Fane, A.G., Fell, C.J.D. and Suzuki, A. (1983) The effect of pH and ionic environment on the ultrafiltration of protein solutions with retentive membranes. *J. Membrane Sci.*, **16**, 195.

Fane, A.G., Chang, S. and Chardon, E. (2002) Submerged hollow fibre membrane module – design options and operational considerations. *Desalination*, **146**, 231–236.

Fang, H.H.P. and Shi, X. (2005) Pore fouling of microfiltration membranes by activated sludge. *J. Membrane Sci.*, **264**, 161–166.

Fawehinmi, F., Lens, P., Stephenson, T., Rogalla, F. and Jefferson, B. (2004) The influence of operating conditions on extracellullar polymeric substances (eps), soluble microbial products (smp) and bio-fouling in anaerobic membrane bioreactors, *Proceedings of Water Environment – Membrane Technology Conference*, Seoul, Korea.

Field, R.W., Wu, D., Howell, J.A. and Gupta, B.B. (1995) Critical flux concept for microfiltration fouling. *J. Membrane Sci.*, **100**, 259–272.

Flemming, H.C. and Wingender, J. (2001) Relevance of microbial extracellular polymeric substances (epss) – part i: Structural and ecological aspects. *Water Sci. Technol.*, **43**, 1–8.

Flemming, H.C., Schaule, G., Griebe, T., Schmitt, J. and Tamachkiarowa, A. (1997) Biofouling – the achilles heel of membrane processes. *Desalination*, **113**, 215–225.

Fonseca, A.D., Crespo, J.G., Almeida, J.S. and Reis, M.A. (2000) Drinking water denitrification using a novel ion-exchange membrane bioreactor. *Environ. Sci. Technol.*, **34**, 1557–1562.

Frederickson, K. and Cicek, N. (2004) Performance comparison of a pilot-scale membrane bioreactor and a full-scale sequencing batch reactor with sand filtration: treatment of low strength wastewater from a northern Canadian aboriginal community, *Proceedings of Water Environment – Membrane Technology Conference*, Seoul, Korea.

Freitas, C. and Teixeira, J.A. (2001) Oxygen mass transfer in a high solids loading three-phase internal-loop airlift reactor. *Chem. Eng. J.*, **84**, 57–61.

Frolund, B., Palmgren, R., Keiding, K. and Nielsen, P.H. (1996) Extraction of extracellular polymers from activated sludge using a cation exchange resin. *Water Res.*, **30**, 1749–1758.

Fuchs, W., Schatzmayr, G. and Braun, R. (1997) Nitrate removal from drinking water using a membrane fixed biofilm reactor. *Appl. Microbiol. Biotechnol.*, **48**, 267–274.

Fuchs, W., Braun, R. and Theiss, M. (2005) Influence of various wastewater parameters on the fouling capacity during membrane filtration, *Proceedings of International Congress on Membranes and Membrane Processes* (ICOM), Seoul, Korea.

Fujie, K., Hu, H.-Y., Ikeda, Y. and Urano, K. (1992) Gas–liquid oxygen transfer characteristics in an aerobic submerged biofilter for the wastewater treatment. *Chem. Eng. Sci.*, **47**, 3745–3752.

Gander, M.A., Jefferson, B. and Judd, S.J. (2000) Membrane bioreactors for use in small wastewater treatment plants: Membrane materials and effluent quality. *Water Sci. Technol.*, **41**, 205–211.

Gao, M., Yang, M., Li, H., Wang, Y. and Pan, F. (2004a) Nitrification and sludge characteristics in a submerged membrane bioreactor on synthetic inorganic wastewater. *Desalination*, **170**, 177–185.

Gao, M., Yang, M., Li, H., Yang, Q. and Zhang, Y. (2004b) Comparison between a submerged membrane bioreactor and a conventional activated sludge system on treating ammonia-bearing inorganic wastewater. *J. Biotechnol.*, **108**, 265–269.

Garcia-Ochoa, F., Castro, E.G. and Santos, V.E. (2000) Oxygen transfer and uptake rates during xanthan gum production. *Enzyme Microbial. Technol.*, **27**, 680–690.

Germain, E. (2004). Biomass effects on Membrane Bioreactor operations, *EngD Thesis*, Cranfield University.

Ghosh, R. and Cui, Z.F. (1999) Mass transfer in gas-sparged ultrafiltration: upward slug flow in tubular membranes. *J. Membrane Sci.*, **162**, 91.

Ghyoot, W.R. and Verstraete, W.H. (1997) Coupling membrane filtration to anaerobic primary sludge digestion. *Environ. Technol.*, **18**, 569.

Ghyoot, W. and Verstraete, W. (2000) Reduced sludge production in a two-stage membrane-assisted bioreactor. *Water Res.*, **34**, 205–215.

Gillot, S., Capela, S. and Heduit, M. (2000) Effect of horizontal flow on oxygen transfer in clean water and in clean water with surfactants. *Water Res.*, **34**, 678–683.

Gorner, T., de Donato, P., Ameil, M.-H., Montarges-Pelletier, E. and Lartiges, B.S. (2003) Activated sludge exopolymers: Separation and identification using size exclusion chromatography and infrared micro-spectroscopy, *Water Res.*, **37**, 2388–2393.

Grace, H.P. (1956) Resistance and compressibility of filter cakes. *Chem. Eng. Prog.*, **49**, 303–318.

Grady, C.P.L., Daigger, G.T., and Lim, H.C. (1998). Biological Wastewater Treatment, 2nd Ed., Marcel Dekker Masoud, A., Sohrabi, M., Vahabzadeh, F., and Bonakdarpour, B. (2001). Hydrodynamic and mass transfer characterization of a down flow jet loop bioreactor, *Biochem. Eng. J.*, **8**(3), 241–250.

Green, G. and Belfort, G. (1980) Fouling of ultrafiltration membranes: lateral migration and the particle trajectory model. *Desalination*, **35**, 129–147.

Grelier, P., Rosenberger, S. and Tazi-Pain, A. (2005) Influence of sludge retention time on membrane bioreactor hydraulic performance, *Proceedings of International Congress on Membranes and Membrane Processes* (ICOM), Seoul, Korea.

Grethlein, H.E. (1978) Anaerobic digestion and membrane separation of domestic wastewater. *J. WPCF*, 754.

Gu, X.-S. (1993) *Mathematical Models in Biological Wastewater Treatment*. Tsinghua University press, Beijing.

Günder, B. (2001) *The Membrane Coupled-Activated Sludge Process in Municipal Wastewater Treatment*. Technomic Publishing Company Inc., Lancaster.

Gunder, B. and Krauth, K. (1998) Replacement of secondary clarification by membrane separation – results with plate and hollow fibre modules. *Water Sci. Technol.*, **38**, 383–393.

Gunder, B. and Krauth, K. (1999) Replacement of secondary clarification by membrane separation-results with tubular, plate and hollow fibre modules. *Water Sci. Technol.*, **40**, 311–320.

Guo, W.S., Vigneswaran, S. and Ngo, H.H. (2004) A rational approach in controlling membrane fouling problems: Pretreatments to a submerged hollow fiber membrane system, *Proceedings of Water Environment – Membrane Technology Conference*, Seoul, Korea.

Hai, F.I., Yamamoto, K. and Fukushi, K. (2005) Different fouling modes of submerged hollow-fiber and FS membranes induced by high strength wastewater with concurrent biofouling. *Desalination*, **180**, 89–97.

Han, S.S., Bae, T.H., Jang, G.G. and Tak, T.M. (2005) Influence of sludge retention time on membrane fouling and bioactivities in membrane bioreactor system. *Proc. Biochem.*, **40**, 2393–2400.

Haugen, K.S., Semmens, M.J. and Novak, P.J. (2002) A novel *in situ* technology for the treatment of nitrate contaminated groundwater. *Water Res.*, **36**, 3497–3506.

He, Y., Li, C., Wu, Z. and Gu, G. (1999) Study on molecular weight cut-off of the anaerobic ultrafiltration membrane bioreactor. *China Water Wastewater*, **15**, 10.

He, Y., Xu, P., Li, C. and Zhang, B. (2005) High-concentration food wastewater treatment by an anaerobic membrane bioreactor. *Water Res.*, **39**, 4110–4118.

Hermanowicz, S.W. (2004) Membrane filtration of biological solids: a unified framework and its applications to membrane bioreactors, *Proceedings of Water Environment – Membrane Technology Conference*, Seoul, Korea.

Hernández, A.E., Belalcazar, L.C., Rodriguez, M.S. and Giraldo, E. (2002) Retention of granular sludge at high hydraulic loading rates in an anaerobic membrane bioreactor with immersed filtration. *Water Sci. Technol.*, **45** (10), 169.

Hernandez Rojas, M.E., Van Kaam, R., Schetrite, S. and Albasi, C. (2005) Role and variations of supernatant compounds in submerged membrane bioreactor fouling. *Desalination*, **179**, 95–107.

Ho, C.M., Tseng, S.K. and Chang, Y.J. (2001) Autotrophic denitrification via a novel membrane-attached biofilm reactor. *Lett. Appl. Microbiol.*, **33**, 201–205.

Holbrook, R.D., Higgins, M.J., Murthy, S.N., Fonseca, A.D., Fleischer, E.J., Daigger, G.T., Grizzard, T.J., Love, N.G. and Novak, J.T. (2004) Effect of alum addition on the performance of submerged membranes for wastewater treatment. *Water Env. Res.*, **76**, 2699–2702.

Hong, S.P., Bae, T.H., Tak, T.M., Hong, S. and Randall, A. (2002) Fouling control in activated sludge submerged hollow fiber membrane bioreactors. *Desalination*, **143**, 219–228.

Howell, J.A. (1995) Subcritical flux operation of microfiltration. *J. Membrane Sci.*, **107**, 165–171.

Howell, J.A., Chua, H.C. and Arnot, T.C. (2004) *In situ* manipulation of critical flux in a submerged membrane bioreactor using variable aeration rates, and effects of membrane history. *J. Membrane Sci.*, **242**, 13–19.

Huang, X., Gui, P. and Qian, Y. (2001) Effect of sludge retention time on microbial behaviour in a submerged membrane bioreactor. *Proc. Biochem.*, **36**, 1001–1006.

Hwang, S.-J. and Lu, W.-J. (1997) Gas–liquid mass transfer in an internal loop airlift reactor with low density particles. *Chem. Eng. Sci.*, **52**, 853–857.

Imasaka, T., Kanekuni, N., So, H. and Yoshino, H. (1989) Cross-flow of methane fermentation broth by ceramic membrane. *J. Ferment. Bioeng.*, **68**, 200.

Iranpour, R., Magallanes, M., Zermeño, M., Varsh, V., Abrishamchi, A. and Stenstrom, M. (2000) Assessment of aeration basin performance efficiency: sampling methods and tank coverage. *Water Res.*, **34**(12), 3137–3152.

Ishiguro, K., Imai, K. and Sawada, S. (1994) Effects of biological treatment conditions on permeate flux of uf membrane in a membrane/activated-sludge wastewater treatment system. *Desalination*, **98**, 119–126.

Itonaga, T., Kimura, K. and Watanabe, Y. (2004) Influence of suspension viscosity and colloidal particles on permeability of membrane used in membrane bioreactor (mbr). *Water Sci. Technol.*, **50**, 301–309.

IUPAC (1985) Reporting physisorption data. *Pure Appl. Chem.*, **57**, 603.

Jang, N.J., Yeo, Y.H., Hwang, M.H., Vigneswaran, S., Cho, J. and Kim, I.S. (2004) The effect of dissolved air on the filtration resistance in the hollow fiber membrane bioreactor, *Proceedings of Water Environment – Membrane Technology Conference*, Seoul, Korea.

Jang, N., Ren, X., Choi, K. and Kim, I.S. (2005a) Comparison of membrane biofouling in nitrification and denitrification for the membrane bio-reactor (mbr), *Proceedings of IWA*, Aspire, Singapore.

Jang, N.J., Trussell, R.S., Merlo, R.P., Jenkins, D., Hermanowicz, S.W. and Kim, I.S. (2005b) Exocellular polymeric substances molecular weight distribution and filtration resistance as a function of food to microorganism ratio in the submerged membrane bioreactor, *Proceedings of International Congress on Membranes and Membrane Processes* (ICOM), Seoul, Korea.

Jefferson, B., Brookes, A., Le Clech, P. and Judd, S.J. (2004) Methods for understanding organic fouling in mbrs. *Water Sci. Technol.*, **49**, 237–244.

Ji, L. and Zhou, J. (2006) Influence of aeration on microbial polymers and membrane fouling in submerged membrane bioreactors. *J. Membrane Sci.*, in press.

Jiang, T., Kennedy, M.D., Guinzbourg, B.F., Vanrolleghem, P.A. and Schippers, J.C. (2005) Optimising the operation of a mbr pilot plant by quantitative analysis of the membrane fouling mechanism. *Water Sci. Technol.*, **51**, 19–25.

Jin, B., Yu, Q., Yan, X.Q. and van Leeuwen, J.H. (2001) Characterization and improvement of oxygen transfer in pilot plant external air-lift bioreactor for mycelial biomass production. *World J. Microbiol. Biotechnol.*, **17**, 265–272.

Jinhua, P., Fukushi, K. and Yamamoto, K. (2004) Structure of microbial communities on membrane surface in a submerged membrane bioreactor, *Proceedings of Water Environment – Membrane Technology Conference*, Seoul, Korea.

Judd, S. (2005) Fouling control in submerged membrane bioreactors. *Water Sci. Technol.*, **51**(6–7), 27–34.

Judd, S. and Jefferson, B. (2003) Membranes for industrial wastewater recovery and re-use, Elsevier, Oxford.

Judd, S.J., Le-Clech, P., Cui, Z. and Taha, T. (2001) Theoretical and experimental representation of a submerged MBR system, *3rd International Meeting on Membrane Bioreactors*, 1–13, Cranfield University, May.

Judd, S.J., Robinson, T., Alvarez-Vazquez, H., Holdner, J. and Jefferson, B. (2004) Impact of membrane material on membrane bioreactor permeability. *Water Environ. – Membrane Technol., (WEMT 2004)*, Seoul, June 2004.

Kang, I.J. (1996) Comparison of fouling characteristics between ceramic and polymeric membranes in membrane coupled anaerobic bioreactors, *MS Thesis*, Department of Chemical Technology, Seoul National University, Korea.

Kang, I.J., Yoon, S.H. and Lee, C.H. (2002) Comparison of the filtration characteristics of organic and inorganic membranes in a membrane-coupled anaerobic bioreactor. *Water Res.*, **36**, 1803.

Kang, I.-J., Lee, C.-H. and Kim, K.-J. (2003) Characteristics of microfiltration membranes in a membrane coupled sequencing batch reactor system. *Water Res.*, **37**, 1192–1197.

Kayawake, E., Narukami, Y. and Yamagata, M. (1991) Anaerobic digestion by a ceramic membrane enclosed reactor. *J. Ferment. Bioeng.*, **71**, 122.

Kim, J. and DiGiano, F.A. (2004) The effects of hollow fiber length and aeration rate on filtration behavior in submerged microfiltration, *Proceedings of Water Environment – Membrane Technology Conference*, Seoul, Korea.

Kim, J.S. and Lee, C.H. (2003) Effect of powdered activated carbon on the performance of an aerobic membrane bioreactor: comparison between cross-flow and submerged membrane systems. *Water Env. Res.*, **75**, 300–307.

Kim, J., Jang, M., Chio, H. and Kim, S. (2004) Characteristics of membrane and module affecting membrane fouling, *Proceedings of Water Environment – Membrane Technology Conference*, Seoul, Korea.

Kimura, K., Nakamura, M. and Yoshimasa, W. (2002) Nitrate removal by a combination of elemental sulphur based denitrification and membrane filtration. *Water Res.*, **36**, 1758–1766.

Kitamura, Y., Maekawa, T., Tagawa, A., Hayashi, H. and Farrell-Poe, K.L. (1996) Treatment of strong organic, nitrogenous wastewater by an anaerobic contact process incorporating ultrafiltration. *Appl. Eng. Agric.*, **12**, 709.

Klein, J., Rosenberg, M., Markos, J., Dolgos, O., Kroslak, M. and Kristofikova, Lu (2002) Biotransformation of glucose to gluconic acid by Aspergillus niger–study of mass transfer in an airlift bioreactor. *Biochem. Eng. J.*, **10**, 197–205.

Koide, K., Shibata, K., Ito, H., Kim, S.Y. and Ohtaguchi, K. (1992) Gas holdup and volumetric liquid-phase mass transfer coefficient in a gel-particle suspended bubble column with draught tube. *J. Chem. Eng. Japan*, **25**, 11–16.

Komaromy, P. and Sisak, C. (1994) Investigation of gas–liquid oxygen-transport 3-phase bioreactor. *Hungarian J. Ind. Chem.*, **22**, 147–151.

Kouakou, E., Salmon, T., Toye, D., Marchot, P. and Crine, M. (2005) Gas–liquid mass transfer in a circulating jet-loop nitrifying MBR. *Chem. Eng. Sci.*, **60**, 6346–6353.

Krampe, J. and Krauth, K. (2003) Oxygen transfer into activated sludge with high MLSS concentrations. *Water Sci. Technol.*, **47**, 297–303.

Krauth, K. and Staab, K.F. (1993) Pressurized bioreactor with membrane filtration for wastewater treatment. *Water Res.*, **27**, 405–411.

Kwon, D.Y. and Vigneswaran, S. (1998) Influence of particle size and surface charge on critical flux of crossflow microfiltration. *Water Sci. Technol.*, **38**, 481–488.

Kwon, D.Y., Vigneswaran, S., Fane, A.G. and Ben Aim, R. (2000) Experimental determination of critical flux in cross-flow microfiltration. *Sep. Purif. Technol.*, **19**, 169–181.

Laspidou, C.S. and Rittmann, B.E. (2002) A unified theory for extracellular polymeric substances, soluble microbial products, and active and inert biomass. *Water Res.*, **36**, 2711–2720.

Lazarova, V., Julián, M., Laurent, D. and Jaques M. (1997) A novel circulating bed reactor: hydrodynamics, mass transfer and nitrification capacity. *Chem. Eng. Sci.*, **52**, (21–22) 3919–3927.

Le-Clech, P., Alvarez-Vazquez, H., Jefferson, B. and Judd, S. (2003a) Fluid hydrodynamics in submerged and sidestream membrane bioreactors. *Water Sci. Technol.*, **48**, 113–119.

Le-Clech, P., Jefferson, B., Chang, I.S. and Judd, S.J. (2003b) Critical flux determination by the flux-step method in a submerged membrane bioreactor. *J. Membrane Sci.*, **227** 81–93.

Le-Clech, P., Jefferson, B. and Judd, S.J. (2003c) Impact of aeration, solids concentration and membrane characteristics on the hydraulic performance of a membrane bioreactor. *J. Membrane Sci.*, **218**, 117–129.

Le-Clech, P., Fane, A., Leslie, G. and Childress, A. (2005a) The operator's perspective. *Filt. Sep.*, **42**, 20–23.

Le-Clech, P., Jefferson, B. and Judd, S.J. (2005b) A comparison of submerged and sidestream tubular membrane bioreactor configurations. *Desalination*, **173**, 113–122.

Le-Clech, P., Cao, Z., Wan, P.Y., Wiley, D.E. and Fane, A.G. (In preparation) Feasibility of using hand-made constant temperature anemometry in spiral wound membrane. *J. Membrane Sci.*

Lee, B.-H. and Choi, Y.-J. (2004) Effect of hrt and cod load on the nitrogen removal in a membrane bioreactor with intermittent feeding and aeration, *Proceedings of Water Environment – Membrane Technology Conference*, Seoul, Korea.

Lee, J., Ahn, W.-Y. and Lee, C.-H. (2001a) Comparison of the filtration characteristics between attached and suspended growth microorganisms in submerged membrane bioreactor. *Water Res.*, **35**, 2435–2445.

Lee, J.C., Kim, J.S., Kang, I.J., Cho, M.H., Park, P.K. and Lee, C.H. (2001b) Potential and limitations of alum or zeolite addition to improve the performance of a submerged membrane bioreactor. *Water Sci. Technol.*, **43**, 59–66.

Lee, K.-C. and Rittmann, B.E. (2002) Applying a novel autohydrogenotrophic hollow fiber biofilm reactor for denitrification of drinking water. *Water Res.*, **36**, 2040–2052.

Lee, K.-C. and Rittmann, B.E. (2003) Effects of pH and precipitation on autohydrogenotrophic denitrification using the hollow-fibre membrane-biofilm reactor. *Water Res.*, **37**, 1551–1556.

Lee, N. and Welander, T. (1994) Influence of predators on Nitrification in aerobic biofilm processes. *Water Sci. Technol.*, **29**(7), 355–363.

Lee, S., Burt, A., Rusoti, G. and Buckland, B. (1999) Microfiltration of yeast cells using a rotary disk dynamic filtration system. *Biotechnol. Bioeng.*, **48**, 386.

Lee, S.M., Jung, J.Y. and Chung, Y.C. (2001c) Novel method for enhancing permeate flux of submerged membrane system in two-phase anaerobic reactor. *Water Res.*, **35**, 471.

Lee, W., Kang, S. and Shin, H. (2003) Sludge characteristics and their contribution to microfiltration in submerged membrane bioreactors. *J. Membrane Sci.*, **216**, 217–227.

Lee, W., Jeon, J.-H., Cho, Y., Chung, K.Y. and Min, B.-R. (2005) Behavior of tmp according to membrane pore size, *Proceedings of International Congress on Membranes and Membrane Processes (ICOM)*, Seoul, Korea.

Lesage, N., Sperandio, M. and Cabassud, C. (2005) Performances of a hybrid adsorption/submerged membrane biological process for toxic waste removal. *Water Sci. Technol.*, **51**, 173–180.

Lesjean, B., Rosenberger, S., Laabs, C., Jekel, M., Gnirss, R. and Amy, G. (2005) Correlation between membrane fouling and soluble/colloidal organic substances in membrane bioreactors for municipal wastewater treatment. *Water Sci. Technol.*, **51**, 1–8.

Lettinga, G. and Vinken, J.N. (1980) Feasibility of the upflow anaerobic sludge blanket (UASB) process for treatment of low strength wastes, *35th Purdue Industrial Waste Conference Proceedings*, pp. 625–634.

Li, H., Yang, M., Zhang, Y., Liu, X., Gao, M. and Kamagata, Y. (2005a) Comparison of nitrification performance and microbial community between submerged membrane bioreactor and conventional activated sludge system. *Water Sci. Technol.*, **51**, 193–200.

Li, X., Gao, F., Hua, Z., Du, G. and Chen, J. (2005b) Treatment of synthetic wastewater by a novel mbr with granular sludge developed for controlling membrane fouling. *Sep. Purif. Technol.*, **46**, 19–25.

Li, Y.Z., He, Y.L., Liu, Y.H., Yang, S.C. and Zhang, G.J. (2005c) Comparison of the filtration characteristics between biological powdered activated carbon sludge and activated sludge in submerged membrane bioreactors. *Desalination*, **174**, 305–314.

Lim, A.L. and Bai, R. (2003) Membrane fouling and cleaning in microfiltration of activated sludge wastewater. *J. Membrane Sci.*, **216**, 279–290.

Lim, B.-R., Ahn, K.-H., Song, K.-G. and Ji-nWoo, C. (2004) Microbial community in biofilm on membrane surface of submerged mbr: Effect of in-line cleaning chemical agent, *Proceedings of Water Environment – Membrane Technology Conference*, Seoul, Korea.

Lindert, M., Kochbeck, B., Pruss, J., Warnecke, H.J. and Hempel, D.C. (1992) Scale-up of airlift-loop bioreactors based on modeling the oxygen mass-transfer. *Chem. Eng. Sci.*, **47**, 2281–2286.

Lipnizki, F. and Field, R.W. (2001) Pervaporation-based hybrid processes in treating phenolic wastewater: Technical aspects and cost engineering. *Sep. Sci. Technol.*, **36**, 3311–3335.

Liu, Y. and Fang, H.H.P. (2003) Influences of extracellular polymeric substances (eps) on flocculation, settling, and dewatering of activated sludge. *Crit. Rev. Env. Sci. Technol.*, **33**, 237–273.

Liu, R., Huang, X., Sun, Y.F. and Qian, Y. (2003) Hydrodynamic effect on sludge accumulation over membrane surfaces in a submerged membrane bioreactor. *Proc. Biochem.*, **39**, 157–163.

Liu, R., Huang, X., Chen, L., Wen, X. and Qian, Y. (2005) Operational performance of a submerged membrane bioreactor for reclamation of bath wastewater. *Proc. Biochem.*, **40**, 125–130.

Livingston, A.G. (1993b) A novel membrane bioreactor for detoxifying industrial wastewaters: II. Biodegradation of 3-chloronitrobenzene in an industrially produced wastewater. *Biotechnol. Bioeng.*, **41**, 927–936.

Livingston, A.G. (1994) Extractive membrane bioreactors: a new process technology for detoxifying chemical industry wastewaters. *J. Chem. Technol. Biotechnol.*, **60**, 117–124.

Lobos, J., Wisniewski, C., Heran, M. and Grasmick, A. (2005) Effects of starvation conditions on biomass behaviour for minimization of sludge production in membrane bioreactors. *Water Sci. Technol.*, **51**, 35–44.

Low, E. and Chase, H. (1999) The effect of maintanence energy requirements on biomass production during wastewater treatment. *Water Res.*, **33**(3), 847–854.

Lowry, O.H., Rosebourgh, N.J., Farr, A.R. and Randall, R.J. (1951) Protein measurement with the folin phenol reagent. *J. Biol Chem.*, **193**, 265–275.

Lubbecke, S., Vogelpohl, A. and Dewjanin, W. (1995) Wastewater treatment in a biological high-performance system with high biomass concentration. *Water Res.*, **29**, 793–802.

Ludzack, F. and Noran, D. (1965) Tolerance of high salinities by conventional wastewater treatment process. *J. Water Pollut. Cont. Fed.*, **37**, 00–00.

Luonsi, A., Laitinen, N., Beyer, K., Levanen, E., Poussade, Y. and Nystrom, M. (2002) Separation of ctmp mill-activated sludge with ceramic membranes. *Desalination*, **146**, 399–404.

Luxmy, B.S., Nakajima, F. and Yamamoto, K. (2000) Analysis of bacterial community in membrane-separation bioreactors by fluorescent *in situ* hybridization (FISH) and denaturing gradient gel electrophoresis (DGGE) techniques. *Water Sci. Technol.*, **41**(10–11), 259–268.

Ma, L., Li, X., Du, G., Chen, J. and Shen, Z. (2005) Influence of the filtration modes on colloid adsorption on the membrane in submerged membrane bioreactor. *Colloid. Surface. A: Physicochem. Eng. Aspect.*, **264**, 120–125.

Madaeni, S.S., Fane, A.G. and Wiley, D.E. (1999) Factors influencing critical flux in membrane filtration of activated sludge. *J. Chem. Technol. Biotechnol.*, **74**, 539–543.

Madoni, P., Davoli, D. and Chierici, E. (1993) Comparative analysis of the activated sludge microfauna in several sewage treatment works. *Water Res.*, 27.

Manem, J. and Sanderson, R. (1996) Membrane bioreactors. In Mallevialle, J., Odendaal, P.E. and Wiesner, M.R. (eds.) *Water Treatment Membrane Processes.* McGraw Hill, New York, pp. 17.1–17.31.

Mansell, B.O. and Schroeder, E.D. (1999) Biological denitrification in a continuous flow membrane bioreactor. *Water Res.*, **33**(8), 1845–1850.

Mansell, B.O. and Schroeder, E.D. (2002) Hydrogenotrophic denitrification in a microporous membrane bioreactor. *Water Res.*, **36**(19), 4683–4690.

Manz, W., Wagner, M., Amann, R. and Schleifer, K.-H. (1994) *In situ* characterisation of the microbial consortia active in two wastewater treatment plants. *Water Res.*, **36**(2), 413–420.

Masselin, I., Chasseray, X., Durand-Bourlier, L., Lainé, J-M., Syzaret, P.-Y. and Lemordant, D. (2001) Effect of sonication on polymeric membranes. *J. Membrane Sci.*, **181**, 213–220.

Matějů, V.,Čižinská, S., Krejčí, J. and Janoch, T. (1992) Biological water denitrification – a review. *Enzyme Microb. Technol.*, **14**, 170–183.

Meng, F., Zhang, H., Yang, F., Li, Y., Xiao, J. and Zhang, X. (2006) Effect of filamentous bacteria on membrane fouling in submerged membrane bioreactor. *J. Membrane Sci.*, in press.

Mercier-Bonin, M., Fonade, C. and Lafforgue-Delmorme, C. (1997) How slug flow can enhance the ultrafiltration flux of mineral tubular membranes. *J. Membrane Sci.*, **128**, 103.

Metcalf, Eddy, (2003) *Wastewater Engineering – Treatment and Reuse* (3rd edn). McGraw-Hill, New York.

Miyashita, S., Honjyo, K., Kato, O., Watari, K., Takashima, T., Itakura, M., Okazaki, H., Kinoshita, I. and Inoue, N. (2000) Gas diffuser for aeration vessel of membrane assembly, US Patent 6,328,886.

Mo, H., Oleszkiewicz, J.A., Cicek, N. and Rezania, B. (2005) Incorporating membrane gas diffusion into a membrane bioreactor for hydrogenotrophic denitrification of groundwater *Water Sci. Technol.*, **51**(6–7), 357–364.

Morgan, J.W., Forster, C.F. and Evison, L. (1990) A comparative study of the nature of biopolymers extracted from anaerobic and activated sludges. *Water Res.*, **24**, 743–750.

Mulder, M. (2000) *Basic Principles of Membrane Technology*. Kluwer Academic Publishers, Dordrecht.

Mueller, J.A., Boyle, W.C. and Popel, H.J. (2002) Aeration: principles and practice. Eckenfelder, W.W., Malina, J.R. and Patterson, J.W. (eds.) *Water Quality Management Library*. CRC Press, Boca Raton.

Muller, E.B., Stouthamer, A.H., Vanverseveld, H.W. and Eikelboom, D.H. (1995) Aerobic domestic waste-water treatment in a pilot-plant with complete sludge retention by cross-flow filtration. *Water Res.*, **29**, 1179–1189.

Nagaoka, H. and Nemoto, H. (2005) Influence of extracellular polymeric substances on nitrogen removal in an intermittently-aerated membrane bioreactor. *Water Sci. Technol.*, **51**, 151–158.

Nagaoka, H., Ueda, S. and Miya, A. (1996) Influence of bacterial extracellular polymers on the membrane separation activated sludge process. *Water Sci. Technol.*, **34**, 165–172.

Nagaoka, H., Yamanishi, S. and Miya, A. (1998) Modeling of biofouling by extracellular polymers in a membrane separation activated sludge system. *Water Sci. Technol.*, **38**, 497–504.

Nakao, K., Harada, T., Furumoto, K., Kiefner, A. and Popovic, M. (1999) Mass transfer properties of bubble columns suspending immobilized glucose oxidase gel beads for gluconic acid production. *Canadian J. Chem. Eng.*, **77**, 816–825.

Ng, C.A., Sun, D., Zhang, J., Chua, H.C., Bing, W., Tay, S. and Fane, A. (2005) Strategies to improve the sustainable operation of membrane bioreactors, *Proceedings of International Desalination Association Conference*, Singapore.

Nielson, P.H. and Jahn, A. (1999) Extraction of eps. In Wingender, J., Neu, T.R. and Flemming, H.C.E. (eds.) *Microbial Extracellular Polymeric Substances*. Springer-Verlag-eds, Berlin.

Nissinen, T.K., Miettinen, I.T., Martikainen, P.J. and Vartiainen, T. (2001) Molecular size distribution of natural organic matter in raw and drinking waters. *Chemosphere*, **45**, 865–873.

Nordkvist, M., Grotkjaer, T., Hummer, J. and Villadsen, J. (2003) Applying rotary jet heads for mixing and mass transfer in a forced recirculation tank reactor system. *Chem. Eng. Sci.*, **58**, 3877–3890.

Nuhoglu, A., Pekdemir, T., Yildiz, E., Keskinler, B. and Akay, G. (2002) Drinking water denitrification by a membrane bioreactor. *Water Res.*, **36**, 1155–1166.

Ognier, S., Wisnieswski, C. and Grasmick, A. (2001) Biofouling in membrane bioreactors: Phenomenon analysis and modelling, *Proceedings of MBR 3*, Cranfield University, UK.

Ognier, S., Wisniewski, C. and Grasmick, A. (2002a) Influence of macromolecule adsorption during filtration of a membrane bioreactor mixed liquor suspension. *J. Membr. Sci.*, **209**, 27–37.

Ognier, S., Wisniewski, C. and Grasmick, A. (2002b) Membrane fouling during constant flux filtration in membrane bioreactors. *Membrane Technol.*, **2002**, 6–10.

Ognier, S., Wisniewski, C. and Grasmick, A. (2004) Membrane bioreactor fouling in sub-critical filtration conditions: a local critical flux concept. *J. Membrane Sci.*, **229**, 171–177.

Oldani, M., Killer, E., Miquel, A. and Schock, G. (1992) On the nitrate and monovalent cation selectivity of ion exchange membranes used in drinking water purification. *J. Membrane Sci.*, **75**, 265–275.

Orantes, J.C., Wisniewski, C., Heran, M. and Grasmick, A. (2004) Influence of total sludge retension on the performance of a submerge membrane bioreactor, *Proceedings of Water Environment – Membrane Technology Conference*, Seoul, Korea.

Ozbek, B. and Gayik, S. (2001) The studies on the oxygen mass transfer coefficient in a bioreactor. *Proc. Biochem.*, **36**, 729–741.

Pankhania, M., Stephenson, T. and Semmens, M.J. (1994) Hollow fibre bioreactor for wastewater treatment using bubbleless membrane aeration. *Water Res.*, **28**, 2233–2236.

Pankhania, M., Brindle, K. and Stephenson, T. (1999) Membrane aeration bioreactors for wastewater treatment: completely mixed and plug-flow operation. *Chem. Eng. J.*, **73**, 131–136.

Park, H., Choo, K.H. and Lee, C.H. (1999) Flux enhancement with powdered activated carbon addition in the membrane anaerobic bioreactor. *Sep. Sci. Technol.*, **34**, 2781.

Park, D., Lee, D.S. and Park, J.M. (2005a) Continuous biological ferrous iron oxidation in a submerged membrane bioreactor. *Water Sci. Technol.*, **51**, 59–68.

Park, J.S., Yeon, K.M. and Lee, C.H. (2005b) Hydrodynamics and microbial physiology affecting performance of a new mbr, membrane-coupled high-performance compact reactor. *Desalination*, **172**, 181–188.

Pollice, A., Brookes, A., Jefferson, B. and Judd, S. (2005) Sub-critical flux fouling in membrane bioreactors – a review of recent literature. *Desalination*, **174**, 221–230.

Prosnansky, M., Sakakibara, Y. and Kuroda, M. (2002) High-rate denitrification and SS rejection by biofilm-electrode reactor (BER) combined with microfiltration. *Water Res.*, **36**, 4801–4810.

Psoch, C. and Schiewer, S. (2005b) Long-term study of an intermittent air sparged mbr for synthetic wastewater treatment. *J. Membrane Sci.*, **260**, 56–65.

Rabie, H.R., Côté, P., Singh, M. and Janson, A. Cyclic aeration system for submerged membrane modules, US Patent 6,881,343.

Rautenbach, R. and Albrecht, R. (1989) *Membrane Processes.* John Wiley and Sons ed., New York.

Reid, E. (2006) Salinity shocking and fouling amelioration in membrane bioreactors, *EngD tThesis*, School of Water Sciences, Cranfield University.

Reid, E. Judd, S. and Churchouse, S. (2004) Long term fouling in membrane bioreactors, *9th World Filtration Congress*, New Orleans, April.

Reising, A.R. and Schroeder, E.D. (1996) Denitrification incorporating microporous membranes. *J. Environ. Eng.*, **122**(7), 599–604.

Ren, N., Chen, Z., Wang, A. and Hu, D. (2005) Removal of organic pollutants and analysis of mlss-cod removal relationship at different hrts in a submerged membrane bioreactor. *Int. Biodeteriorat. Biodegrad.*, **55**, 279–284.

Rinzema, A. (1988) Anaerobic treatment of wastewater with high concentration of lipids or sulphate, *PhD Thesis*, Wageningen Agricultural University, Wageningen, The Netherlands.

Roggy, D.K., Novak., P.J., Hozalski, R.M., Clapp, L.W. and Semmens, M.J. (2002) Membrane gas transfer for groundwater remediation: Chemical and biological fouling. *Environ. Eng. Sci.*, **19**(6), 563–574.

Romero, C.A. and Davis, R.H. (1991) Experimental verification of the shear-induced hydrodynamic diffusion model of crossflow microfiltration. *J. Membrane Sci.*, **62**, 249–273.

Rosenberger, S. and Kraume, M. (2002) Filterability of activated sludge in membrane bioreactors. *Desalination*, **146**, 373–379.

Rosenburger, S., Kraume, M. and Szewzyk, U. (1999) Operation of different membrane bioreactors experimental results and physiological state of the microorganisms, *Proceedings IWA Conference Membrane Technology in Environmental Management*, Tokyo 310–316.

Rosenberger, S., Kruger, U., Witzig, R., Manz, W., Szewzyk, U. and Kraume, M. (2002) Performance of a bioreactor with submerged membranes for aerobic treatment of municipal waste water. *Water Res.*, **36**, 413–420.

Rosenberger, S., Evenblij, H., te Poele, S., Wintgens, T. and Laabs, C. (2005) The importance of liquid phase analyses to understand fouling in membrane assisted activated sludge processes -six case studies of different European research groups. *J. Membrane Sci.*, **263**, 113–126.

Rosso, D. and Stenstrom, M. (2005) Comparative economic analysis of the impact of mean cell retention time and denitrification on aeration systems. *Water Res.*, **39**, 3773–3780.

Saba, E., Kumazawa, H., Lee, C.H. and Narukawa, H. (1987) Gas–liquid interfacial area and liquid-side mass-transfer coefficient in a slurry bubble column. *Ind. Eng. Chem. Res.*, **26**, 112–116.

Sato, T. and Ishii, Y. (1991) Effects of activated-sludge properties on water flux of ultrafiltration membrane used for human excrement treatment. *Water Sci. Technol.*, **23**, 1601–1608.

Saw, C.B., Anderson, G.K., James, A. and Le, M.S. (1986) A membrane technique for biomass retention in anaerobic waste treatment, *Proceedings 40th Industrial Waste Conference*, Purdue University, West Lafayette, Indiana, Ann Arbor Science, Ann Arbor, Michigan, 805.

Schiener, P., Nachaiyasit, S. and Stuckey, D.C. (1998) Production of soluble microbial products (SMP) in an anaerobic baffled reactor: composition, biodegradability, and the effect of process parameters. *Environ. Technol.*, **19**, 391.

Schoeberl, P., Brik, M., Bertoni, M., Braun, R. and Fuchs, W. (2005) Optimization of operational parameters for a submerged membrane bioreactor treating dyehouse wastewater. *Sep. Purif. Technol.*, **44**, 61–68.

Schrader, G.A., Zwijnenburg, A. and Wessling, M. (2005) The effect of wwtp effluent zeta-potential on direct nanofiltration performance. *J. Membrane Sci.*, **266**, 80–93.

Scott, J.A., Neilson, D.J., Liu, W. and Boon, P.N. (1998) A dual function membrane bioreactor system for enhanced aerobic remediation of high-strength industrial waste. *Water Sci. Technol.*, **38**, 413–420.

Seghezzo, L., Zeeman, G., Van Lier, J.B., Hamelers, H.V.M. and Lettinga, G. (1998) A review: the anaerobic treatment of sewage in UASB and EGSB reactors. *Bioresource Technol.*, **65**, 175–190.

Seghezzo, L., Guerra, R.G., Gonzalez, S.M., Trupiano, A.P., Figueroa, M.E., Cuevas, C.M., Zeeman, G. and Lettinga, G. (2002) Removal efficiency and methanogenic activity profiles in a pilot scale UASB reactor treating settled sewage at moderate temperatures. *Water Sci. Technol.*, **45**(10), 243–248.

Shimizu, Y., Okuno, Y.-I., Uryu, K., Ohtsubo, S. and Watanabe, A. (1996) Filtration characteristics of hollow fiber microfiltration membranes used in membrane bioreactor for domestic wastewater treatment. *Water Res.*, **30**, 2385–2392.

Shin, H.-S. and Kang, S.-T. (2003) Characteristics and fates of soluble microbial products in ceramic membrane bioreactor at various sludge retention times. *Water Res.*, **37**, 121–127.

Singh, K.S. and Viraraghavan, T. (2003) Impact of temperature on performance, microbial, and hydrodynamic aspect of USAB reactors treating municipal wastewater. *Water Sci. Technol.*, **48**(6), 211.

Smith, B.C. and Skidmore, D.R. (1990) Mass-transfer phenomena in an airlift reactor – effects of solids loading and temperature. *Biotech. Bioeng.*, **35**, 483–491.

Smith, S., Judd, S., Stephenson, T. and Jefferson, B. (2003) Membrane bioreactors – hybrid activated sludge or a new process? *Membrane Tech.*, 5–8.

Smith, P.J., Vigneswaran, S., Ngo, H.H., Ben-Aim, R. and Nguyen, H. (2005) Design of a generic control system for optimising back flush durations in a submerged membrane hybrid reactor. *J. Membrane Sci.*, **255**, 99–106.

Smith, R. (2006). The MBR at Buxton wastewater plant, presented at *The Use and Practice of Membranes in Water and Wastewater in the UK*, Meeting of the Chartered Institution of Water and Envionmental Management (CIWEM), Glasgow, January 17, 2006.

Soares, M. (2000) Biological denitrification of groundwater. *Water Air Soil Poll.*, **123**, 183–193.

Sofia, A., Ng, W.J. and Ong, S.L. (2004) Engineering design approaches for minimum fouling in submerged MBR. *Desalination*, **160**, 67–74.

Stenstrom, M.K. and Redmon, D.T. (1996) Oxygen transfer performance of fine pore aeration in ASBs – a full scale review *TAPPI International Environmental Conference*, Orlando, FL., May 6.

Stephenson, T., Judd, S., Jefferson, B. and Brindle, K. (2000) *Membrane Bioreactors for Wastewater Treatment*, IWA Publishing, London.

Stewart, V. (1988) Nitrate respiration in relation to facultative metabolism. *Enterobacteria Microbiol. Rev.*, **52**, 190–232.

Stuckey, D.C. (2003) The submerged anaerobic membrane bioreactor (SAMBR): an intensification of anaerobic wastewater treatment, *Abstract of Presentation to the Department Civil Engineering at the University of Minnesota*, September 10, Minneapolis, USA.

Stuckey, D.C. and Hu, A. (2003) The submerged anaerobic membrane bioreactor (SAMBR): an intensification of anaerobic wastewater treatment, *Presented at the IWA Leading Edge Conference on Drinking Water and Wastewater Treatment Technologies*, Noordwijk/Amsterdam, The Netherlands.

Sun, Y., Huang, X., Chen, F. and Wen, X. (2004) A dual functional filtration/aeration membrane bioreactor for domestic wastewater treatment, *Proceedings of Water Environment – Membrane Technology Conference*, Seoul, Korea.

Suzuki, Y., Miyahara, S. and Tokeishi, K. (1993) Oxygen supply method using gas permeable film for wastewater treatment. *Water Sci. Technol.*, **28**(7), 243–250.

Tao, G., Kekre, K., Wei, Z., Lee, T.C., Viswanath, B. and Seah, H. (2005) Membrane bioreactors for water reclamation. *Water Sci. Technol.*, **51**, 431–440.

Tardieu, E., Grasmick, A., Geaugey, V. and Manem, J. (1999) Influence of hydrodynamics on fouling velocity in a recirculated mbr for wastewater treatment. *J. Membrane Sci.*, **156**, 131–140.

Tarnacki, K., Lyko, S., Wintgens, T., Melin, T. and Natau, F. (2005) Impact of extra-cellular polymeric substances on the filterability of activated sludge in membrane bioreactors for landfill leachate treatment. *Desalination*, **179**, 181–190.

Tokuz, R. and Eckenfelder, W. (1979) The effect of inorganic salts on the activated sludge process performance. *Water Res.*, **13**, 99–104.

Tsai, H.-H., Ravindran, V., Williams, M.D. and Pirbazari, M. (2004) Forecasting the performance of membrane bioreactor process for groundwater denitrification. *J. Environ. Eng. Sci.*, **3**, 507–521.

Ueda, T., Hata, K., Kikuoka, Y. and Seino, O. (1997) Effects of aeration on suction pressure in a submerged membrane bioreactor. *Water Res.*, **31**, 489–494.

Urbain, V., Benoit, R. and Manem, J. (1996) Membrane bioreactor: a new treatment tool. *J. Am. Water Works Assoc.*, **88**(5), 75–86.

Vallero, M.V.G., Lettinga, G. and Lens, P.N.L. (2005) High rate sulfate reduction in a submerged anaerobic membrane bioreactor (sambar) at high salinity. *J. Membrane Sci.*, **253**, 217–232.

Van Dijk, L. and Roncken, G.G. (1997) Membrane bioreactors for wastewater treatment: the state of the art and new developments. *Water Sci. Technol.*, **35**(10), 35–41.

Van Lier, J.B. (1995) Thermophillic anaerobic wastewater treatment, temperature aspects and process stability, *PhD Thesis*, Department of Environmental Technology, Wageningen Agricultural University, The Netherlands.

Van Loosdrecht, M.C.M. and Henze, M. (1999) Maintanence, endogenous respiration, lysis, decay and preditation. *Water Sci. Technol.*, **39**(1), 107–117.

Verlaan, P. and Tramper, J. (1987) Hydrodynamics, axial dispersion and gas–liquid oxygen transfer in an airlift-loop bioreactor with three-phase flow, *International Conference on Bioreactors and Biotransfomations*, Scotland, pp. 363–373.

Velizarov, S., Rodrigues, C.M., Reis, M.A. and Crespo, J.G. (2000/2001) Mechanism of charged pollutants removal in an ion exchange membrane bioreactor: drinking water denitrification. *Biotechnol. Bioeng.*, **71**(4), 245–254.

Velizarov, S., Rodrigues, C.M., Reis, M.A. and Crespo, J.G., (2003) Removal of trace mono-valent inorganic pollutants in an ion exchange membrane bioreactor: analysis of transport rate in a denitrification process. *J. Membrane Sci.*, **217**, 269–284.

Vigneswaran, S., Shim, W.G., Chaudhary, D.S. and Ben Aim, R. (2004) Mathematical modelling of submerged membrane bioreactor (smbr) system used in wastewater treatment, *Proceedings of Water Environment – Membrane Technology Conference*, Seoul, Korea.

Visvanathan, C. and Ben Aim, R. (1989) Studies on colloidal membrane fouling mechanisms in crossflow microfiltration. *J. Membrane Sci.*, **45**, 3–15.

Visvanathan, C., Yang, B.S., Muttamara, S. and Maythanukhraw, R. (1997) Application of air backflushing technique in membrane bioreactor. *Water Sci. Technol.*, **36**, 259–266.

Wagner, M., Rath, G., Amann, R., Koops, H.-P. and Schleifer, K.-H. (1996) *In situ* analysis fo nitrifying bacteria in sewage treatment plants. *Water Sci. Technol.*, **34**(1/2), 237–244.

Wagner, M., Noguera, D., Jutretschko, S., Rath, G., Koop, H.-P. and Schleifer, K.-H. (1998) Combining fluorescent *in situ* hybridization (FISH) with cultivation and mathematical modelling to study population structure and function of ammonia-oxidising bacteria in activated sludge. *Water Sci. Technol.*, **37**(4/5), 441–449.

Wagner, M., Cornel, P. and Krause, S. (2002) Efficiency of different aeration systems in full scale membrane bioreactors, *75th Annual Technical Conference WEFTEC*, Chicago, USA.

Wen, C., Huang, X. and Qian, Y. (1999) Domestic wastewater treatment using an anaerobic bioreactor coupled with membrane filtration. *Proc. Biochem.*, **35**, 335.

Wen, X., Bu, Q. and Huang, X. (2004) Study on fouling characteristic of an axial hollow fibers cross-flow microfiltration under different flux operations, *Proceedings of Water Environment – Membrane Technology Conference*, Seoul, Korea.

WERF (2005) www.werf.org/products/MembraneTool/home/default.asp, accessed: November 2005.

Wicaksana, F., Fane, A.G. and Chen, V. (2006) Fibre movement induced by bubbling using submerged hollow fibre membranes. *J. Membrane Sci.*, in press.

Wisniewski, C. and Grasmick, A. (1998) Floc size distribution in a membrane bioreactor and consequences for membrane fouling. *Colloid. Surface. A: Physicochem. Eng. Aspect.*, **138**, 403–411.

Witzig, R., Manz, W., Rosenburger, S., Krüger, U., Kraume, M. and Szewzyk, U. (2002) Microbiological aspects of a bioreactor with submerged membranes for aerobic treatment of municipal wastewater. *Water Res.*, **36**, 394–402.

Wong, H.M., Shang, C. and Chen, G. (2004) Factors affecting virus removal in a membrane bioreactor, *Proceedings of Water Environment – Membrane Technology Conference*, Seoul, Korea.

Wu, Y., Huang, X., Wen, X. and Chen, F. (2004) Function of dynamic membrane in self-forming dynamic membrane coupled bioreactor, *Proceedings of Water Environment – Membrane Technology Conference*, Seoul, Korea.

Xing, C.-H, Wu, W.-Z. and Tardieu, E. (2003) Excess sludge production in membrane bioreactors: a theoretical investigation. *J. Env. Engng.*, **129**(4), 291–297.

Yamagiwa, K. and Ohkawa, A. (1994) Simultaneous organic carbon removal and nitrification by biofilm formed on oxygen enrichment membrane. *J. Chem. Eng. (Japan)*, **27**, 638–643.

Ye, Y., Le-Clech, P., Chen, V. and Fane, A.G. (2005) Evolution of fouling during crossflow filtration of model eps solutions, *J. Membrane Sci.*, **264**, 190–199.

Yeo, A. and Fane, A.G. (2004) Performance of individual fibers in a submerged hollow fiber bundle, *Proceedings of Water Environment – Membrane Technology Conference*, Seoul, Korea.

Yeom, I.T., Nah, Y.M. and Ahn, K.H. (1999) Treatment of household wastewater using an intermittently aerated membrane bioreactor. *Desalination*, **124**, 193–203.

Yeom, I.T., Lee, K.R., Choi, Y.G., Kim, H.S. and Lee, Y. (2004) Evaluation of a membrane bioreactor system coupled with sludge pretreatment for aerobic sludge digestion, *Proceedings of Water Environment – Membrane Technology Conference*, Seoul, Korea.

Yeoman, S., Stephenson, T., Lester, J., and Perry, R. (1986). Biotechnology for Phosphorous removal during wastewater treatment. *Biotech. Adv.* **4**(1), 13–26.

Yin, X., Han, P.F., Lu, X.P. and Wang, Y.R. (2004) A review on the dewaterability of bio-sludge and ultrasound pretreatment. *Ultrason. Sonochem.*, **11**, 337–348.

Yoon, S.H, Kang, I.J. and Lee, C.H. (1999) Fouling of inorganic membrane and flux enhancement in membrane-coupled anaerobic bioreactor. *Sep. Sci. Technol.*, **34**, 709.

Yoon, S.H., Collins, J.H., Musale, D., Sundararajan, S., Tsai, S.P., Hallsby, G.A., Kong, J.F., Koppes, J. and Cachia, P. (2005) Effects of flux enhancing polymer on the characteristics of sludge in membrane bioreactor process. *Water Sci. Technol.*, **51**, 151–157.

Yu, H.-Y., Xie, Y.-J., Hu, M.-X., Wang, J.-L., Wang, S.-Y. and Xu, Z.-K. (2005a) Surface modification of polypropylene microporous membrane to improve its antifouling property in mbr: Co2 plasma treatment. *J. Membrane Sci.*, **254**, 219–227.

Yu, H.Y., Hu, M.X., Xu, Z.K., Wang, J.L. and Wang, S.Y. (2005b) Surface modification of polypropylene microporous membranes to improve their antifouling property in mbr: Nh3 plasma treatment. *Sep. Purif. Technol.*, **45**, 8–15.

Zhang, X.Q., Bishop, P.L. and Kinkle, B.K. (1999) Comparison of extraction methods for quantifying extracellular polymers in biofilms. *Water Sci. Technol.*, **39**, 211–218.

Zhang, Y., Bu, D., Liu, C.-G., Luo, X. and Gu, P. (2004) Study on retarding membrane fouling by ferric salts dosing in membrane bioreactors, *Proceedings of Water Environment – Membrane Technology Conference*, Seoul, Korea.

Zhang, S.T., Qu, Y.B., Liu, Y.H., Yang, F.L., Zhang, X.W., Furukawa, K. and Yamada, Y. (2005) Experimental study of domestic sewage treatment with a metal membrane bioreactor. *Desalination*, **177**, 83–93.

Zhang, J., Chua, H.C., Zhou, J. and Fane, A.G. (2006) Factors affecting the membrane performance in submerged membrane bioreactors. *J. Membrane Sci.*, (Submitted).

Zheng, X., Fan, Y.B. and Wei, Y.S. (2003) A pilot scale anoxic/oxic membrane bioreactor (a/o mbr) for woolen mill dyeing wastewater treatment. *J. Environ. Sci. – China*, **15**, 449–455.

Zhongwei, D., Liying, L. and Runyu, M. (2003) Study on the effect of flow maldistribution on the performance of the hollow fiber modules used in membrane distillation. *J. Membrane Sci.*, **215**, 11–23.

Zumft, W.G. (1992) The denitrifying prokaryotes. In Balows, A., Trüper, H.G., Dworkin, M., Harder, W. and Schleifer, K.H. (eds.) *The Prokaryotes*, Springer Verlag, New York, pp. 482–554.

Chapter 3

Design

With acknowledgements to:

Section 3.1	Harriet Fletcher	School of Water Sciences, Cranfield University, UK
Section 3.2.4	Kiran Kekre, Tao Guihe	Centre for Advanced Water Technology, Singapore
	Harry Seah	Public Utilities Board, Singapore
Section 3.2.5	Guiseppe Guglielmi, Gianni Andreottola	Department of. Civil and Environmental Engineering, University of Trento, Italy
Section 3.2.6	Adriano Joss, Hansruedi Siegrist	Engineering Department, Eawag, Switzerland
Section 3.3	Harriet Fletcher	School of Water Sciences, Cranfield University, UK

3.1 Membrane bioreactor system operational parameters

There are essentially three main elements of a membrane bioreactor (MBR) contributing to operating costs, ignoring membrane replacement. These are:

1. liquid pumping
2. membrane maintenance
3. aeration

Of these by far the most significant, especially for immersed technologies, is aeration. As already discussed in Chapter 2, aeration is used both for scouring an immersed membrane and for suspending and maintaining a viable biomass. Design of an MBR therefore demands knowledge both of the feedwater quality, which principally determines the oxygen demand for biotreatment, and the aeration demand for fouling control, which relates to a number system characteristics as summarised in Fig. 2.18. The various governing principles have been discussed in Chapter 2. Design equations are summarised below.

3.1.1 Liquid pumping

The key hydraulic operating parameters for MBR operation are obviously flux J litres per m^2 per hour (LMH) and transmembrane pressure (TMP) ΔP_m (bar or Pa), from which the permeability K ($J/\Delta P_m$ in LMH/bar), kg/(m^2s) is obtained. In sidestream MBRs the specific energy demand (the energy demanded per unit permeate product volume) for continuous membrane permeation relates primarily to the pressure and conversion, and is roughly given by:

$$W_h = \frac{1}{\rho\xi\Theta}\sum\Delta P \qquad (3.1)$$

where ρ is the density (kg/m^3), Θ is the conversion achieved by a single passage of fluid along the length of the module (nominally 100% for submerged systems) and ξ is the pumping energy efficiency. In the above equation $\Sigma\Delta P$ refers to the sum of all individual pressure losses which, for a sidestream system, comprises ΔP_m and the pressure losses in the retentate channels, ΔP_r, all pressure values being in SI units. As explained in Section 2.3.1, maximising Θ to reduce W_h relies on increasing the total length of the flow path through the membrane channel and increasing the shear rate, but this has the effect of increasing ΔP_r. The optimum operating conditions for a given channel dimension (tube diameter or parallel flow channel separation) is therefore determined by relatively straight-forward hydraulic considerations, perhaps complicated slightly by the non-Newtonian nature of the sludge (Section 2.2.5.3).

For submerged MBRs, the total pressure loss is of lesser importance, since it is necessarily very low to sustain operation. In this configuration the pressure loss, principally the TMP, reflects state of membrane fouling, through K, but contributes very little to the energy demand. Instead, the specific aeration energy demand is mainly governed by the aeration intensity (Section 3.1.3.2).

Other contributions to pumping energy demand include the various transfer operations, such as from the balancing tank to the bioreactor, the bioreactor to the membrane tank and the bioreactor or membrane tank to the denitrification tank or zone. All of these operations are secondary contributors to energy demand. However, given that it provides the highest flow, the recycle for denitrification (Section 2.2.6), if used, would normally incur the greatest energy demand as determined by the recycle ratio Q_{dn}/Q_p, where Q_{dn} is the recycle flow for denitrification and Q_p is the feed flow. In full-scale systems this value (r) normally takes values between 1 and 3 (WEF, 1996) but can be up to 4 (Metcalf and Eddy, 2003) depending on the feedwater quality. Pumping power is given by:

$$W_p = \frac{\rho g H}{1000\xi} Q_{pump} / Q_p \quad (3.2)$$

where W_p is the power input to the pump (kW), H is the pump head including the system losses (m of water), ρ is the density of the pumped fluid (kg/m^3) and ξ is the pump efficiency. This relationship yields the theoretical power requirement for all liquid pumping operations, such as transfer from the bioreactor tank to the membrane tank. In the case of biomass recycle the term Q_{pump}/Q_p approximates to r. Note that ξ is highly variable and a more accurate measure of pumping power requirements can be found by consulting data sheets produced by the pump manufacturers.

3.1.2 Membrane maintenance

Other factors impacting on operating cost relate to physical and chemical membrane cleaning, which incur process downtime, loss of permeate product (in the case of backflushing) and membrane replacement. The latter can be accounted for simply by amortisation, although actual data on membrane life is scarce since for most plants the start-up date is recent enough for the plants still to be operating with their original membranes. Membrane replacement costs are potentially very significant, but data from some of the more established plant are somewhat encouraging in this regard (Table 5.2).

Physical and chemical backwashing requirements are dependent primarily on the membrane and process configurations and the feedwater quality. Thus far only rules of thumb are available for relationships between feedwater quality and membrane operation and maintenance (O&M); O&M protocols for specific technologies are normally recommended by the membrane and/or process suppliers and sometimes further adapted for specific applications. Fundamental relationships between cleaning requirements and operating conditions, usually flux and aeration for submerged systems, have been generated from scientific studies of fouling (Section 2.3.7.1). However, arguably the most useful sources of information for membrane cleaning requirements are comparative pilot trials – the assessment of different MBR technologies challenged with the same feedwater (Section 3.2) – and full-scale reference sites (Chapter 5).

Key design parameters relating to membrane cleaning (Sections 2.1.4.3 and 2.3.9.2) are:

- period between physical cleans (t_p), where the physical clean may be either backflushing or relaxation;
- duration of the physical clean (τ_p);
- period between chemical cleans (t_c);
- duration of the chemical clean (τ_c);
- backflush flux (J_b);
- cleaning reagent concentration (c_c) and volume (v_c) normalised to membrane area.

If it can then be assumed that a complete chemical cleaning cycle, which will contain a number of physical cleaning cycles (Fig. 2.11), restores membrane permeability to a sustainable level then the net flux J_{net} can be calculated:

$$J_{net} = \frac{n(Jt_p - J_b\tau_p)}{t_c + \tau_c} \qquad (3.3)$$

where n is the number of physical cleaning cycles per chemical clean:

$$n = \frac{t_c}{t_p + \tau_p} \qquad (3.4)$$

t_c and t_p may be determined by threshold parameter values, specifically the maximum operating pressure or the minimum membrane permeability. Note that it is normally the net flux which is most appropriate to use for energy demand calculations, and that the pumping energy in Equation (3.1) can be modified for chemical cleaning down time:

$$W_{h,net} = W_h \frac{t_c}{t_c + \tau_c} \qquad (3.5)$$

Other costs of chemical cleaning relate to the cost of the chemical reagent itself. The total mass of cleaning of chemical cleaning reagent is simply the product of volume and concentration. For a periodic chemical clean in place (CIP), either maintenance or recovery (Section 2.3.9.2), the specific mass per unit permeate product is simply:

$$M_c = \frac{c_c v_c}{J_{net} A_m (t_c + \tau_c)} \qquad (3.6)$$

where A_m is the membrane area. If the cleaning reagent is flushed through the membrane *in situ* then the volume of cleaning reagent used can be found from:

$$v_c = J_c A_m \tau_c \qquad (3.7)$$

where J_c is the cleaning flux. From the above two equations:

$$M_c = c_c \frac{J_c}{J'_{net}} \frac{\tau_c}{(t_c + \tau_c)} \tag{3.8}$$

Equation (3.8) is applicable to both a chemically enhanced backwashing (CEB) and a CIP, the values for c_c, t_c and τ_c being much lower for a CEB.

The frequency of chemical cleaning can, if necessary, be determined from the rate of decline of permeability, where permeability decline rate at constant flux is determined from the pressure p and time t at two points (subscripted 1 and 2) within the cleaning cycle:

$$\frac{\Delta K}{\Delta t} = \frac{J_{net}}{t_2 - t_1}\left(\frac{p_2 - p_1}{p_1 p_2}\right) \tag{3.9}$$

where p_1 refers to the recovered permeability after cleaning and p_2 to the minimum acceptable permeability.

3.1.3 Aeration

3.1.3.1 Biomass aeration

The first component of aeration concerns the bioreactor and, specifically, the demand of the mixed liquor for air required for (a) solids agitation, and (b) dissolved oxygen (DO) for maintaining a viable micro-organism population for biotreatment. In biotreatment DO is normally the key design parameter. The oxygen requirement (m_O, kg/day) for a biological system relates to the flow rate (Q, m^3/day), substrate degradation ($S_0 - S$, Section 2.1.4.1), sludge production (P_x, Section 2.1.4.2) and concentration of total kjeldahl nitrogen (TKN) that is oxidized to form products (NO_x, Section 2.1.4.4). This relationship is derived from a mass balance on the system (Section 2.2.5.1):

$$m_O = Q(S - S_e) - 1.42P_x + 4.33Q(NO_x) - 2.83Q(NO_x) \tag{2.17}$$

The oxygen is most commonly transferred to the biomass by bubbling air, or in some cases pure oxygen, into the system through diffusers. Only a portion of the air, or oxygen, which is fed to the system is transferred into the biomass. This is quantified by the oxygen transfer efficiency (OTE). The transfer efficiency is dependent on the type of diffuser used and the specific system design. Manufacturers provide an OTE for their diffuser in clean water at 20°C, three correction factors α, β and φ (Section 2.2.5.3) can be used to convert this to process water (i.e. the mixed liquor in the case of MBRs). The airflow through the blower required to maintain the biological community ($m_{A,b}$, kg/h) can be calculated from:

$$Q_{A,b} = \frac{R_O x OTE}{24 \rho_A \alpha \beta \varphi} \tag{3.10}$$

However, to simplify calculations in process design the three correction factors are often combined into a single α factor.

Unlike other flows $m_{A,b}$ is calculated as a mass flow as opposed to a volumetric flowrate. It is more practically useful to consider the volumetric biomass aeration demand, for which $m_{A,b}$ must be converted to a volumetric flow rate, normalised to some selected temperature (usually 20°C) by the air density, to provide air flow rate per unit permeate flow:

$$R_b = \frac{Q_{A,b}}{J'_{net}A_m} \quad (3.11)$$

where J'_{net} is the net flux normalised to 20°C.

3.1.3.2 Membrane aeration

It is necessary to aerate the membrane unit in an MBR to scour solids from the membrane surface (Section 2.3.7.1). In practice the membrane aeration value is not defined theoretically since the relationship between aeration and flux decline is not well understood at present. Membrane aeration values are based on previous experience, and in many cases the suppliers recommend an appropriate aeration rate. As is the case for membrane cleaning regimes the most valuable data are that which are collected from pilot trials and full-scale case studies. The key contributing factor to energy demand in submerged systems is the specific aeration demand, the ratio of Q_A either to membrane area (SAD_m) or permeate volume (SAD_p):

$$SAD_m = \frac{Q_{A,m}}{A_m} \quad (3.12)$$

$$SAD_p = \frac{Q_{A,m}}{JA_m} \quad (3.13)$$

where the subscript m refers specifically to the individual membrane element or module. Whereas SAD_m has units of m^3 air per unit m^2 per unit time, and thus nominally m/h, SAD_p is unitless. It should be noted that both are temperature dependent, since gas volume (and thus flow) increases with temperature as does flux, through decreasing viscosity. If normalised to 20°C the air flow is demoted units of Nm^3/h.

3.1.3.3 Diffuser type

There are three types of aeration used in MBR plants: coarse bubble aeration, fine bubble aeration and, less commonly, jet aeration. The principal differences between the two main aerator types are given in Table 3.1. Traditionally, fine bubble diffusion has been used for biomass aeration and a separate coarse bubble aeration system applied for membrane scouring. In many proprietary systems separate tanks are

Table 3.1 Main features of aeration systems

	Fine bubble	Coarse bubble
Bubble size	2–5 mm[a]	6–10 m[a]
OTE (percentage of O_2 transfer per m depth)	3–10%[b]	1–3%[b]
Mechanical component	Air blower	Air blower
Diffuser type	Ceramic or membrane diffuser disk, dome or tube	Steel or plastic disk or tube
Shear rate[c]	Bubble velocity $\propto d^2$ (from Stokes Law). The small bubble sizes provide lower velocity and hence smaller shear forces.	Bubble velocity, and so shear, is higher than fine bubble aeration since the larger bubbles rise faster than small bubbles.
Diffuser cost[d]	Approximately £40 per diffuser	Approximately £15 per diffuser

[a]EPA, 1989.
[b]Data from survey of manufacturers and from literature study. The large variation in the efficiency data for fine bubble aeration is attributed to changes in the distribution of the diffuser nozzles over the tank floor and diffuser age. Fine bubble diffusers are susceptible to fouling and oxygen transfer efficiency can decrease up to 19% over 2 years of operation.
[c]Shear rate is a measure of propensity to ameliorate membrane fouling (Section 2.3.7.1).
[d]Data obtained from manufacturer quotes for use as a guideline only.

provided for the membrane to simplify membrane cleaning operations. During the process of scouring the membrane, if air is used, there is some transfer of oxygen into the biomass which raises its DO level in the biomass. Membrane aeration is usually carried out using coarse bubble aeration because of the increased turbulence and hence shear forces created (Section 2.3.7.1), whilst biomass aeration is usually performed using fine bubble devices because of the enhanced oxygen transfer (Table 3.1).

Oxygen transfer is corrected from clean water to process water (i.e. mixed liquor) conditions by the α factor. Several relationships have been developed that relate mass transfer to the mixed liquor suspended solids (MLSS) or viscosity of a system (Section 2.2.5.3), most of them taking the form:

$$\alpha = a.e^{-b.X} \tag{3.14}$$

where a and b are constants and X is the MLSS concentration in g/m^3; a varies from 1 to 4.3 and b from 0.04 to 0.23 (Fig. 2.19).

3.1.3.4 Blower power consumption

The theoretical power consumption for a blower is given by (Appendix A):

$$Power = \frac{P_{A,1} T_{K,1} \lambda}{2.73 \times 10^5 \, \xi (\lambda - 1)} \left[\left(\frac{P_{A,2}}{P_{A,1}} \right)^{1-\frac{1}{\lambda}} - 1 \right] Q_A \tag{3.15a}$$

where $P_{A,1}$ and $P_{A,2}$ are the inlet (normally atmospheric) and outlet absolute pressures (Pa) respectively, ξ is the blower efficiency, λ is the ratio of specific heat capacity at constant pressure to constant volume and takes a value of 1.4 for air, $T_{K,1}$ is the inlet temperature (K), and Q_A is the volumetric flow rate of air. The blower efficiency (ξ) varies significantly according to the type of blower and its mechanical design, the rotational speed, and the blower rating for the task. As with pumping operations more accurate data can be found from manufacturers' data sheets. If Equation (3.15) is divided by the total membrane module area A_m to which the flow rate Q_A applies, then one obtains an expression in kW/m^2 for power per unit membrane area:

$$W_{b,m} = \frac{P_{A,1}T_{k,1}\lambda}{2.73 \times 10^5\, \xi(\lambda - 1)}\left[\left(\frac{P_{A,2}}{P_{A,1}}\right)^{1-\frac{1}{\lambda}} - 1\right]\frac{Q_A}{A_m} = \frac{kQ_A}{A_m} \tag{3.15b}$$

where k is therefore a function of pressure, and thus the hydrostatic head ΔP_h determined by the depth at which the aerator is placed in the tank, as well as temperature. If Equation (3.15) is divided by the permeate flow rate Q_P, that is $J_{net}A_m$, where J_{net} is the net flux ($m^3/(m^2\,h)$), then one obtains an expression for specific energy demand in kWh/m^3:

$$W_{b,V} = \frac{kQ_A}{JA_m} \tag{3.16}$$

Hence, whilst the hydrostatic head wholly or partly determines the flux, depending on whether suction is applied, its main contribution to energy demand does not relate to membrane permeation directly but to its impact on aeration energy. If the flux and flow are normalised to 20°C and 101.3 kPa (i.e.J'_{net} and Q'_A respectively), then k in Equation (3.15b) simplifies to:

$$k = \frac{108.748\lambda}{\xi(\lambda - 1)}\left[\left(\frac{P_{A,2}}{101.325}\right)^{1-\frac{1}{\lambda}} - 1\right] \tag{3.17}$$

As already stated in most hollow fibre (HF) MBR technologies the membrane aeration is carried out separately from the bioreactor aeration and does not necessarily employ the same type of aerator. Some of the oxygen transfer is achieved by the membrane aeration, and the extent to which this occurs depends mainly on the type of membrane aerator employed.

3.1.4 Design calculation: summary

Inter-relationships within an MBR process are complex (Fig. 2.26), but the most crucial relationships with respect to operating costs are those associated with aeration since this provides the largest component of the process operating cost. The impacts of aeration on the various operating parameters have already been discussed and depicted in Fig. 2.18, and biological and physical parameters used for the determination of operating costs are listed in Tables 3.2 and 3.3 respectively. Note that it is most consistent to normalise against permeate product volume to produce specific energy

Table 3.2 Biological operating parameters

Raw data		Calculated data	
Average flow (m^3/h)	Q	Maximum specific substrate utilisation rate (k g/(g/day))	$\frac{\mu_m}{Y}$
Peak flow (m^3/h)	Q_{peak}	Effluent BOD (S g/m^3)	$\frac{K_s(1-k_e\theta_x)}{\theta_x(Yk-k_e)-1}$
BOD influent (g/m^3)	S_0	Specific growth of nitrifying bacteria, μ_n g/(g/day)	$\frac{1}{\theta_x}$
Biodegradable COD (g/m^3) (often taken as 1.6 BOD)	$bCOD$	Effluent nitrogen (N_e g/m^3)	$\frac{K_n(\mu_n+k_{e,n})}{\mu_{n,m}-k_{e,n}-\mu_n}$
Total suspended solids influent (g/m^3)	TSS	Cell debris (f_d g/g)	$\approx \frac{VSS}{S_0}$
TSS effluent (g/m^3) (normally negligible)	TSS_e	Sludge yield (P_x g/day)	$\frac{QY(S_0-S)}{1+k_e\theta_x}+\frac{f_a k_e QY\theta_x(S_0-S)}{1+k_e\theta_x}+\frac{QY_n NO_x}{1+k_{e,n}\theta_x}$
Volatile suspended solids (g/m^3)	VSS	Non-biodegradable solids (X_0 g/day)	$(TSS - VSS)Q$
Nitrogen influent (g/m^3)	TKN	Concentration of ammonia oxidised to form nitrate, NO_x (g/m^3)	$N - N_e + 0.12P_x$
Wastewater temperature (°C)	T_w	Sludge wastage flow (Q_w m^3/day)	$\frac{V}{\theta_x}$
SRT (day)	θ_x	Observed yield, Y_{obs} mg VSS/(mg BOD day)	$\frac{Y}{1+k_e\theta_x}+\frac{f_d k_e Y\theta_x}{1+k_e\theta_x}$

(*Table to be continued on next page*)

Table 3.2 *(continued)*

Raw data		Calculated data	
Design mixed liquor suspended solids (g/m^3)	$MLSS$	Aerobic tank volume V m^3	$\frac{P_x \theta_x}{X}$
Anoxic tank as percentage of aerobic tank size	V_a/V_{an}	Food to micro-organisms ratio, $F{:}M$ kg BOD/(k g TSS/day)	$\frac{SQ}{VX}$
Maximum specific growth rate (heterotrophic) (g VSS/(g VSS/day))	μ_m	Oxygen requirement, m_0 kg/day	$Q[(S_0 - S) - 1.42P_x/Q + 4.33NO_x - 2.83NO_x$
Saturation coefficient (heterotrophic) (g/m^3 BOD)	K_s	Required airflow to meet biological requirements, $Q_{A,b}$ kg/day	$\frac{R_0 xOTE}{\rho_A 0.21\alpha\beta\tau}$
Endogenous decay coefficient (heterotrophic), g VSS/(g VSS/day)	k_e	α factor	$e^{-0.084.X}$
Yield coefficient (heterotrophic) g VSS/(g BOD)	Y	Sludge wastage, Q_w m^3/day	$\frac{V}{\theta_x}$
Maximum specific growth rate (nitrification), (g VSS/(g VSS/day))	$\mu_{m,n}$	Sludge waste per unit permeate ($Q_{w,V}$ $m^3/(m^3$ day)	$\frac{Q_w}{J'_{net} A_m}$
Endogenous decay coefficient (nitrification) (g VSS/(g VSS/day))	$k_{e,n}$	Airflow per unit permeate (R_b $m^3/(m^3$/day))	$\frac{Q_{A,b}}{\rho_A J'_{net} A_m}$
Yield coefficient (nitrification), g VSS/(g BOD)	Y_n		

Table 3.3 Physical operating parameters

Raw data		Normalised or derived data	
Mean flux (LMH)	J	Temperature-corrected flux, J', LMH	$J/1.024^{(T-20)}$
Mean transmembrane pressure (bar)	ΔP_{m}	Temperature-corrected mean permeability (LMH/bar)	$J'/\Delta P_{\mathrm{m}}$
Aeration rate (m^3/h)	$Q_{\mathrm{A,m}}$	Temperature, pressure-corrected aeration rate, $Q'_{\mathrm{A,m}}$ (m^3/h)	$Q_{\mathrm{A,m}}\left(\frac{293}{T_{\mathrm{a},K}}\right)\left(\frac{P_{\mathrm{a,1}}}{101.325}\right)$
Temperature of wastewater (K)	$T_{\mathrm{w,K}}$	Temperature-corrected net flux, J'_{net} (LMH)	$n\frac{(J't_{\mathrm{b}} - J'_{\mathrm{b}}\tau_{\mathrm{b}})}{t_{\mathrm{c}} + \tau_{\mathrm{c}}}$
Inlet air temperature (K)	$T_{\mathrm{a,K}}$	Membrane aeration demand per unit membrane area, SAD$_{\mathrm{m}}$ (N m^3/(h m^2))	$\frac{Q'_{\mathrm{A,m}}}{A'_{\mathrm{m}}}$
Inlet air pressure (kPa)	$P_{\mathrm{a,1}}$	Membrane aeration demand per unit permeate flow, SAD$_{\mathrm{p}}$	$\frac{Q'_{\mathrm{A,m}}}{J'_{\mathrm{net}}A_{\mathrm{m}}}$
Membrane area (m^2)	A_{m}	Specific membrane aeration energy demand, $W_{\mathrm{b,V}}$ (kWh/m^3)	$\frac{kQ'_{\mathrm{A,m}}}{\rho_{\mathrm{a}}J'_{\mathrm{net}}A_{\mathrm{m}}}$
Physical cleaning (backflush) interval (h)	t_{b}	Specific hydraulic energy demand for membrane permeation, W_{h} (kWh/m^3)	$\frac{\Delta P}{\rho_{\mathrm{p}}\xi\Theta}$
Physical cleaning (backflush) duration (h)	τ_{b}	Specific recirculation energy demand per unit permeate volume, W_{p} (kWh/m^3)	$\frac{r\rho_{b}gH}{1000\xi}$
Backflush flux (LMH)	J_{b}	Mass of chemical reagent per unit permeate volume, M_{c} (kg/m^3)	$\frac{c_{\mathrm{c}}v_{\mathrm{c}}}{J'_{\mathrm{net}}A_{\mathrm{m}}(t_{\mathrm{c}} + \tau_{\mathrm{c}})}$
Cleaning interval (h)	t_{c}		
Cleaning duration (h)	τ_{c}		
Cleaning reagent strength (kg/m^3)	c_{c}		
Cleaning reagent volume (m^3)	v_{c}		
Conversion	Θ		
Density (kg/m^3) $_{\mathrm{p}}$ relates to permeate, $_{\mathrm{b}}$ to biomass and $_{\mathrm{a}}$ to air	ρ		
Pumping efficiency	ξ		
Blower outlet pressure (m)	$P_{\mathrm{a,2}}$		

$_{\mathrm{p}}$ relates to permeate, $_{\mathrm{a}}$ relates to air.

demand components W. Since energy demand relates to pumping, a knowledge of the pumping energy efficiency is required as well as the cost per unit of electrical energy. Chemical reagent costs are also a contributing factor but these (represented by M_{c}) are normally very small compared to energy demand and membrane replacement. The latter (F) is obviously a key parameter and relates to irrecoverable fouling. Unfortunately, there is insufficient historical data to enable F to be determined, but it is normal for suppliers to give a guarantee with their membrane products which in effect provide F for costing purposes. As with all biological processes sludge disposal costs contribute to the MBR operating costs. However, since the sludge yield is reduced in MBR processes, the quantity of sludge generated is relatively low compared with a conventional activated sludge process (ASP) and the disposal cost correspondingly low.

For a given comprehensive set of data, the determination of energy demand proceeds through the calculation of:

- oxygen requirements for the biomass,
- oxygen transfer coefficient from the aerator characteristics,
- alpha factor from empirical correlations with MLSS,
- specific aeration demand from the aeration rate and the net flux,
- air flow rate through the blower,
- blower power requirement,
- pumping energy for both permeate extraction and recirculation.

Design is thus critically dependent on the selected operating (gross) or net flux (J or J_{net} respectively) chosen and the aeration demand required to maintain this flux, as quantified by R_v, under the conditions of physical and chemical cleaning employed. Total aeration demand will then also depend on the extent to which membrane aeration generates dissolved oxygen which is subsequently used for sustaining the biomass. Since there is currently no generally-applicable first-principles or empirical relationship proposed between flux and membrane aeration for three-phase intermediate flow, such as persists in an immersed MBR, it is necessary to review heuristic information on this topic to identify appropriate relationships which can subsequently be used for design purposes.

3.2 Data for technology comparison, immersed systems

3.2.1 Introduction

Since MBR performance is highly dependent upon feedwater quality, true comparison of the performance of different MBR technologies can only be achieved when they are tested against the same feedwater matrix. Two options exist for such comparisons:

(a) the use of analogues of precisely controlled composition, and
(b) the simultaneous trialling of technologies challenged with the same feedwater.

The dichotomy over the respective benefits of analogue and real matrices for research and development (R&D) purposes is not limited to MBRs and applies across many sectors. The use of analogues for feedwaters in water and wastewater treatment R&D allows total control of feed quality and allows tests to be conducted sequentially or simultaneously without detracting from the performance comparison. On the other

Table 3.4 Comparative pilot plant trials

Technology tested	Reference						
	Adham *et al.* (2005)	van der Roest *et al.* (2002)	Tao *et al.* (2005)	Honolulu	Lawrence *et al.* (2005)	Trento	Eawag
Zenon	×	×	×	×		×	×
Kubota	×	×	×	×		×	×
Mitsubishi Rayon	×	×	×	×			×
Norit X-Flow	–	×	–				
Huber				(×)	×		
Memcor	×			×		×	
Toray					×		

hand it is widely recognised that analogues can never satisfactorily represent real sewage, particularly so in the case of such crucially important parameters as fouling propensity, and can be extremely expensive to produce. For pilot trials of MBR technologies of a reasonable scale (i.e. based on a small number of full-scale membrane modules), conducting trials based on real feedwaters is always preferred. A number of such trials have been carried out since around the turn of the millennium which permit a useful technology comparison (Table 3.4), albeit with certain caveats. The studies identified in Table 3.2 all employ at least one full-scale membrane module per bioreactor and at least three different technologies. Not all of these studies have been published, however.

3.2.2 Beverwijk wastewater treatment plant, the Netherlands

An extensive comparative pilot trial was carried out at Beverwijk–Zaanstreek wastewater treatment works between 2000 and 2004, with a substantial body of work published in 2002 (van der Roest *et al.*, 2002). The work represents one of the earliest large-scale comparative pilot trials and was conducted by DHV in collaboration with the Dutch Foundation of Applied Water Research (STOWA). The ultimate goal of the work was the construction of a number of full-scale plants of 60–240 megalitres per day (MLD) capacity in the Netherlands, scheduled for installation between 2003 and 2006. Results from trials on four MBR systems of 24–120 m^3/day capacity were published in the 2002 report. The four technologies originally tested were Kubota, Norit X-Flow, Mitsubishi Rayon and Zenon. Subsequent reports by this group (Lawrence *et al.*, 2005; Schyns *et al.*, 2003) have not contained the same level of technical detail regarding operation and maintenance.

The reported trials (van der Roest *et al.*, 2002) were conducted in four phases:

I. Primary clarification with ferric dosing prior to screening
II. Primary clarification with simultaneous ferric dosing (i.e. downstream of screening)
III. Raw wastewater with simultaneous precipitation
IV. Raw wastewater with bio-P removal.

Table 3.5 Feedwater quality

Parameter (mg/L)	Raw sewage	Advanced primary treated sewage
COD	548–621	297–422
TKN	57–61	39–67
Total phosphate-P	9.3–12.1	7.1–8.3

Table 3.6 Design information

	Kubota	Mitsubishi Rayon	Zenon	Norit X-Flow
Flow (m^3/day)				
Peak	190–240	154–230		30–43
Normal (design)	48–72	32–48	44–74	9
Screening	2.0 mm basket filter, SD; 2.5–1 mm rotary drum filter, DD	0.75 mm parabolic sieve	0.75 mm static half drum with brush	0.5 mm rotating drum
Tank sizes (m^3)				
Denitrification	12.0	15.6	12.0	3.6
Nitrification	–	7.8	7.7	2.1
Membrane	18.8	10.8	3.9	~0.1 (ALSS)
Recycle ratio	~8	~12	~8	~5
Membrane				
Type	510	SUR 234	ZW500, a or c	F-4385
Configuration	150 panels per deck, SD/DD	Triple deck	4 modules	8 modules, 2 × 4
Total membrane area (m^2)	240	315	184	240

SD = single deck; DD = double deck, ALSS-airlift sidestream: no membrane tank.

All MBRs were configured with a denitrification zone, with an anaerobic tank added for Phase IV. The Kubota system was initially fitted with a single deck; a double deck was fitted mid-way through Phase II. Data presented in the following tables refer only to the first three phases. Water quality data (Table 3.5) relates to the range of mean values (i.e. not the total range of values recorded) reported for the four technologies tested. The process design and pre-treatment (i.e. screening) were all as specified by the supplier. A noticeable impact on alpha factor was recorded independent of the membrane type, with the value decreasing from 0.78–0.79 for the conventional ASP in operation at the works to 0.43–0.54 across all the MBR technologies.

Design (Table 3.6) and operation and maintenance (Table 3.7) data are reported below. Comparison of process performance is somewhat difficult from the reported data, since the process operating conditions were changed frequently throughout the trials. Variation in hydraulic retention time (HRT) meant that the organic loading rate varied significantly between the different technologies tested, from 0.043 to 0.059 kg biological oxygen demand (BOD)/m^3 for the Mitsubishi Rayon module to

Table 3.7 O&M data

	Kubota	Mitsubishi Rayon	Zenon	Norit X-Flow
Membrane aeration capacity (Nm^3/h)	90–180	75–120	100 (cycled)	140
Cycle (min)	8 on/2 relax[a]	20 h on/4 h off[c]	bflsh[f]	20 h on/4 h off[h]
Net flux (LMH)				
• Normal	8.3–12.5	5–8	20[e]	15–20, 37[i]
• Peak	32.5–42	20.3–30.6	35[e]	50
Biological aeration capacity (Nm^3/h)	160	160	100	140
F/M ratio	0.04–0.18	0.02–0.14	0.04–0.18	0.04–0.12
HRT (h)	10.2–15.4	15–22	7.6–12.3	15.2
SRT (day)	27–70	31–87	26–51	42–66
MLSS (g/L)	10.5–12	8.9–11.6	10.4–11.2	
Chemical cleaning reagents	NaOCl, 0.5% Oxalic acid, 1%	NaOCl, 0.5% followed by acid	NaOCl, 1%, followed by 0.3% citric acid	NaOCl, 0.5%, followed by *Ultrasil*[j]
Derived data				
SAD_m[a] ($Nm^3/(h\,m^2)$)	0.75	0.28–0.38	0.54[g]	0.33–0.6[h]
SAD_p (m^3 air/m^3 permeate)	60–90 normal; 18–23 peak	48–56 normal 12–14 peak	27 15	30–40 normal 12–16 peak
Mean permeability, LMH/bar	200–250 w/o r[a] 500–800 w r[a] 350 peak w r[a]	200 normal[d] 140–150 peak[d]	200–250 320–350 after clean	250 normal 75–200 peak
Permeability decline $\Delta K \Delta t$, LMH/(bar h)	1.5[b]	0.39[b]	20[b]	

[a]Relaxation introduced mid-way through Phase I; permeability data refers to without (w/o.) and with (w.) relaxation.
[b]Refers to peak flux operation: for the Zenon membrane this was 60 LMH.
[c]"Night-time" relaxation introduced during Phase III, along with backflushing at 20 LMH.
[d]Assumed to be with relaxation.
[e]Refers to 500 c module.
[f]Authors state "ratio of net to gross flux was 83–85%"; bflsh = backflushed.
[g]Intermittent operation.
[h]Night-time relaxation introduced during Phase II.
[i]With weekly maintenance clean.
[j]Combination of sulphuric and phosphoric acid.

0.075–0.11 for the Zenon. Ferric dosing was applied at different concentrations for the different technologies.

The protocol adopted was for operation under flow conditions set by the flow to the sewage treatment works: a fixed proportion of the flow was directed to the MBR pilot plants. This led to rather conservative average fluxes. Operating (or gross) fluxes ranged between 10 and 20 LMH for most of the time, leading to even lower net fluxes of 8–15 LMH. As a result, high permeabilities were generally recorded, particularly for the Kubota membrane and especially following the introduction of cleaning strategies such as relaxation. Commensurately high SAD numbers (SAD_p = 60–90, m^3 air/m^3 permeate in the case of Kubota) resulted. Peak flow tests conducted on each of the technologies always resulted in a significant drop in permeability.

For each of the trials operational problems were encountered which led to irreversible, and sometimes irrecoverable, fouling. This included partial aerator blockage, membrane tube blockage and partial dewatering (i.e. sludge thickening due to failure of the feed flow). Repeated chemical cleans were thus employed over the course of the trials, both for maintenance and recovery of membrane permeability, and relaxation was only introduced mid-way through the trials. A strong temperature dependence of permeability was noted for all membranes tested at low temperature, with the sustainable permeability decreasing markedly at temperatures below 10°C.

It was concluded from this trial that the Kubota could not be operates routinely at net fluxes of 38.5 LMH, although the fouling rate was decreased when relaxation was introduced. Peak flux operation of both the Zenon and Kubota systems at 41–43 LMH for 100 h led to a manageable decline in permeability, however, with permeability recovering to normal levels on reverting to low flux operation. Under the most optimal conditions of a regular maintenance clean and backflushing at 20.8 LMH, the permeability of the Mitsubishi Rayon membrane was maintained at between 200 and 300 LMH/bar at some unspecified flux. A net flux of 20.8 LMH was achieved for a period of 5 days, but peak fluxes of 28 LMH or more led to a rapid decline in permeability. Enhanced cleaning at higher reagent concentrations (1 wt% NaOCl) and temperatures (40°C) may have led to the noted deterioration of the membranes. The Norit X-Flow membrane was apparently, eventually, successfully operated at 37 LMH with weekly maintenance cleaning and night-time relaxation (4 h every 24 h). However, there were problems maintaining this flux due to tube blockage. The Zenon plant, fitted with the 500 c module, appears to have been operated under conditions whereby a reasonable net flux of around 20 LMH with an accompanying permeability of 200–250 LMH was sustained through maintenance cleaning twice weekly over an extended time period. It is possible that the other technologies could have also been sustained higher fluxes in operated under more optimal conditions.

3.2.3 Point Loma Wastewater Treatment Plant, San Diego

This trial resulted from an award made to the City of San Diego and Montgomery Watson Harza (MWH) from the Bureau of Reclamation to evaluate MBR + reverse osmosis (RO) technology for wastewater reclamation (Adham *et al.*, 2004). A 16-month study was conducted at the Point Loma Wastewater Treatment Plant (PLWTP) at San Diego, CA on four MBR systems: Mitsubishi Rayon, Zenon, Kubota and Memcor. The technologies were each challenged first with raw wastewater (Part 1) and then with advanced primary treated effluent containing polymer and coagulant residual (Part 2), arising from clarification with ferric chloride (27 mg/L, average dose) and a long chain, high-molecular-weight anionic polymer. Details of feedwater quality (Table 3.8), MBR technology design (Table 3.9) and operation (Table 3.10) are given below.

In this trial only the advanced primary treated effluent feedwater was tested on all four technologies. Results from untreated primary sewage were inconclusive since the feed water evidently contained unclarified ferric coagulant. All four technologies performed satisfactorily with respect to nitrification, disinfection and BOD removal. Only the Kubota was operated with denitrification and consequently demanded less

Table 3.8 Feedwater quality, PLWTP

Parameter (mg/L)	Raw sewage	Advanced primary treated sewage
Ammonia-N	27.3	26.6
BOD_5	213	97
COD	463	216
TOC	40	44
TKN	42.9	44.8
Ortho-phosphate-P	0.61	0.46

Table 3.9 Design information, PLWTP

	Kubota	USFilter	Zenon	Mitsubishi Rayon
Screening	3.2 mm travelling band screen	1.0 mm wedge wire slotted rotary screen	0.75 mm perforated rotary drum screen	0.75 mm perforated rotary drum screen
Tank sizes (m^3)				
Denitrification	6.42	3.79*	–	–
Nitrification	12.5	5.7	4.92	6.06
Membrane	–	0.34	0.7	0.95
Recycle ratio	4 (303 l/min)	–	–	–
Membrane				
Type	510	B10 R	ZW500d	Sterapore SUR
Configuration	100 panels/deck, double deck	4 modules	3 modules	50 modules/ bank, two banks
Total membrane area (m^2)	160	37	69	100

*Not used in trial.

oxygen from fine bubble aeration than the other technologies (Table 3.10). This being said, the data for specific aeration demand appear to demonstrate that, for this trial:

(a) The Mitsubishi Rayon system had a significantly higher specific aeration demand with respect to membrane aeration (SAD_m and SAD_p), reflecting the somewhat lower permeability of the membrane material.
(b) There was little difference in the figures for the total specific aeration demand ($TSAD_p$) of all those technologies operating without denitrification under the most optimum conditions, the figures ranging from 51 to 62.

The latter point is interesting in that it suggests one of two things: (a) that the air from fine bubble aeration somehow ameliorated membrane fouling, or (b) that the coarse bubble aeration contributed substantially to the DO.

The permeability data also provides interesting information on system operation and maintenance. The Zenon system was the only one operated with regular (three times a week) maintenance cleans by backpulsing with chlorine (250 ppm) or citric

Table 3.10 O&M data, PLWTP

	Kubota	USFilter Memcor	Zenon	Mitsubishi Rayon
Membrane aeration rate (Nm^3/h)	96	14.4	36[c]	90–114[e]
Cycle (min)	9 on/1 relax	12 on/0.75 relax/ 0.25 bflsh	10 on/0.5 relax	12 on/2 relax
Net flux (LMH)	25	19.2–24.2, pt 1 24.2[e]–40, pt 2	37.2[d]	20–25[e]
Fine bubble aeration rate (Nm^3/h)	<18	42, pt 1 78, pt 2	96	17–27[e]
Target (actual) DO (mg/L)	2 (3–5)	>1 (2–4)	1 (0.5–1.5)	>1 (1–2)
HRT (h)	5.1	6–11.5, pt 1 6, pt 2	2–5	3–4
SRT (day)	10–40	12–25	15–20	25–40
MLSS (g/L)	10–14, pt 1 9–12, pt 2	8–10, pt 1 6–8, pt 2	8–10	8–14
Derived data				
SAD_m[a] ($Nm^3/(hm^2)$)	0.60	0.39	0.52	0.9–1.14[e]
SAD_p (m^3 air/m^3 permeate)	24	16–20, pt 1 9.8[e]–16, pt 2	14[e]	45
$TSAD_p$ (m^3 total air/m^3) permeate	<29	63–79, pt 1 62[e]–103, pt 2	51	52–66[e]
Mean permeability[e] (LMH/bar)	250	150	270	140
Permeability decline[e] $\Delta K/\Delta t$, LMH/(bar h)	0	0.12	0.21	0.17
Cleaning cycle time[b], day	–	67	59	49

pt 1 = part 1: raw sewage containing ferric matter; pt 2 = part 2: advanced primary treated (clarified) sewage.
[a]Specific aeration demand.
[b]Based on a cycle between 0.1 and 0.5 bar at the given flux J and permeability decline $\Delta K/\Delta t$, hence $t_c = 8J/(\Delta K/\Delta t)$.
[c]Intermittent: 10 s on, 10 s off.
[d]Sustained by maintenance cleaning by backpulsing w. 250 ppm NaOCl 3 times a week.
[e]Data refer to that recorded for the latter part of the trial: advanced primary feed, most optimal conditions.

acid (2%) four times with a 30 s soak time between each cycle. This incurred downtime but allowed the system to operate at a relatively high net flux (37.2 LMH, significantly higher than the more commonly applied value of 25 LMH). The projected cleaning cycle time, calculated from the permeability decline value and TMP boundary values of 0.1 and 0.5 bar, was similar for the three HF technologies. The cleaning cycle time of the Kubota technology, on the other hand, could not be predicted since, under the conditions employed, there was no noticeable permeability decline. It is possible that the inclusion of denitrification for this process treatment scheme may have ameliorated fouling in some way. Cleaning protocols employed are summarised in Table 3.11.

The MBR effluent was very low in turbidity (<0.1 NTU on average, *cf.* 36–210 NTU for the feedwater) across all technologies tested, and average effluent BOD, total organic

Table 3.11 Cleaning protocols for the four MBR technologies, PLWTP

	Kubota	Mitsubishi Rayon	Zenon	Memcor
Recovery				
Reagents used	NaOCl, 0.5% Oxalic acid, 1%	NaOCl, 0.3% Citric acid, 0.2%	NaOCl, 0.2% Citric acid, 0.2–0.3% (pH 2–3)	NaOCl, 100 mg/L as Cl_2
Protocol	CIP by soaking for 2 h; reagents applied consecutively	CIP by flushing through for 2 h and soaking for 2 more hours	Tank drained and membranes hosed down with water; membranes backpulsed with reagent until flux stable	Tank drained and membranes rinsed with water for 10 min; membranes flushed through with reagent then soaked for 4 min
Maintenance				
	Reagent	Concentration	Duration	
Zenon	NaOCl	250 mg/L	10–15 s × 3 backpulses at 55 LMH, 30 s relaxation between pulses	

carbon (TOC) and chemical oxygen demand (COD) concentrations were respectively ≤2 mg/L, ≤9 mg/L and ≤31 mg/L. Silt density index (SDI) values measured for the Kubota MBR effluent during Phase I ranged from 0.9 to 1.1. Effluent from the Kubota MBR was treated downstream using Saehan BL and Hydranautics LFC3 reverse osmosis membranes. These were operated with minimal fouling during operation on raw wastewater and advanced primary effluent. The average net operating pressure of the Saehan 4040 BL (low-pressure) RO membranes measured during testing was 3. 1 bar, and that of the Hydranautics LFC3 (fouling resistant) RO membranes measured during testing was 8.3 bar. A 1–2 mg/L dose of chloramine in the RO feed was effective at mitigating membrane biofouling.

3.2.4 Bedok Water Reclamation Plant, Singapore

This trial arose from the Singapore NEWater project under the auspices of the Centre for Advanced Water Technology of the Singapore Utilities International Private Limited, a wholly owned subsidiary of Singapore Public Utilities Board. The work is intended to produce high grade water of drinking water standards, called NEWater, from municipal effluent to ensure a diversified and sustainable water supply. The project began in 2000 with the installation of ultrafiltration/microfiltration (UF/MF) plants for polishing secondary sewage prior to RO treatment. The MBR/RO option has been explored with trials of three MBR pilot plants operating simultaneously. The three technologies are unspecified in the report produced (Tao *et al.*, 2005) but have membrane properties similar to those of the Kubota, Zenon and Mitsubishi Rayon products. The plants were commissioned in March 2003 and fed throughout with primary settled sewage (Table 3.12).

Table 3.12 Feedwater quality, Bedok WRP

Parameter (mg/L)	Concentration
Ammonia-N	33
BOD5	99
COD	265, of which 109 soluble
TKN	33
Total P	9.2

Table 3.13 Design and O&M data, Bedok WRP

Parameter	MBR A	MBR B	MBR C
Membrane	0.4 μm FS, 0.8 m^2 panel area	0.4 μm HF, 280 m^2 element area	0.035 μm HF, 31.5 m^2 element area
Probable technology	*Kubota, double deck*	*Mitsubishi Rayon*	*Zenon (500d)*
Membrane area (m^2)	480	1120	1008
Tank sizes			
Anoxic tank volume (m^2)	30.8	37.5	25.2
Aerobic tank volume (m^2)	11.4	–	27.9
Memb. tank volume (m^2)	32.8	37.5	21.8
O&M			
MLSS g/L	6–12	6–14	4–13
Net flux (LMH)	13–28.4 (26)	16–24 (24)	6.2–29.3 (12.4)
Initial TMP	4	17	10
Cycle (min)	9 on/1 relax	13 on/2 relax	12 on/0.5 backflush +relax
Cleaning cycle time (day)	90	120	3.5*
Chemical cleaning reagents	0.6% NaOCl 1% oxalic acid	0.3% NaOCl 2% citric acid	NaOCl (citric acid)*
Derived data			
SAD_p (m^3 air/m^3) permeate	28–50 (50)	16–24 (24)	20–30 (30)
Initial permeability (LMH/bar)	650	66	124
SAD_p *(m^3 air/m^3 permeate) (energy demand, kWh/m^3)*			
Baseline	50 (1.4)	24 (1.3)	30 (1.7)
High flux	34 (1.2)	21 (1.0)	25 (1.3)
Low aeration	28 (1.0)	16 (0.8)	20 (1.1)

*Maintenance cleaning employed; citric acid cleaning suspended after 11 months.

The MBR plants are described in Table 3.13. The plants were all of around 75 m^3 total tank and all were submerged MBRs with a separate membrane tank. The baseline HRT and solids retention time (SRT) values were respectively 6 h (hence 300 m^3/day flow) and 21 days for all tanks, and the target MLSS was 10 g/L. An *ex situ* clean was only carried out once on MBR A and only due to failure of the aeration system. Following operation for 2–3 months under baseline conditions the flux was increased for an unspecified period and following this the aeration reduced for another unspecified period. It is not clear whether the conditions identified were optimal. According to the analysis conducted, MBR B (assumed to be Mitsubishi Rayon) was

Figure 3.1 The planned location of the new demonstration MBR at Ulu Pandan

lower in energy demand than the other two MBRs. However, MBR B was also the only one configured without a separate aeration tank. Contrary to previous reported data (Fig. 2.19), these authors reported a linear relationship between viscosity and MLSS between 4 and 14 g/L for all three systems.

The mean product water quality from each of the MBRs tested was found to be broadly similar at <0.2 NTU, <2 mg/L TKN, <1 mg/L NH_4^+-N and <5 mg/L TOC, and slightly higher than product from UF polishing of secondary effluent (7 mg/L TOC, 4.5 mg/L TKN and 3 mg/L NH_4^+-N). However, for MBRs A and B the permeate TOC was found to rise significantly – to as high as 100 mg/L – following chemical clean. As the authors suggested, this was likely to be due to the removal of the protective gel layer by chemical cleaning, this layer taking around 36 h of filtration to be reformed. For citric or oxalic acid cleaning, TOC in the permeate may also arise from the chemical cleaning reagent.

The pilot testing has shown the MBR-RO option to produce a slightly superior quality product water than the conventional approach of secondary treatment followed by UF/MF + RO, specifically with respect to TOC, nitrate and ammonia (Qin *et al.*, 2006), and is also lower in cost. Future expansion of wastewater treatment plants in Singapore is therefore likely to be based on MBR technology. A 23 MLD MBR demonstration plant at Ulu Pandan (Fig. 3.1) for testing at municipal scale is planned for completion in 2006.

3.2.5 Pietramurata, University of Trento

An extensive comparative pilot trial has been conducted by the University of Trento at Pietramurata wastewater treatment plant in Italy to evaluate different innovative technologies for the upgrading of existing plants. The plants tested were originally

Zenon and Kubota, but more recently a Memcor plant has also been studied. The first plant was installed in June 2001. This consisted of two separate MBRs whose biotreatment tanks were provided by dividing a single rectangular stainless steel tank into two separated treatment lines for a Zenon and Kubota MBR, both configured with pre-denitrification. All experimental activities on site have been performed between March and December, since the whole system is outdoors and the ambient temperature is too low to operate during the winter months.

The biological volume was calculated according to a rather conservative flux value of 15 LMH resulting in volumes of 5.14 and 4.23 m^3 for the aerobic tanks, and 2.76 and 2.27 m^3 for the anoxic tanks for the respective Zenon and Kubota processes. The permeate is removed by suction, originally using progressing cavity (*Mono*) pumps but a self-priming pump was installed in 2003. A progressing cavity pump is also used for recirculation from the aerobic to the anoxic tank at recycle ratios between 3 and 5. Aeration for the biological process is supplied to both aerobic tanks from the main aeration pipe of the full-scale plant and is measured by specific airflow meters. The Zenon/Memcor and Kubota systems were operated under different SRTs, imposed by daily sludge wasting. All the usual operating parameters have been monitored (TMP, DO, etc.) and, since this effluent is re-used for agriculture, from 2004 onwards the effluent nitrate concentration from the Zenon plant has been measured.

Membrane modules originally installed comprised a Zenon 500 c (66.6 m^2 membrane surface area) and a Kubota E50 (40 m^2 of membrane surface area). In November 2003 a new hollow fibre module (Memcor) was installed as a sidestream to the Zenon MBR tank. The module, 40 m^2, was immersed in a tank (0.3 m^3) and fed with sludge from the aerobic tank and fed back to it under gravity. Due to this special combination, the permeate flow suctioned from both HF lines was often reduced to avoid excessive organic loading of the biological system. Simultaneous short- and long-term flux-step tests on both systems were likewise also avoided. Due to some technical and budgeting problems, operation of some modules was stopped for long periods; the Zenon module was not used during 2003 and the Kubota module was not used during 2005. In 2005 the Memcor module was operated for a few weeks on biomass previously acclimatised by the Zenon module. The aerator depth was 1.9–2.0 m for the Zenon and Memcor modules and 2.4 m for the Kubota stack.

During 2002 the Zenon module was cleaned monthly using 300 mg/L NaOCl as Cl_2. From 2004 onwards the reagent concentration for the monthly clean was reduced to 200 mg/L. From 2002 to 2003 the Kubota module was cleaned twice yearly with 3000 mg/L (i.e. 0.3 wt%) NaOCl as Cl_2 and once a year with a 10 000 mg/L (1 wt%) solution of oxalic acid. From 2004 the module was cleaned monthly with a 200 mg/L solution of NaOCl with no supplementary acid cleaning. The Memcor module was cleaned monthly using 300 mg/L NaOCl. For all three modules the membranes were additionally cleaned using 100 mg/L NaOCl immediately before each flux-step test.

The mean feed COD and N-NH_4^+ levels were 575–988 and 23–33 mg/L respectively, with the BOD/COD ratio generally being <0.5. The corresponding outlet concentrations were <22 and <5 respectively, with the total organic nitrogen (TON) being below 11 mg/L at all times. Summary operation and maintenance data are summarised in Table 3.14, and mean and apparent optimum values in Table 3.15. These are discussed in the context of other data in Section 3.3.

Table 3.14 O&M data, Pietramurata WWTP

Parameter	No[a].	Zenon		Kubota		Memcor	
Membrane aeration rate (Nm^3/h)	1–2	16.7	25.0	38	46	8	
	3	25.0	25.0	38	46	8	
	4	–	–	38	46	–	–
	5	–	–	28		–	–
Cycle (min)	1–2	9 on/1 off		9 on/1 off		9 on/1 off	
	3	9 on/1 off		8 on/3 off		–	–
	4	–	–	9 on/1 off		–	
	5	–	–	continuous		–	–
Net flux (LMH)[b]	1	7.2	28.8	6.3	28.4	9	31.5
	2	9	30.6	7.6	11.9	21.6	
	3	8.6	14	6.9	16.7	–	–
	4	–	–	8.6	9.5	–	–
	5	–	–	16	37.5	–	–
HRT (h)	1	4.1	16.5	10.7	42.9	4.5	14
	2	4.5	14	18.1	18.1	11	
	3	10	13.4	10	22	–	–
	4	–	–	18		–	–
	5	–	–	5.5	8	–	–
SRT (day)	1	35		40		8	20
	2	8	20	8		15	
	3	15		20		–	–
	4	–	–	12		–	–
	5	–	–	9	12	–	–
Derived data							
Permeability LMH/bar	1	144	144	630	81	250	58
	2	138	61	167	264	300	200
	3	123	108	276	239	–	–
	4	–	–	264	271	–	–
	5	–	–	320	100	–	–
SAD_m ($Nm^3/(h\,m^2)$)	1	0.25	0.375	0.95	1.2	0.2	
	2	0.25	0.375	0.95	1.2	0.2	
	3	0.375	0.375	0.95	1.2	–	–
	4	–	–	0.95	1.2	–	–
	5	–	–	0.7		–	–
SAD_p (m^3 air/m^3 permeate)	1	35	13	37	136	5.7	20.0
	2	28	12	87	113	8.3	
	3	44	27	50	100	–	–
	4	–	–	110	100	–	–
	5	–	–	19	44	–	–

[a]Trial number; trials range between 2 and 9 months in duration.
[b]Temperature-corrected to 20°C.

3.2.6 Eawag pilot plant MBR, Kloten/Opfikon, Switzerland

Eawag (The Swiss Federal Institute of Aquatic Science and Technology) is a Swiss national research organisation committed to ecological, economical and socially responsible management of water. A pilot scale MBR was set up and operated

Table 3.15 O&M data: most optimum and mean, Pietramurata WWTP

Parameter	Zenon		Kubota		Memcor	
	Optimum[a]	**Average**	Opt[a]	**Average**	Opt[a]	**Average**
HRT (h)	14	**10**	6	**17**	11	**10**
SRT (day)	20	**21**	9	**17**	15	**14**
Membrane aeration rate (Nm^3/h)	25	**22**	28	**39**	8	**8**
Cycle (min)	**9 on / 1 off**			**–[b]**	**9 on / 1 off**	
Flux (LMH)	31	**16**	16	**15**	22	**21**
Permeability (LMH/bar)	61	**120**	320	**261**	270	**182**
SAD_m ($Nm^3/(hm^2)$)	0.38	**0.33**	0.70	**0.98**	0.20	**0.20**
SAD_p (m^3 air/m^3) permeate	12	**27**	19	**79**	22	**17**

[a]Conditions employed under which the highest fluxes and lowest aeration rates were sustained.
[b]9 on/1 off routinely, continuous operation for final trial.

Table 3.16 Feedwater quality, Kloten/Opfikon WWTP (value ± standard deviation)

Parameter mg/L	MBR feed (after primary clarification)
Ammonia-N	21 ± 5 mg NH_4^+-N L^{-1}
BOD5	not available
COD	282 ± 91mg COD L^{-1}
TOC	not available
TKN	38 ± 9 mg NL^{-1}
Ortho-phosphate-P	2.1 ± 0.7 mg PL^{-1}
P total	4.2 ± 1.3 mg PL^{-1}

continuously for 4 years by the Engineering Department of Eawag to gain hands-on experience of running such plants. The test protocol incorporated the study of the impact of membrane maintenance protocols, SRT and chemical flocculants on performance with reference to three different MBR technologies. The pilot plant was also used for studying the fate of micropollutants in an MBR compared with conventional technologies (Göbel *et al.*, 2005; Huber *et al.*, 2005; Joss *et al.*, 2005, 2006; Ternes *et al.*, 2004).

The pilot plant is located on the municipal sewage treatment plant of the works at Kloten/Opfikon, a conventional activated sludge plant with an average dry weather flow of 17 ± 4.3 MLD (55 000 population equivalent (p.e.); maximum flow 645 l/s). This plant receives sewage from Zürich airport (~50% of the influent flow) and from the nearby towns of Kloten and Opfikon. The pilot plant is fed with primary effluent (Table 3.16) from the full-scale plant following primary treatment, comprising screening at 10 mm, a sand/oil trap and primary clarification at 2h HRT). The flow rate was proportional (~0.2%) to that of the full-scale facility (similar to Beverwijk trial, Section 3.2.1). The volume of wastewater treated was 30 ± 12 m^3/day, with a maximum daily flux of 74 m^3/day.

The pilot comprised a cascade of 2–6 tanks, each of 2 m^3 volume, providing anaerobic and anoxic treatment at recycle rates of 80 ± 30 and 50 ± 23 m^3/day respectively, and at mean sludge concentrations of around 4.8 g/L. Aerobic treatment at mean

Table 3.17 Design information, Kloten/Opfikon WWTP (value ± standard deviation)

	Kubota	Zenon	Mitsubishi Rayon
Tank size (m^3)	2.6	1.6	1.9
Sludge recycle ratio	1.3 ± 0.5	1.2 ± 0.5	1.3 ± 0.5
Sludge content (gSS/L)	8.3 ± 3.4	8.2 ± 3.5	7.0 ± 2.9
Membrane			
Type	0.8 m^2 panels	ZW500a	Sterapore SUR
Configuration	50 panels	1 module	80 × 1 m^2
Total membrane area (m^2)	40	46	80
Net permeate production (m^3/day)	9.1 ± 3.9	11 ± 4.5	944 ± 3.8
Maximum permeate production (m^3/day)	25.4	28.7	24.2

Table 3.18 O&M data, Kloten/Opfikon WWTP

	Kubota	Zenon	Mitsubishi Rayon
Membrane aeration rate (Nm^3/h)	60 ± 4	50 ± 5, air cycling on/off 10 s/10 s	29 ± 4
Cycle (min)	8 on/2 relax	5 on/0.5 bflsh.	8 on/0.5 bflsh. +1.5 relax
Average net flux (LMH)	9.5 ± 4	10 ± 4	4.8 ± 2
Gross flux (LMH)	17 ± 4	19.5 ± 3	10 ± 2
Fine bubble aeration rate (Nm^3/h)	0–10	0–10	0–10
HRT (h)*	3 ± 1.3	1.9 ± 0.8	1.9 ± 1.1
SRT (day)	15–67	15–67	15–67
MLSS (g/L)	8.3 ± 3.4	8.2 ± 3.5	7 ± 2.9
Derived data			
SAD_m ($Nm^3/(h\,m^2)$)	1.5	0.54	0.37
SAD_p (m^3 air/m^3 permeate)	88	28	38
$TSAD_p$ (m^3 total air/m^3 permeate)	88–90	28–38	38–48
Mean permeability (LMH/bar)	200 ± 160	200 ± 83	90 ± 70
Cleaning cycle time[2], day	variable	variable, 14 day	variable

*HRT in the membrane compartment only.

MLSS concentrations of between 7 and 8.3 g/L was coupled with membrane separation in discrete compartments, operated in parallel and providing a membrane area of between 40 and 80 m^2 (Table 3.17), all modules being single deck. Four phases of work can be identified:

- 1st year (day 1 to 385): 17 ± 2 days sludge age, 8 m^3 anaerobic volume.
- 2nd year (day 410 to 745): 32 ± 3 days sludge age, 6 m^3 anaerobic volume.
- 3rd year (day 770 to 1090): 58 ± 10 days sludge age, 6 m^3 anaerobic volume.
- 4th year (day 1090 to 1510): 56 ± 11 days sludge age, no anaerobic volume, Fe^{3+} addition.

Data for the three technologies (Table 3.18) indicate considerable changes in permeability (Fig. 3.2), particularly for the Kubota module. Permeability fluctuations for this membrane did not appear to correlate with either operational changes such

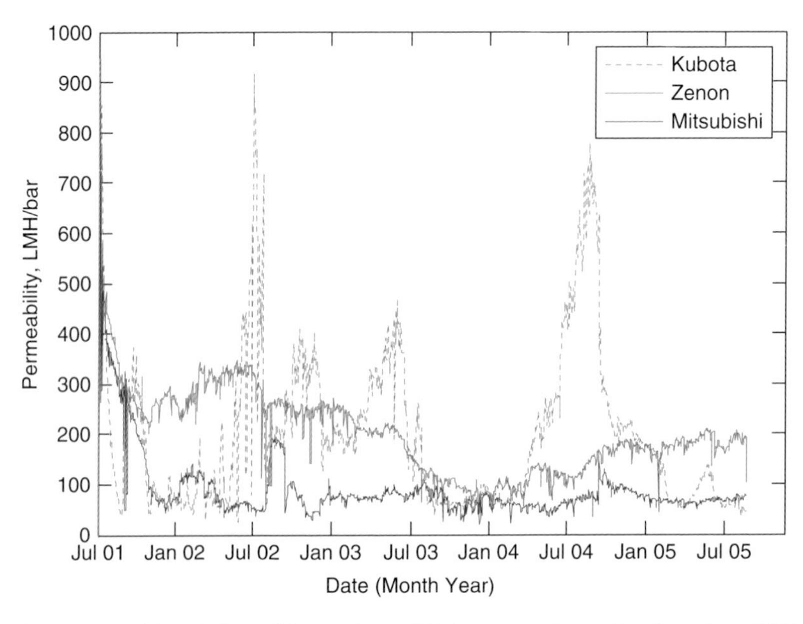

Figure 3.2 Permeability of the modules as observed during the total operation time of the MBR pilot. Data corrected to 20°C according to viscosity correction

Table 3.19 MBR cleaning protocols, Kloten/Opfikon WWTP

	Kubota	Mitsubishi Rayon	Zenon
Reagents used	NaOCl, 150–2000 mg/L Citric acid pH 2.5	NaOCl, 150–1000 mg/L Citric acid pH 2.5	NaOCl, 150–1000 mg/L Citric acid pH 2.5
Established protocol	–*	–*	Maintenance clean, 150 mg/L NaOCl and citric acid clean every 14 days.

*No protocol tested yielded satisfactorily reproducible results.

as cleaning protocols. For the Zenon module, however, permeability correlated roughly with MLSS concentration in the membrane compartment from the start of operation until January 2004. The permeability increase from January 2004 onwards correlated with the introduction of regular maintenance CIPs (Table 3.19) every 14 days with 150 ppm NaOCl followed by a citric acid clean pH 2.5.

The Mitsubishi Rayon unit provided a low permeability throughout the study due to clogging of the hollow fibres by dehydrated sludge. Several measures taken to ameliorate this problem, such as overnight relaxation and intensive or regular chemical cleaning, were unsuccessful. Regular backflushing for 30 s during each permeate production cycle, as applied to the Zenon module, provided a stable low permeability. No difference in efficacy of maintenance cleaning at low hypochlorite concentrations "in air" (Section 2.3.9.2) rather than in place was noted.

Maintenance CIP was conducted through backflushing for 20–30 min using 150 mg/L NaOCl followed by an acid clean at pH 2.5 ± 0.5. For hollow fibre membranes the backflush (1.5 times the maximum operating flux) was applied intermittently with 20–30 s pulses at applied at 5 min intervals. Flat sheet (FS) membranes where backflushed continuously under gravity at a maximum hydrostatic head of 0.1 bar (the limit recommended by the supplier is 0.2 bar). Chemical usage per backflush was between 2 L/m^2 for the HF membranes to 5 L/m^2 for the FS ones. Intensive (recovery) cleaning was through soaking the membranes for 3–6 h in a high-strength (0.1–0.5 wt%) NaOCl solution with pulsed aeration combined with suction for 5–10 s every 20 minutes. Hypochlorite cleaning led to significant foaming due to formation took place due to the organic matter present. The amount of chemical solution required depended on the packing density of the membrane, but appeared to be in the range of 20–100 L/m^2. Care was taken to drain the membrane of hypochlorite before applying the acid wash of 0.5 mM sulphuric acid together with a 5 mM citric acid buffer, the mineral acid being used to counter the alkalinity, the buffered acid pH being 2.8.

3.3 MBR design and operation

3.3.1 Reference data

3.3.1.1 Aeration

Comparison of the pilot study data in Sections 3.2.1–3.2.6 (Table 3.20) demonstrates the difficulty in attempting to generalise with design and operation of MBRs. A summary of the data reveals substantial variation in some of the key performance parameters, possibly because the conditions were not uniformly optimised. Extending the analysis to available full-scale data from Chapter 5 (Table 3.21) provides a more reasonable basis for an analysis and establishing appropriate operating conditions.

It has generally been observed from laboratory-scale studies that the attainable flux increases with increasing aeration rate due to increased scouring. This is manifested as an increase in the critical or sustainable flux (Section 2.3.7.1). In full-scale plants increased aeration would be expected to produce an increase in sustainable net permeability. Figure 3.3, based on the available data, would appear to indicate that this is indeed the case. According to this figure there is a general tendency for increasing permeability with increasing SAD_m, though the data are very highly scattered. Some of this data scatter can be attributed to obvious outliers, namely either plant operating under sub-optimal conditions and/or small unstaffed plant where blowers are more likely to be oversized to maintain permeability and so limit maintenance requirements. Sustainable permeability also changes according to the nature of aeration, that is the specifications of the aerator itself and the mode of application (continuous or intermittent), and the cleaning protocol. Although physical cleaning is to some extent accounted for by using net rather than gross flux in calculating permeability, maintenance cleaning with hypochlorite permits higher permeabilities to be sustained. A study of the data for flux for the two main technologies (Fig. 3.4,

Table 3.20 Summary of pilot plant SAD data, four common MBR technologies

Technology	J (LMH)	K (LMH/bar)	SAD_m	SAD_p	Source
Kubota	8.3–12.5	500–500	0.75	60–90	DHV
	32.5–42[a]	350[a]	0.75[a]	60–90[a]	DHV[a]
	25	250	0.6	24	MWH
	26	650[b]	0.67–1.2	28–50	PUB
	18–25	200–500	0.67–1.2	28–50	PUB
	15–16	261–320	0.7–0.98	19–79	Trento[c]
	9.5	200	1.5	88	Eawag
Mitsubishi Rayon	5–8	200	0.28–0.38	48–56	DHV
	20[a]	140–150[a]	0.28–0.38[a]	12–14[a]	DHV[a]
	20–25	140	0.9–1.14	45	MWH
	16–24	66[b]	0.38–0.58	16–24	PUB
	4.8	90	0.37	38	Eawag
Zenon	20	200–250	0.54	27	DHV
	35[a]	200–250[a]	0.54[a]	15[a]	DHV[a]
	37.2	270	0.52	14	MWH
	6.2–29.6	124[b]	0.25–0.37	20–30	PUB
	16–31	61–120	0.33–0.38	12–27	Trento[c]
	10	200	0.54	28	Eawag
Memcor	20–40	150	0.39	10–20	MWH
	21–22	182–270	0.2	17–22	Trento

[a]Peak loading conditions; [b]initial permeability; [c]range refers to mean optimum.

Table 3.21 Summary of full-scale plant specific aeration demand data

Technology	Capacity MLD	Flux LMH	K LMH/bar	SAD_m [2], Nm/h	SAD_p [2] –	MLSS g/L
FS						
Kubota	1.9	20	350	0.75	32	12–18
	13	33	330	1.06	32	8–12
	4.3	25	680	0.56	23	na
Brightwater	1.2	27	150	1.28	47	12–15
Toray	0.53	25	208	0.54	22	6–18
	1.1,i	21.6	1500	0.4	19	22
Huber	0.11	24	250	0.35	22	ns
Colloide	0.29	25	62.5	0.5	20	ns
HF						
Zenon	2	18	95	1	56	15
	48*	18	144	0.29	16	8–10
	0.15*,i	12	71	0.65	54	10–15
	50*	25	175	0.4	17	12
Mitsubishi Rayon	0.38	10	30	0.65	65	12
Memcor	0.61	16	150	0.18	11	12
Asahi-kasei	0.9, i	16	80	0.24	15	8
KMS Puron	0.63	25	160	0.25	10	ns

*Intermittent aeration.
i – Industrial effluent feedwater.
na – Not applicable.
ns – Not specified.

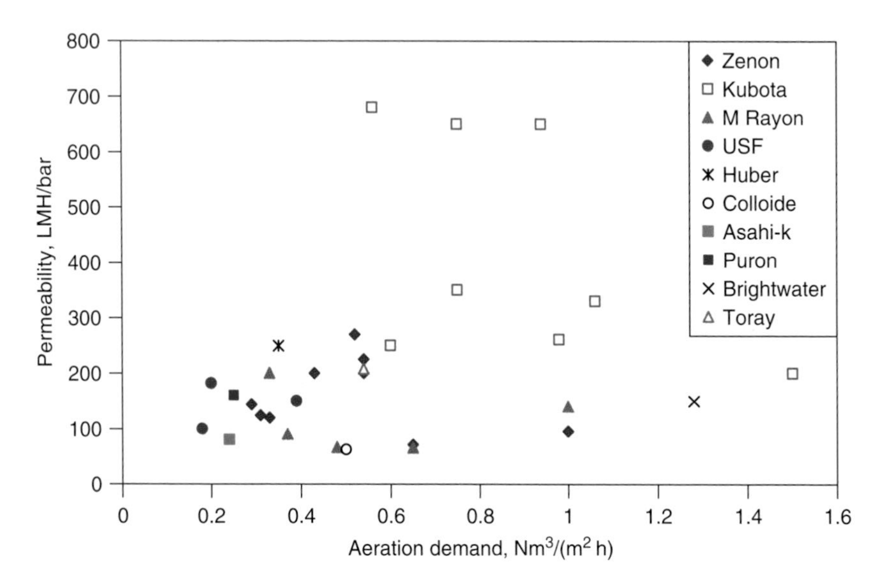

Figure 3.3 Permeability vs. specific aeration demand for available data provided in pilot plant and full-scale data from Chapters 4 and 5 respectively

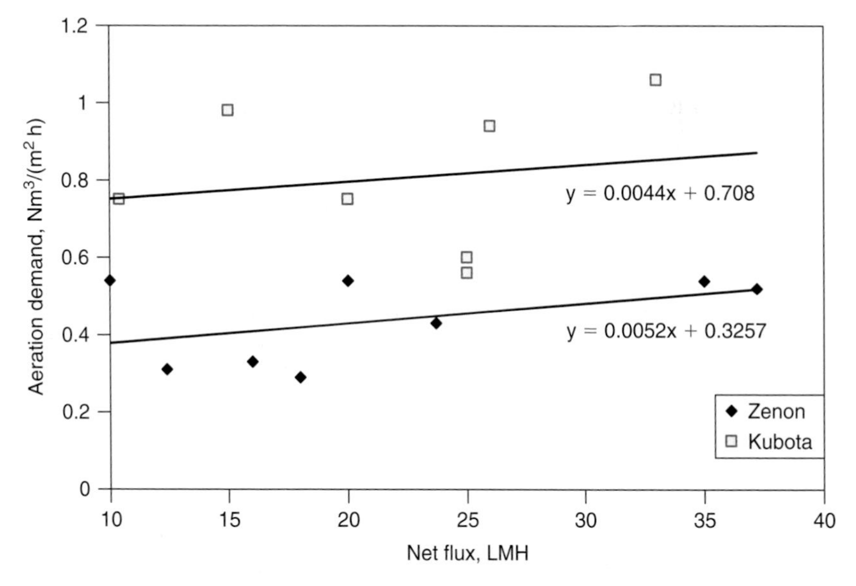

Figure 3.4 Specific aeration demand vs. flux for Kubota and Zenon data

1–2 outliers for each of the two technologies have been omitted), on the other hand, reveals that the relationship between flux and aeration demand in practice is weak.

If the more obvious outliers are ignored then some general trends can be identified:

(a) FS systems tend to operate at high permeabilities (generally >200 LMH/bar) and are associated with high aeration demands, both as SAD_m and SAD_p.

No trend is evident in this data subset, though all but the highest (and probably non-optimal) SAD_p values lie within the range 20–40.

(b) HF systems tend to operate at lower permeabilities (generally $<$200 LMH/bar) and are associated with lower aeration demands, sometimes achieved by employing aeration intermittently. For these systems the permeability attained is roughly linearly related to aeration demand (Fig. 3.2), with around 0.4 m^3 permeate generated per Nm^3 air per bar, and SAD_p values are generally between 15 and 30.

Conventionally MBRs, in common with all membrane processes, are designed on the basis of operating flux. However, permeability more accurately reflects the fouling condition of the membrane. Although the permeability of the virgin membrane material varies somewhat from one membrane product to another, a low operating permeability is generally indicative of a high degree of fouling. Given that permeability is the ratio of flux to TMP, then operation within given boundaries of permeability will necessarily determine the range of operational TMP values and vice versa. According to Fig. 3.3 it follows that, for HF systems at least, the required aeration rate for maintaining a given permeability (or TMP for a given target flux) can be estimated from the permeability itself. Alternatively, flux can be correlated with aeration rate from data depicted in Fig. 3.4; that is:

$$SAD_m = 0.0044J_{net} + 0.708 \quad \text{for FS module} \tag{3.18}$$

$$SAD_p = 0.0052J_{net} + 0.326 \quad \text{for HF module} \tag{3.19}$$

although clearly more data is required to ratify such correlations.

3.3.1.2 Cleaning

A summary of physical and chemical cleaning protocols for the pilot plant study data (Table 3.22) and full-scale installation data (Table 3.23) reveals that physical cleaning is predominantly by relaxation rather than backflushing. Pilot plant data indicate downtime for physical cleaning to account for between 4% and 20% of the operating time, with no profound difference between the two configurations. On the other hand, maintenance cleaning every 3–14 days is routinely employed for some HF technologies whereas chemical cleaning is usually limited to recovery cleans alone for FS systems.

For both FS and HF systems recovery cleans are generally applied at intervals of 6–18 months depending on flux. Recovery cleans generally employ hypochlorite concentrations between 0.2 and 1 wt% NaOCl (Table 3.9), usually adjusted to a pH of ~12. Both maintenance and recovery cleaning are brief and/or infrequent enough to add little to the percentage downtime. For example recovery cleans at Porlock and Brescia have been applied, on average, once every 9 and 15 months respectively. Even for an overnight soak of 16 h, the total downtime for cleaning amounts to $<$0.25% in such cases. For Zenon sewage treatment plants the maintenance cleaning cycle normally incurs no more than 2 h of downtime and is employed once every one or

Table 3.22 Summary of pilot plant physical cleaning protocol data

Technology	Filtration cycle (min)		percentage on	Source
	on	off		
Kubota	8	2r	80	DHV
	9	1r	90	MWH
	9	1r	90	PUB
	9/8	1/3r	73–90	Trento
Mitsubishi Rayon	1200	240r	83	DHV
	12	2r	86	MWH
	13	2r	87	PUB
Zenon	ns	ns; b,r	85	DHV
	10	0.5r	95	MWH
	12	0.5b,r	96	PUB
	9	1r	90	Trento

r – Relaxation; b – backflushing.

Table 3.23 Summary of full-scale plant cleaning protocol data

Technology	Cap[a] MLD	Filtration cycle (min)		Cleaning cycle	
		on	r or b	Interval	Type
FS					
Kubota	1.9	1380	60 r	8–9 m	CIP
	13	1380	60 r	6 m	CIP
	4.3	10–20	1–2	6 m	CIP
Brightwater	1.2	55	5 r	>18 m	CIP
Toray	0.53	ns	ns	none[b]	
	1.1	8	2	12 m	CIP
Huber	0.11	9	1	none	–
Colloide	0.29	6	2	ns	–
HF					
Zenon	2	10	0.75 b	1 w/6 m	mCIP/R
	48[b]	10.5	1.5 r	1 w/15 m	mCIP/R
	0.15[b],i	10	0.5 b	0.5 w	mCIP
	48[b]	7	1	2 w	mCIA
Mitsubishi Rayon	0.38	12–13	2–3	ns	ns
Memcor	0.61	12	1	3 m	CIA
Asahi-kasei	0.9,i	9	1	1–3 m	CEB
KMS Puron	0.63*	5	ns	3 w	mCIP

w – Weeks; m – months.
mCIP – maintenance clean in place; mCIA – maintenance clean in air.
CEB – Chemically enhanced backflush; R – recovery clean.
*Moderate backflush used combined with air scouring.

two weeks, amounting to <1.2% downtime. Thus, for the majority of the most recent plant, the ratio of the net to gross flux is determined primarily by period of relaxation or backflushing alone. The most onerous impact of chemical cleaning appears to be chemicals usage and chemical waste discharge, both of which are minor.

Table 3.24 Biokinetic constants

Constant	Range	Unit	Constant	Range	Unit
K_s	6–192	g/m^3	K_n	0.01–0.1	$g\ NH_4\text{-}N/m^3$
k_e	0.023–0.2	per day	$k_{e,n}$	0.025–0.15	per day
μ_m	3–13.2	per day	$\mu_{m,n}$	0.2–2.21	per day
Y	0.28–0.67	kg VSS/kg COD	Y_n	0.1–0.15	kg VSS/kg NH_4-N

3.3.2 Biokinetic constants

The design of an MBR from biokinetic considerations relies on assumptions regarding values of the biokinetic constants, as is the case with other aerobic treatment processes. Ranges of values are summarised in Table 3.24 below. A more comprehensive listing of biokinetic constants, along with references identifying their origins is provided in Appendix B. Typical values of $K_s = 60$, $k_e = 0.08$, $\mu_m = 8$, $Y = 0.4$, $K_n = 0.07$, $k_{e,n} = 0.1$, $\mu_{m,n} = 0.7$, $Y_n = 0.13$ have been used in the example design that follows in Section 3.3.3. Note that the values selected are not necessarily the most appropriate for an MBR.

3.3.3 Design calculation

A complete design for an immersed membrane bioreactor (iMBR) can be carried out on the basis of the information presented, provided the nature of the interrelationship can be determined between aeration and:

(a) permeability and cleaning protocol for the membrane permeation component,
(b) feed water quality, flows and biokinetics for the biological component.

Whilst it is possible to base the latter on biokinetics, the former cannot reasonably be calculated from first principles. The calculation for an HF technology based on a flow of 5 MLD is detailed below. Equation (3.19) has been used to correlate aeration with flux. Assumptions concerning the organic content of the feed (Table 3.25), values for biokinetic constants (Table 3.26), biotreatment operation (Table 3.27) membrane operation (Table 3.28) and aeration operation (Table 3.29) lead to the system design an operating parameters detailed in subsequent tables (Tables 3.30–3.36). Shaded cells refer to calculated values. Sludge disposal costs are excluded since costs associated with this are process dependent post-treatment of the sludge, such as thickening, has a large impact on disposal costs. However, in large full-scale plants sludge treatment and disposal can be assumed to be small compared with power costs.

The design of a pumped sidestream membrane bioreactor (sMBR) is, in principle, somewhat more straightforward than an iMBR since membrane permeation is carried out by crossflow filtration without air. In this case a correlation of permeability and flux with crossflow velocity can provide the information needed to calculate energy demand for pumping, based on Equation (3.1) along with the classical biokinetic expressions for oxygen demand by the biomass.

A much discussed issue in iMBR technology is the relative costs of processes based on FS and HF module configurations. Considerable effort has been devoted to increasing

Table 3.25 Feedwater specification

Parameter	Value	Unit	Parameter	Value	Unit
Q	5000	m^3/day	T_w	12	°C
Q_p	203384	L/h	$T_{w,K}$	285	°K
Q_{peak}	1000	m^3/h	T_a	15	°C
BOD S_0	220	g/m^3	$T_{a,K}$	288	°K
COD	430	g/m^3	$P_{a,1}$	101	kPa
BOD:COD ratio	1:2		ρ_a	1.23	kg/m^3
TSS	250	g/m^3	ρ_b	1003	kg/m^3
VSS	190	g/m^3	ρ_p	996	kg/m^3
N	40	g/m^3			
Non-biodegradable VSS	0.24	g/gTSS			

Table 3.26 Kinetic constants

BOD			Ammonia		
Parameter	Value	Unit	Parameter	Value	Unit
T = 20°C					
K_s	60	g/m^3	$\mu_{n,m}$	0.41	g/(g/day)
k_e	0.06	per day	K_n	0.05	g/m^3
Y	0.4	g/gBOD	k_{dn}	0.07	g/(g/day)
μ_{max}	4.7	g/(gVSS/day)	Y_n	0.13	g/(g/day)
f_d	0.86	gnbSS/g feed	μ_n	0.040	g/(g/day)
Y_{obs}	0.64	g/gBOD			

*Adjusted according to Metcalfe and Eddy (2003), pp. 709.

Table 3.27 Biological operating parameters

Parameter	Value	Unit	Parameter	Value
X	8000	g/m^3	V_{an}/V	0.3
θ_x	25	day	r	2

Table 3.28 Membrane operating data

Parameter	Value	Unit	Parameter	Value	Unit
K	100	LMH/bar	ΔP_m	0.25	bar
J	25	LMH	SAD_m	0.92	$m^3/(m^2 h)$
t_b	0.167	h	t_c	168	h
τ_b	0.013	h	τ_c	2	h
J_b	35	LMH	c_c	0.25	kg/m^3
n	938		J_{CIP}	45	LMH

membrane aeration efficiency for both membrane configurations, yet evidence suggests that the aeration energy demand for the FS configuration is inherently higher than that of the HF technologies (Figs. 3.3–3.4). Applying Equations (3.18) and (3.19) for SAD_m vs. J_{net} allows the specific cost, in \$/m^3 permeate, to be computed for

Table 3.29 Aeration operating data

Parameter	Biology	Membrane
Diffuser type	Fine bubble	Coarse bubble
OTE per m	0.05	0.02
α		0.49
β		0.95
ϕ		0.89

Table 3.30 Biological parameters

Parameter	Value	Unit	Parameter	Value	Unit
V	2980	m^3	Aerobic HRT (average)	14.3	h
V_{an}	894	m^3	Aerobic HRT (peak)	3	h
S_e	0.81	g/m^3	Anoxic HRT (ave)	4.3	h
N_e	0.018	g/m^3	Anoxic HRT (peak)	0.9	h
R_0	1395	kg/day	R'_0	1434	N kg/day

Table 3.31 Sludge yield

Parameter	Value	Unit
NO_x	40	g/m^3
P_x	654	kg/day
X_0	300	kg/day
$P_x + X_0$	954	kg/day
Q_w	119	m^3/day
$Q_w/(J'_{net}A_m)$	0.02	m^3/m^3

Table 3.32 Membrane calculations

Parameter	Value	Unit	Parameter	Value	Unit
A_m	9887	m^2	J_{net}	21	LMH
Area per element	250	m^2	$J_{net,acual}$	20	LMH
Actual membrane area	10 000	m^2	$J'_{net,acual}$	25	N LMH
Number of elements	40		J'_b	42	N LMH
K'	120	N LMH/bar	J'_{actual}	30	N LMH

Table 3.33 Membrane Operation

Parameter	Value	Unit	Parameter	Value	Unit
Aeration time	0.5	h/h	$M_{A,m}$	139110	kg/day
$Q_{A,m}$	4600	m^3/h	v_c	225	kg
$Q'_{A,m}$	4729	Nm^3/h	M_c	0.0065	kg/m^3
SAD_m	0.47	$Nm^3/(h\,m^2)$			
SAD_p	19.2	Nm^3/m^3			

Table 3.34 Aeration design

Parameter	Value	Unit
Tank depth	3	m
$P_{a,2}$	131	kPa
OTE membrane	0.06	%
OTE biological	0.14	%
$Q'_{A,m}$	1.61	Nm^3/s
O_2 transferred by membrane aeration	24.87	kg/day
O_2 required to maintain biology	1409	kg/day
$Q'_{A,b}$	1.38	Nm^3/s

Table 3.35 Power requirements

Air blower parameter	Value	Unit	Liquid pump parameter	Value	Unit
ξ_{blower}	0.50		ξ_{pump}	0.45	
Power (biological)	80.70	kW	Power (permeate)	3.44	kW
W_b	0.40	kWh/m^3	W_h	0.02	kWh/m^3
Power (membrane)	94.27	kW	Power (recycle)	9.11	kW
$W_{b,V}$	0.46	kWh/m^3	W_p	0.04	kWh/m^3

Table 3.36 Costs

Costs	Value
Power cost/kWh	\$0.10
Total energy/unit permeate	0.922 kWh/m^3
Total energy cost/Unit permeate	\$0.09
Chemical cost/kg	\$0.50
Chemical Usage/Unit permeate	0.0065 kWh/m^3
Total chemical cost/unit permeate	<\$0.01
Total cost/unit permeate	\$0.10
Estimated annual OPEX	\$170,066

the conditions given in Tables 3.25–3.27. According to this correlation (Fig. 3.5), the specific cost per unit volume permeate generated decreases with flux, as would be expected, and is around 33% lower for the HF than for the FS module.

However, a degree of caution is required in interpreting these data in that:

(a) The calculation does not account for the impact either of membrane life, which may possibly relate to permeability (which is generally higher for the FS modules), or sludge dry solids concentration on sludge disposal costs.

(b) There is considerable scatter in the data from which the J_{net}:SAD_m correlations were obtained. Over the range of SAD_m values applied in practice for the same SAD_m value the operating cost of the HF module is ~19% lower than that of the FS module (Fig. 3.6).

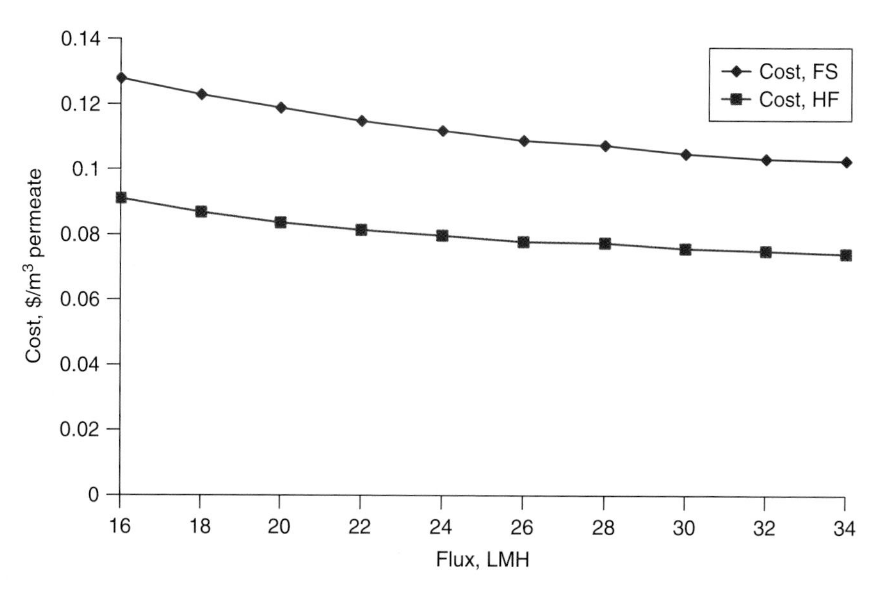

Figure 3.5 Specific cost vs. flux for HS and FS technologies, flux to be related to aeration demand according to Equations (3.18) and (3.19)

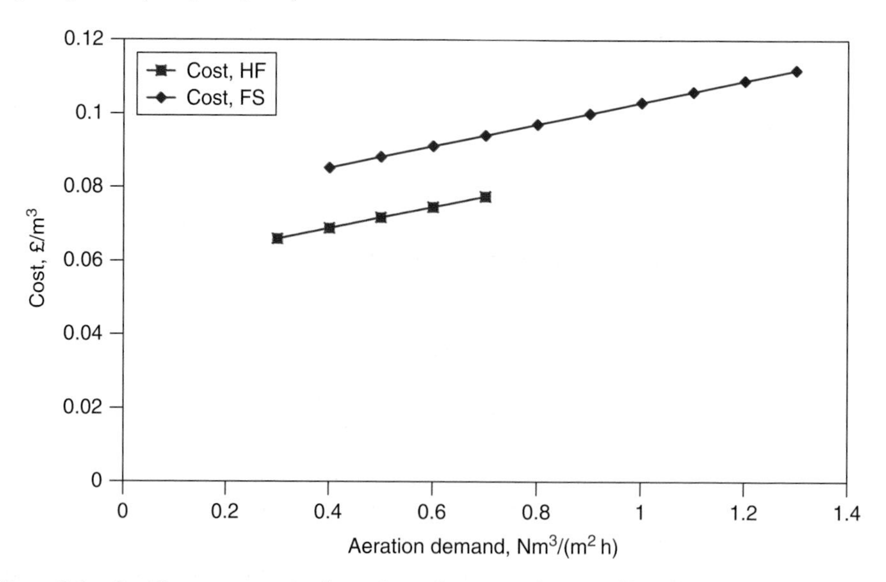

Figure 3.6 Specific cost vs. aeration demand over the ranges of aeration demand observed in practice for HF and FS technologies

(c) Many of the FS MBRs listed are operated without supplementary fine bubble aeration, such that the membrane aeration also provides air for biotreatment. This means that threes membranes may be over-aerated, which could also explain the higher permeabilities recorded compared with those of the HF MBRs. The one FS MBR operated with segregated membrane aeration yielded the lowest SAD_m value, the second lowest being that recorded for the double-deck FS MBR (Fig. 3.4).

In short, the design calculation demonstrates the principle of MBR process design and can provide trends in operating costs without necessarily providing absolutely correct values.

3.3.4 Design and O&M facets

As well as those facets already discussed, a number of practical aspects of MBR plant design and operation impact upon performance. Some of these are listed below, and further details can be found in WEF (2006).

3.3.4.1 Process control and software

- As with other process plant for water and wastewater treatment, feedback control and alarm triggering relies on monitoring of key parameters such as TMP (for indicating membrane fouling condition and triggering a cleaning cycle), DO (for biological process control) and turbidity (for membrane integrity).
- Unless they are specifically designed to be free from clogging, it is essential that the maintenance schedule includes cleaning of the aerators. This can be achieved by flushing with a dilute hypochlorite solution. For some technologies the aerator flush is scheduled at the same time as the maintenance clean.
- The principle impact of added process complexity is on the software and programmable logic controllers (PLCs), and on ancillary hardware such as pumps, valves and actuators. For example, intermittent aeration at 10 s means that each valve is actuated pneumatically over 3 million times per year.
- Foaming control and abatement procedures are particularly important in an MBR since aeration is more intense than for an ASP. Designs should include surface wasting or spraying.
- Sludge wasting for SRT control can be based on on-line MLSS measurement, although instruments have only recently developed for this.

3.3.4.2 Pre-treatment and residuals management

- The basic process can be modified in the same way as a conventional ASP to achieve denitrification, chemical phosphorous removal (CPR) and biological nutrient removal (BNR), as indicated in Section 2.2.6. Additional tanks (or tank volume) and sludge transfer pumps must then be sized accordingly, based on similar biokinetic principles as those used for the design for the core aerobic process.
- Pre-clarification can be used to reduce aeration demand (for agitation of biomass) by reducing solids ("trash") loading, but this significantly adds to footprint.
- Upgrading screens is essential, especially for HF iMBRs (Section 2.3.9.1). Grit removal is desirable for smaller plants if no capacity is available to allow the grit to settle out before the membrane tank.
- MBR sludge is generally less settleable than ASP sludge, with floc sizes generally being smaller (Section 2.2.6) and sludge volume index (SVI) values higher. Conventional gravity thickening is therefore less effective for MBR sludges. Membrane thickening can and has been used for this duty, albeit operating at necessarily low fluxes.

3.3.4.3 Tank sizing and redundancy

- DO levels are <0.5 mg/L in the anoxic zone, typically 1.0–2.5 mg/L in the aerobic zone and relatively high in the membrane region (2–6 mg/L) where aeration is intensive. If recycling for denitrification takes place from the membrane aeration tank to the anoxic zone then the anoxic zone becomes slightly aerobic at the sludge inlets, reducing denitrification efficiency. To compensate for this the anoxic zone must be increased to extend the HRT in this region. Alternatively, sludge can be recycled from the aerobic tank.
- Retrofitting places additional constraints on design of iMBRs, since the tank size determines the HRT and the shape and placement of immersed membrane modules. Retrofitting of sidestream modules, on the other hand, is not constrained by the aeration tank dimensions.
- Spare capacity is required for membrane cleaning, which either involves draining of the tank and cleaning in air (Section 2.3.9.2) or CIP. In the former case the biomass in the membrane train being cleaned has to be transferred either to a holding tank or to adjacent membrane tanks. In either case sufficient installed capacity is needed to contend with the total volume entering the works while the membrane train is being cleaned. For large plants with a large number of trains (membrane tanks) the cleaning can be sequenced to avoid large hydraulic shocks on the remaining in-service modules. For small plants a buffer tank may be required.

3.4 Summary

The selection of appropriate design and operating parameter values for an iMBR centres on:

1. choice of membrane module;
2. choice of membrane aerator, if the membrane module is not provided with an integral aerator;
3. membrane aeration rate.

Many, if not all, of these facets are stipulated by the technology provider, and the choice of technology itself (including that between iMBR and sMBR) will be strongly influenced by the duty to which it is being put. However, broadly speaking the mean permeability sustained in an iMBR is dependent on the aeration rate. Failure of membrane surface fouling, which can be irrecoverable, and clogging of the membrane channels. It is therefore essential that the maintenance schedule includes cleaning of the aerators, normally achieved by flushing with a water or hypochlorite solution.

Design of iMBRs relies on accurate information regarding oxygen transfer to the biomass and maintenance of permeability by membrane aeration. The design proceeds via calculation of the oxygen demand of the biomass, which relates primarily to the feedwater composition, the solids retention time and the oxygen transfer efficiency. This procedure applies to all biotreatment processes. However, for an MBR complications arise when estimating the oxygen transfer and hydraulic performance (flux and permeability) provided by the membrane aeration. The correlation of membrane flux and permeability with membrane aeration rate and the applied cleaning protocol relies entirely on heuristically-derived information. Continuing improvements in MBR process design mean that the boundary conditions for appropriate operation of specific technologies are changing, with the process costs tending to decrease year on year as a result. It remains to be seen whether further incremental improvements can further significantly decrease operating costs in the future.

References

Adham, S., DeCarolis, J.F. and Pearce, W. (2004) Optimisation of various MBR systems for water reclamation – phase III, *Desalination and Water Purification Research and Development Program Final Report*, No. 103.

Babcock, R. (2005) www.wrrc.hawaii.edu/research/project_babcock/Babcock-membrane. htm (Accessed December 2005).

EPA (1989) EPA/ASCE Design manual on fine pore aeration. Cincinnati, Ohio.

Göbel, A., Thomsen, A., McArdell, C.S., Joss, A. and Giger, W. (2005) Occurrence and sorption behavior of sulfonamides, macrolides and trimethoprim in conventional activated sludge treatment including sorption to sewage sludge. *Environ. Sci. Technol.*, **39**, 3981–3989.

Huber, M.M., Goebel, A., Joss A., Hermann N., Kampmann M., Löffler D., McArdell, C.S., Ried A., Ternes, T.A. and von Gunten, U. (2005) Oxidation of pharmaceuticals during ozonation of municipal wastewater effluents: a pilot study. *Environ. Sci. Technol.*, **39**, 4290–4299.

Joss, A., Andersen, H., Ternes, T., Richle, P.R. and Siegrist, H. (2004) Removal of estrogens in municipal wastewater treatment under aerobic and anaerobic conditions: consequences for plant optimisation. *Environ. Sci. Technol.*, **38**(11), 3047–3055.

Joss, A., Keller, E., Alder, A., Göbel, A., McArdell, C.S., Ternes, T. and Siegrist, H. (2005) Removal of pharmaceuticals and fragrances in biological wastewater treatment. *Water Res.*, **39**(14), 3139–3152.

Joss, A., Zabczynski, S., Göbel, A., Hoffmann, B., Löffler, D., McArdell, C.S., Ternes, T.A., Thomsen, A. and Siegrist, H. (2006) Biological degradation of pharmaceuticals in municipal wastewater treatment: proposing a classification scheme. *Water Res.*, **40**(8), 1686–1696.

Lawrence, D., Ruiken, C., Piron, D., Kiestra, F. and Schemen, R. (2005) *Dutch MBR Development: Reminiscing the Past Five Years, H_2O*, 36–29.

Metcalf, Eddy. (2003) *Wastewater Engineering – Treatment and Reuse* (3rd edn). McGraw-Hill, New York.

Qin, J.-J., Kekre, K.A., Guihe, T., Ooa, M.-H., Wai, M.-N., Lee, T.C., Viswanath, B. and Seah, H. (2005) New option of MBR-RO process for production of NEWater from domestic sewage. *J. Membrane Sci.*

Schyns, P., Petri, C., van Bentem, A. and Kox, L. (2003) *MBR Varsseveld, a Demonstration of Progression, H_2O*, 10–12.

Tao, G., Kekre, K., Wei, Z., Lee, T.C., Viswanath, B. and Seah, H. (2005) Membrane bioreactors for water reclamation. *Water Sci. Technol.*, **51**(6–7), 431–440.

Ternes, T.A., Joss, A. and Siegrist, H. (2004) Scrutinizing pharmaceuticals and personal care products in wastewater treatment. *Environ. Sci. Technol.*, **38**(20), 393A–399A.

van der Roest, H.F., Lawrence, D.P. and van Bentem, A.G.N. (2002) *Membrane Bioreactors for Municipal Wastewater Treatment*. IWA Publishing.

WEF (1996) **??**

WEF (2006) *Membrane systems for wastewater treatment.* Water Environment Foundation, WEFPress/McGraw-Hill, New York.

Chapter 4

Commercial Technologies

With acknowledgements to:

Section 4.2.1	Ryosuke (Djo) Maekawa	Kubota Membrane Europe Ltd, UK
Section 4.2.2	Paul Zuber	Brightwater Engineering, UK
Section 4.2.3	Paddy McGuinness	Colloide Engineering Systems, Northern Ireland
Section 4.2.4	Torsten Hackner	Hans Huber AG, Germany
	Jason Sims	Huber Technology, UK
Section 4.2.5	Shanshan Chou	Industrial Technology Research Institute, Taiwan
	Wang-Kuan Chang Chen-Hung Ni	Green Environmental Technology Co. Ltd, Taiwan
Section 4.2.6	Nobuyuki Matsuka	Toray Industries Inc., Japan
Section 4.3.1	Enrico Vonghia Jeff Peters	Zenon Environmental Inc., Canada
	Luca Belli Sandro Monti	Zenon Environmental Inc., Italy
Section 4.3.2	Noriaki Fukushima	Mitsubishi Rayon Engineering Co. Ltd, Japan
Section 4.3.3	Ed Jordan	Siemens Water Technologies – Memcor Products, USA
	Nathan Haralson Scott Pallwitz Fufang Zha	Siemens Water Technologies – Memcor Products, Australia
Section 4.3.4	Klaus Vossenkaul	Koch Membrane Systems GmbH, Germany
Section 4.3.5	Atsuo Kubota	Microza Division, Asahi Kasei Chemicals Corporation, Japan

Section 4.3.6	Michael Dimitriou	ITT Advanced Water Treatment, USA
Section 4.4.1	Eric Wildeboer	Berghof Membrane Technology, The Netherlands
Section 4.4.2	Ronald van't Oever	Norit X-Flow BV, The Netherlands
Section 4.4.3	Gunter Gehlert	Wehrle Werk AG, Germany
Section 4.4.4	Steve Wilkes	Millenniumpore, UK
Section 4.5.1	Marine Bence, Sylvie Fraval	Novasep Process, Orelis, France
Section 4.5.2	Olivier Lorain	Polymem, France
Section 4.6	Stefan Krause, Margot Goerzel	Microdyn-Nadir GmbH, Germany
	Phoebe Lam	Lam Environmental Services Ltd and Motimo Membrane Technology Ltd, China

4.1 Introduction

Available and developing commercial membrane bioreactor (MBR) technologies employed for wastewater treatment can be classified according to membrane configuration: flat sheet (FS), hollow fibre (HF) and multitube (MT) (Section 2.1.3). Many such products exist and many more are being developed, and a comprehensive description of all technologies available globally is not possible. In the sections that follow, the leading technologies are described. Other technologies of interest are also outlined, for which most have reference sites (Chapter 5).

4.2 Immersed FS technologies

4.2.1 Kubota

The Kubota membrane module was developed in the late 1980s by the Kubota Corporation, a diversified Japanese engineering company originally best known for agricultural machinery. The development was in response to a Japanese Government initiative to encourage a new generation of a compact wastewater treatment process producing high-quality treated water. The first pilot plant demonstration of the Kubota membranes was conducted in 1990, prior to the first commercial installation soon after. There are now over 2200 Kubota MBRs worldwide, with about 10% of these installed in Europe.

The original FS microfiltration (MF) membrane, the type 510 which is still widely used, comprises a 0.5 m × 1 m flat panel, 6 mm thick, providing an effective membrane area of 0.8 m^2. The membrane itself is a hydrophilicised, chlorinated polyethylene (PE) membrane, supported by a very robust non-woven substrate (Fig. 4.1), which is ultrasonically welded on each side to an acrylonitrile butadiene styrene (ABS) resin plate with a felt spacer material between the membrane and plate. The plate contains a number of narrow channels for collecting the permeate. The nominal pore size is 0.4 μm, but, due to the formation of the dynamic layer on

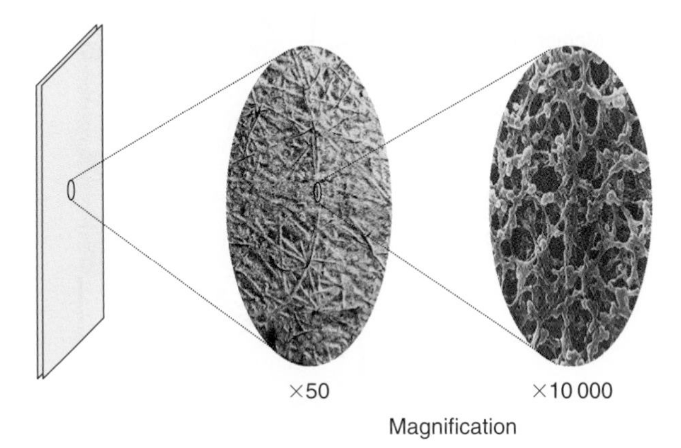

Figure 4.1 Kubota membrane showing substrate and membrane surface

the membrane surface, the effective pore size in operation is considerably lower than this and can be in the ultrafiltration (UF) range.

The membrane panels (Fig. 4.2) are securely fitted into a cassette to form a stack (Fig. 4.3), providing a 7–8 mm membrane separation. This has been found to be sufficient to largely prevent clogging provided pre-treatment using a 3 mm-rated bi-directional screen is employed. Flow from outside to inside the panel is either by

Figure 4.2 Kubota 510 membrane panel (a) schematic and (b) photograph of used panel extracted from cassette

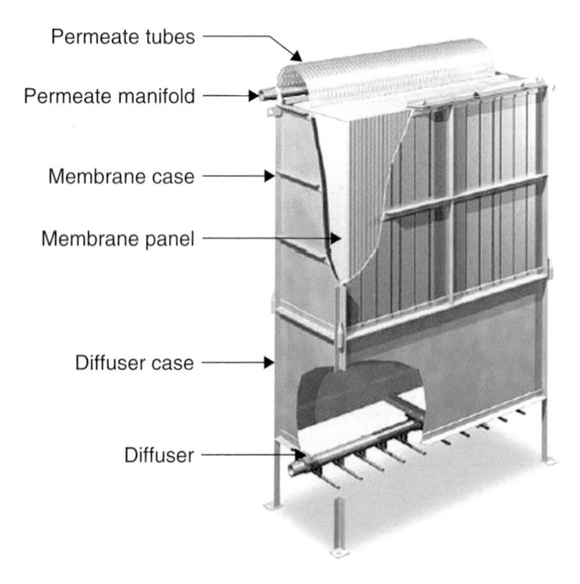

Figure 4.3 Kubota (ES Model) stack, based on the 510 membrane panels

suction or under gravity, routinely between 0.5 and 1.3 m hydrostatic head for gravity-fed operation. Permeate is extracted from a single point at the top of each membrane panel via a polyurethane tube. Aeration via coarse bubble aerators is applied at the base of the tank so as to provide some oxygenation of the biomass, in addition to aerating the membrane stack. The original loop-type aeration pipe containing 8 or 10 mm holes has largely been superseded by a patented sludge flushing aerator. This aerator comprises a central pipe with smaller open-ended lateral branch pipes at regular intervals (Fig. 4.3). Each lateral pipe has 4 mm holes on the top surface. Cleaning of the aerator is achieved by briefly opening an external valve connected via a manifold to the ends of the central pipe(s). This allows vigorous backflow of sludge and air back into the tank. This backflow clears sludge from within the aerator and helps to prevent clogging of the aeration system. To prevent air bubbles from escaping without passing through the stack, a diffuser case is fitted, in effect providing a skirt at the base of the module. Membrane modules contain up to 200 panels in a single-deck configuration.

A step improvement in efficiency was made with the introduction of the double-deck EK design in 2002 (Fig. 4.4). In this design, capital costs are decreased because the available membrane surface area/m^2 of plant footprint is doubled, the number of diffuser cases required is halved and a single cassette base is used for mounting two banks of panels. In addition, the operational costs are reduced because the specific membrane aeration rate (specific aeration demand with respect to membrane area (SAD_m, synonymous with R_M in Equation 3.12) in $Nm^3/(m^2\,h)$, Equation (3.6)) is decreased. In principle, this halves the aeration energy demand, provided the aerator depth in the tank remains unaltered. By June 2005, there were over 40 double-deck

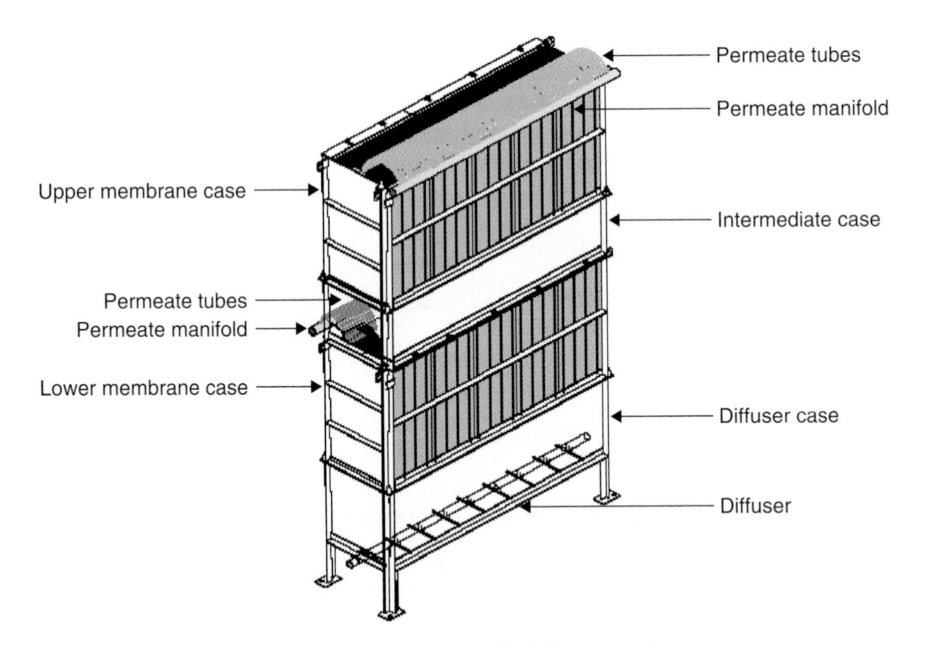

Figure 4.4 Kubota double-deck (EK) stack

membrane plants operating or under construction worldwide, treating a total flow of >250,000 m^3/day (Kennedy and Churchouse, 2005).

A logical alternative to the double-deck design is the creation of a single elongated membrane panel, therefore having a longer airflow path and a larger membrane area. Kubota have recently developed a larger module (type 515) in which the membrane panels interlock to create the module without the need for a separate housing to slot the panels into. Each membrane panel (1.5 m × 0.55 m, providing an effective membrane area of 1.25 m^2) contains internal channels connecting to a moulded permeate manifold section. When the panels are joined together, these sections form the permeate manifold. This design removes the need for separate manifolds and permeate tubes and reduces the complexity of the housing (Fig. 4.5). In addition the greater surface area per panel leads to reduced power consumption for aeration, SAD_m being 0.34 $Nm^3 h^{-1}$ air/m^2 membrane area. This represents a significant reduction in aeration demand, and thus energy consumption, from the original single-deck ES module for which the figure is 0.75 (Table 4.1). This then leads to a lower SAD per unit permeate volume (SAD_p, volume of air per volume of permeate,

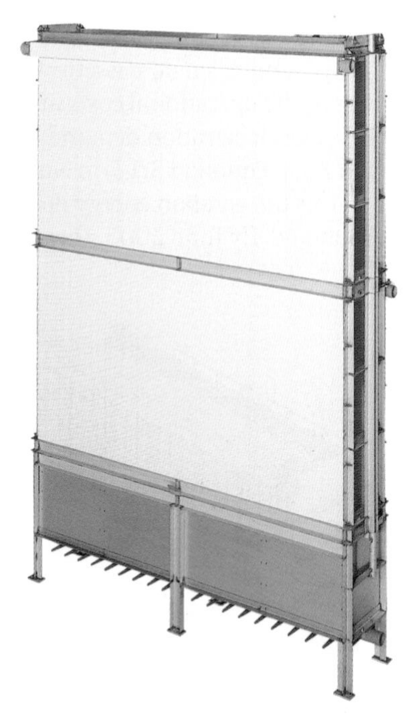

Figure 4.5 The Kubota EW module

Table 4.1 SAD_m figures in $Nm^3/(h\ m^2)$ for Kubota modules

	EW	EM	EK	ES
Standard	0.34	0.48	0.53	0.75

SAD is given as airflow per unit membrane area (Equation (3.6)).

synonymous with R_V Equation (3.13)) Equation (3.7)). Values below 15 can be expected for the double-deck unit for operation at standard flux rates for sewage treatment. The EM (single-deck) and EW modules are expected to be in commercial production by 2006 and are already being designed into future treatment plants for delivery in that year.

4.2.2 Brightwater Engineering

Brightwater Engineering Ltd has been established since 1990. The company designs, supplies and commissions plants for the treatment of sewage, industrial wastes, water and sludges. Brightwater is a member of the Bord na Móna Environmental (UK) Ltd group of companies, which acquired Brightwater in July 2000.

The Brightwater MEMBRIGHT® system (Fig. 4.6) is an FS immersed MBR (iMBR) with 150 kDa polyethylsulphone (PES) membranes mounted on a rigid polypropylene (PP) support. The module is 1120 mm in length, 1215 mm in width for a 50-panel unit (715 mm for 25-panel unit) and 1450 mm high. The respective membrane area provided by the two sizes is 46 and 92 m^2, with each square panel, 950 mm in length, providing an area of 1.84 m^2. The membrane spacing is ~9 mm, the panel support spacing being 10 mm and the panels are clamped in place within a stainless steel frame to form the module. The module is fitted with an integral aerator which ensures even distribution of air across the module.

4.2.3 Colloide Engineering Systems

Colloide Engineering Systems (CES) are a small-to-medium sized enterprises (SME) who provide various treatment technologies for water and wastewater in both the

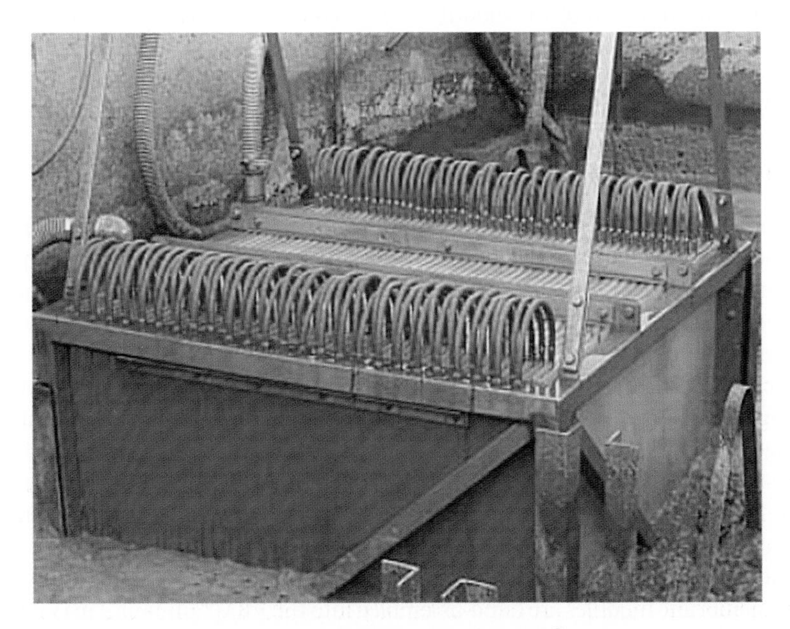

Figure 4.6 The Brightwater MEMBRIGHT® module

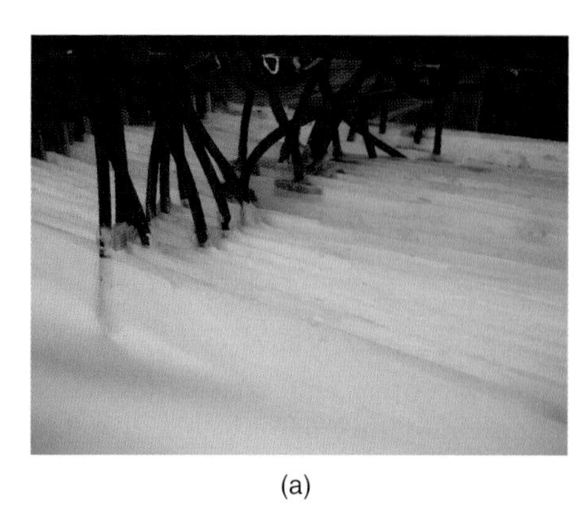

(a)

(b)

Figure 4.7 The CES Sub Snake system: (a) FS elements and (b) permeate extraction and module ends

industrial and municipal sectors, concentrating mainly on small- to medium-scale plants. The company has been developing the MBR technology since 2001. The CES Sub Snake system is unusual in that the membrane modules are bespoke and fabricated from a continuous 0.04 μm PES membrane sheet which is cut to size and then glued at the edges to form a FS module. The membrane is then wrapped, snake-like, around a purpose-built steel or plastic frame comprising a number of rigid vertical poles at each end to make a multiple FS module with a membrane sheet separation of 10 mm. A single tube is inserted into the permeate channel of each FS element for permeate extraction under suction into a common manifold (Fig. 4.7b).

The maximum depth of the module is dictated by the width of the sheeting, and the total membrane area of the module by its overall length. Thus far, the largest modules offered by the company are 10 m^2, provided by 10 elements of 1 m depth and 0.5 m width (double-sided), the total length of the FS being 5 m in this case.

4.2.4 Huber Technology

The Huber VRM® (Vacuum Rotation Membrane) product is differentiated from all the other iMBR systems by having a moving membrane module, which rotates at a frequency of 1–2 rpm. The small shear created by this, combined with scouring of the membrane surface by air from the central coarse bubble aeration, apparently obviates all cleaning; the rotary action means that solids do not collect in any region of the module. The membrane elements themselves comprise a four-plate segment of a hexagon or octagon (Fig. 4.8a), thereby making up one sixth or one eighth of a complete plate. The membrane material itself is based on 0.038 μm pore-size PES material of around 300 μm in thickness (Fig. 4.8b). The individual elements are thus relatively small (0.75 m^2 for four parallel plates, hence 0.19 m^2/element) and, since each are fitted with a permeate extraction tube, the permeate flow path is relatively short. The plates themselves are 6 mm thick and separated by a 6 mm channel.

The membrane modules are hand-assembled into the VRM® units of 2 m (VRM® 20) or 3 m (VRM® 30) diameter. Each module is positioned and fixed in a drum and

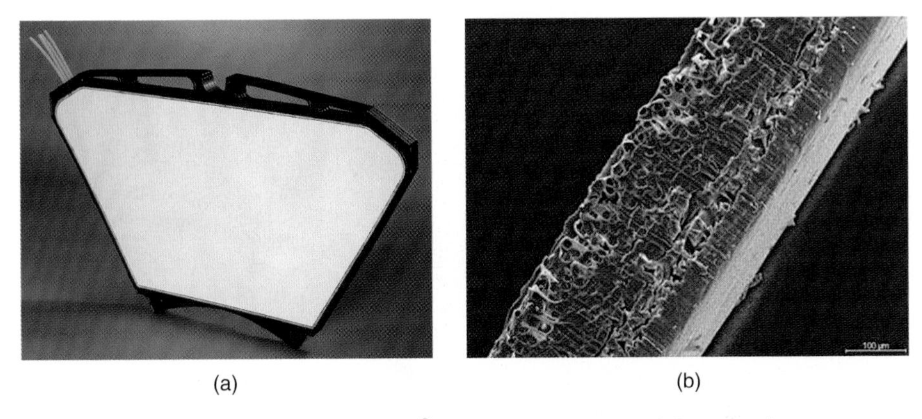

Figure 4.8 *The Huber VRM® membrane (a) element and (b) material*

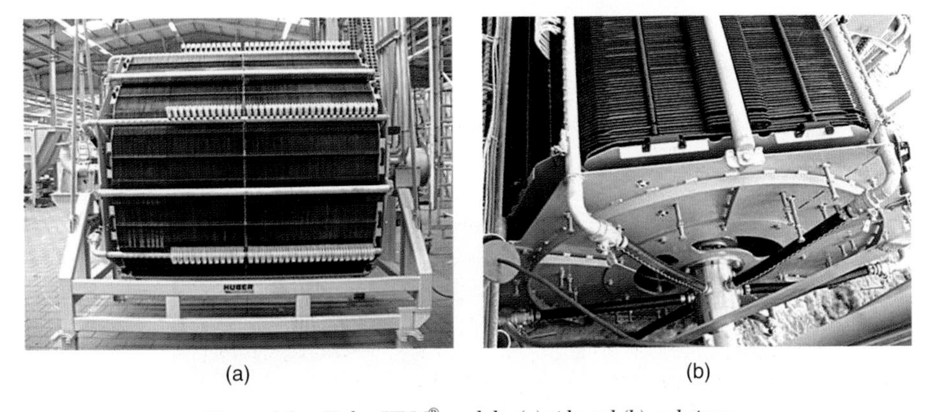

Figure 4.9 *Huber VRM® module, (a) side and (b) end views*

then connected to the collection tubing. The mixed liquor is taken from the aeration tank and is circulated around the VRM® unit and between the plates where permeate water is drawn. The water flows via the tubing into the collecting pipes (Fig. 4.9a). The pipes are joined to the central collection manifold, and the product water is discharged from this central pipe (Fig. 4.9b). The whole module is aerated by two coarse bubble aerators which are placed in the middle of the membrane unit (Fig. 4.10).

4.2.5 The Industrial Technology Research Institute non-woven fabric-based MBR

This is not a commercial product, but a recent development based on a low-rejection FS membrane module using a PP non-woven fabric or felt (NWF) which has reached a technical scale. A number of studies have been performed by various research groups at the bench scale (Alavi Moghaddam *et al.*, 2002; Chang *et al.*, 2001; Chang *et al.*, 2003; Fuchs *et al.*, 2005; Meng *et al.*, 2005; Seo *et al.*, 2002). However, the first

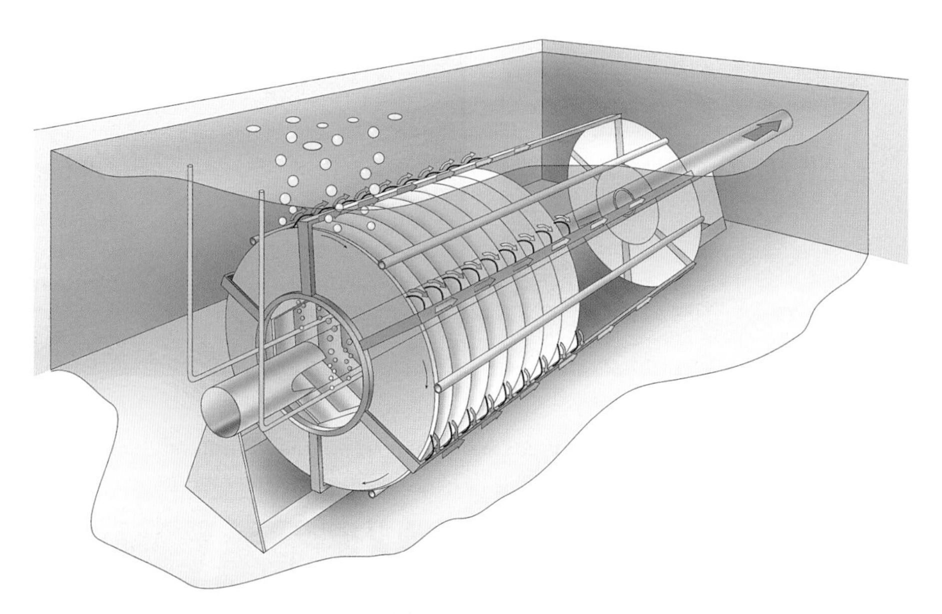

Figure 4.10 The Huber VRM® MBR, schematic

Figure 4.11 The NWF MBR module

pilot-scale demonstration of an NWF MBR appears to have been conducted by the Industrial Technology Research Institute (ITRI) in Taiwan. This centre's work is based on a 0.6 mm-thick mesh of 20 μm nominal pore size, created by 10 μm filaments, supplied as a FS by KNH Enterprise Co. Ltd, Taiwan. The module (Fig. 4.11) contains a number of these sheets glued onto a separator and spaced ~10 mm apart. The membrane material is low in cost, compared with a MF/UF membrane module,

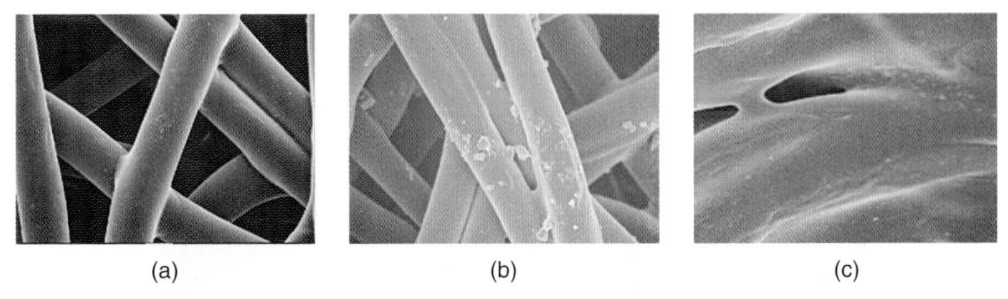

Figure 4.12 NWF during MBR operation (1000 × magnification): (a) originally, (b) after 1 day and (c) after 3 days of operation challenged with a feedwater of 300 mg/L COD and operating at an MLSS of 2.2–3 g/L (Seo et al., *2002)*

and is very robust. However, the large pore size inevitably leads to significant internal fouling with organic material over the course of a few days (Fig. 4.12), as demonstrated from bench-scale tests, leading to a drop in permeability to very low values. On the other hand, the material is sufficiently low in cost for the module to be competitive, provided disinfection is not required (as is the case for many industrial effluent duties).

The process relies on the formation of a perm-selective dynamic layer onto the porous non-woven substrate. As such, there is a certain amount of downtime associated with the generation of this dynamic layer, during which the permeability decreases along with a concomitant increase in selectivity. The use of low-cost porous substrates for dynamic membrane formation has its roots in the South African Renovexx process (Section 5.2.1.1) from the mid 1980s, which was actually based on a woven, rather than non-woven, material. Ongoing research into these materials for MBR duties suggests that commercial MBR technologies based on non-woven substrates may appear in the future.

4.2.6 Toray Industries

Toray is another established Japanese membrane manufacturer of some 25 years standing, specialising principally in reverse osmosis (RO) membranes for pure water applications. The company launched its FS MBR MF membrane product in 2004. The membrane material used is 0.08 μm-rated polyvinylidene difluoride (PVDF) with a standard deviation of 0.03 μm. It is reinforced with a polyethylene terephthalate (PET) non-woven fibre and mounted on an ABS support, into which a number of 1–2 mm permeate channels are cut. Permeate is extracted via a single outlet tube. The element (TPS-50150, Fig. 4.13a) has dimensions of 515 mm by 1608 mm, providing a membrane area of 1.4 m^2, and is 13.5 mm thick (including the membrane separation of 6–7 mm). A smaller element of 0.5 m × 0.5 m also exists, its use being apparently limited to applications where height is constrained (such as on board ships).

Elements are assembled in a stainless steel frame to form modules ranging from 70 m^2 total membrane area (50 elements, TRM140-050S module) or 140 m^2 (100 elements, TRM140-100S module), which can then be either doubled in width

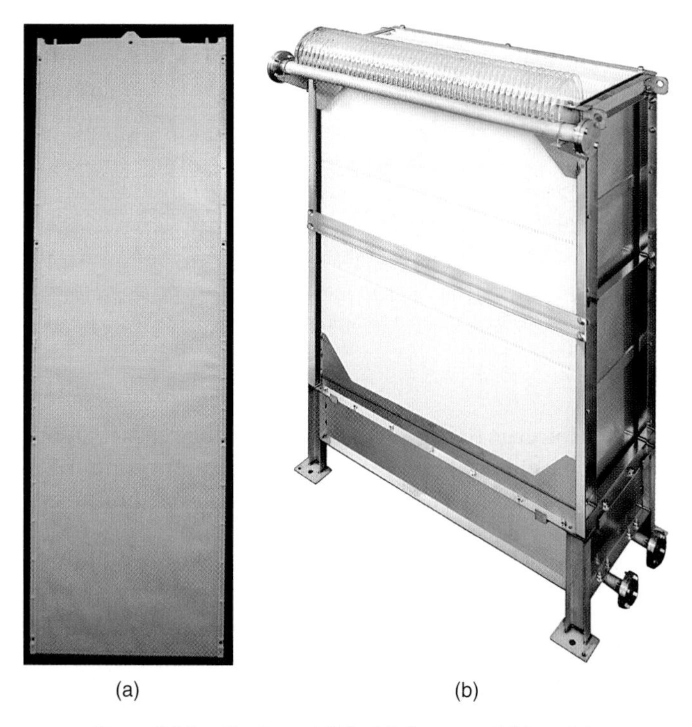

(a) (b)

Figure 4.13 The Toray MBR: (a) element and (b) module

(TRM140-200W module Fig. 4.13(b)) or stacked (TRM140-200D module) to form larger modules. The 100S module has dimensions of 1620 mm long, 810 mm wide and 2100 mm high. A design flux of 33 litres per m^2 per hour (LMH) is assumed (though the quoted range is between 8.3 and 62.6 LMH), along with a maximum transmembrane pressure (TMP) of 0.2 bar. The recommended aeration rate is 0.15–0.25 Nm^3/h for the single-deck unit, half these values for the double-deck module, and pre-screening to 1 mm is stipulated. As of December 2005 there were around 25 operating plant worldwide, all but three being below 0.5 megalitres/day (MLD).

4.3 Immersed HF technologies

4.3.1 Zenon Environmental

Zenon are currently the largest MBR process technology company, operating in over 45 countries across the world with a staff of 1470 and a turnover of $CAD 233.8 m (2004 figures). The company's first HF module (the Moustic, Fig. 4.14a) was developed in 1989 and consisted of sheets of small, extruded HF membranes oriented orthogonal to the flow and with each sheet at 90 degrees to the previous one. The

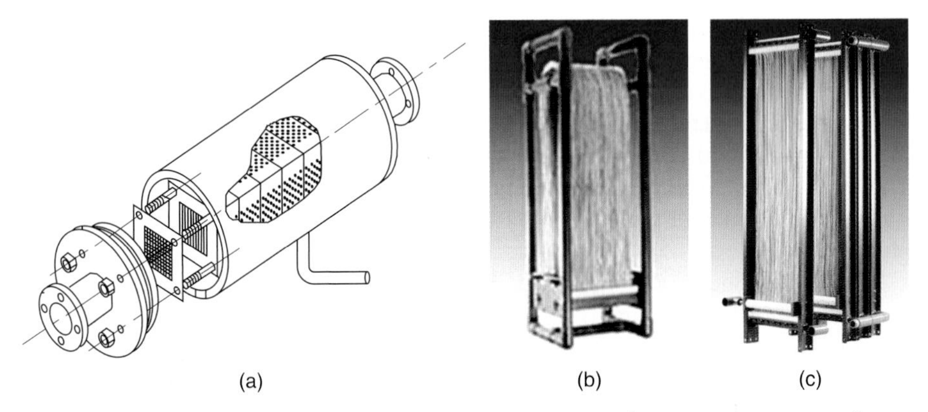

Figure 4.14 Early Zenon modules: (a) the Moustic, (b) the ZeeWeed® 145 and (c) the ZeeWeed® 150

transverse feed flow was applied tube-side of the module. The manufacturing complexity of the module prevented it from being commercialised, but the Moustic represented an important stage in the development of the subsequent ZW series.

Following early trials with the Moustic, a reinforced HF membrane was developed in 1991–1992. The early ZeeWeed® membrane, providing 145 ft^2 (13.5 m^2) of membrane area and thus named the ZeeWeed® 145 (Fig. 4.14b), was an ultrafilter with maximum 0.1 μm pore size, and comprised an inner reinforcing polyester braid on which the membrane layer was cast. Bundles of fibres were potted at both ends into permeating headers to make a prototype module, which was arranged horizontally over an aeration assembly in the tank. Further testing led the development team to move the two headers together and loop the fibre bundle over a raised section of the PVC frame, allowing vertical bubble flow to better mechanically scour the fibres. The ZeeWeed® 145 was first commercially installed in 1993 for industrial wastewater applications at Orlick Industries (Stoney Creek, Ontario, Canada) and GM Sandusky. Module headers consisted of two pieces, one containing the potted fibres and another completing the permeate cavity when both were bolted together. Permeation occurred through both headers, and module aeration was continuous, with aerators located between both headers within the module and on either side of the module at the header level.

It was recognised that the looping of fibres over the PVC frame led to solids build-up and abrasion at the top of the fibre bundle. The problem was substantially ameliorated in the subsequent design, the ZeeWeed® 150 (i.e. 150 ft^2, 13.9 m^2 membrane area, Fig. 4.14c), which was the first self-supporting and stand-alone immersed vertical HF module. The fibres in this module were potted at both ends, with regular spacing between them to ensure better solids removal from the bundle. Modules were custom-manufactured and could be interconnected in any number depending on the application. Permeation occurred through both headers, and module aeration was continuous. The aerators were bolted on to the bottom header where they could introduce air directly into the base of the fibre bundle. The ZeeWeed® 150 was first tested in an MBR application at a land development site called Skyland Baseball Park, in New Jersey in June 1994, and the product subsequently sold in several small land development MBRs, such as Thetis Lake Campground and Trailer Park, BC (1995) – which, as of July 2005, was still operating with its original membranes. ZeeWeed® modules were also used for the first time in potable and industrial pure water treatment applications from 1995 onwards. The early experiences in continuous aeration also led to the development in 1998–1999 (and later patenting) of two-phase aerator flushing to prevent aerator plugging.

Zenon introduced the ZeeWeed® 500a for wastewater treatment and the ZeeWeed® 500b for drinking water treatment in 1997. The new membrane modules had significantly higher surface area per plant footprint than the ZeeWeed® 150, with the drinking water modules constructed with more fibres, and hence a greater surface area (650 ft^2, 60.4 m^2), than those for wastewater (500 ft^2, 46.5 m^2). These modules not only provided a greater membrane area through employing a higher, wider and slightly more densely packed fibre bundle, but could also be housed within a single, standardised cassette of eight modules (Fig. 4.15a). The membrane material was also made more selective than those used previously, with these membranes, the nominal pore size being reduced to 0.04 μm.

Two further vertical HF modules, the ZeeWeed® 500c (23.2 m^2 membrane area) and the ZeeWeed® 500d (340 ft^2, 31.6 m^2) were introduced in 2001 and 2002 respectively. These modules (Fig. 4.15b–c) provided further improvements in clogging amelioration and ease of element removal and refitting over the earlier modules. In addition to further increasing the membrane surface area per module footprint, the ZeeWeed® 500c is more flexible, being no longer self-supporting but designed to be applied to either drinking water or wastewater simply by varying the number of modules within the same, standard-size cassette. Aerators are incorporated into the bottom of each cassette, and permeation is no longer from both ends, but from the top only. This design capitalises on the fact that the highest shear from the air scour occurs at the top section of the membrane fibre in an MBR. When permeation is from the top only, the top section of the membrane is also the point of the greatest flux. The configuration allows the effectiveness of the scouring to match the flux magnitude across the length of the membrane fibre, maximising the treatment capacity of the MBR system. The ZeeWeed® 500c module includes a top permeate header (complete with o-ring seals) that interlocks with other modules to form an integrated permeate collection header at the top of each cassette. Since permeation does not occur at the bottom header, a more streamlined design results with better airflow into the membrane fibre bundle.

The ZeeWeed® 500d, commercially released in 2002, is geared towards even larger scale potable water and wastewater treatment plants. Modules are wider and higher than the ZeeWeed® 500c and can be easily isolated and removed from the cassette. Like the ZeeWeed® 500c, the cassette incorporates aerators and a central permeate header. ZeeWeed® 500d modules are arranged in two parallel rows within the cassette, with central permeate collector pipes between them. To simplify manufacturing, the top and bottom headers are identical, and because these membranes are more frequently used in potable water projects, permeation occurs through both headers. This product is employed in the world's largest MBR plants, currently at the planning

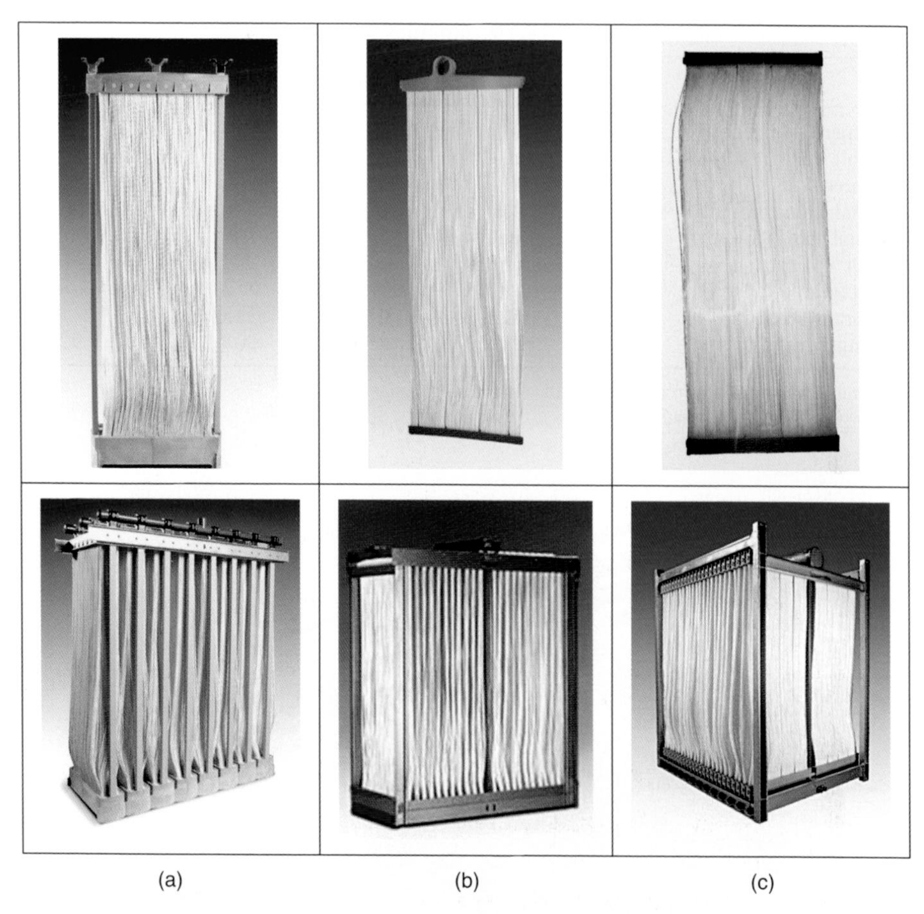

Figure 4.15 Zenon MBR membrane elements and stacks: (a) ZeeWeed® 500a, (b) ZeeWeed® 500c and (c) ZeeWeed® 500d (see also Table 4.2)

or construction stage. These include plants such as John's Creek Environmental Campus in Georgia, USA (41 MLD average day flow; 94 MLD peak hourly flow) and Brightwater, Washington, USA (117 MLD average day flow; 144 MLD peak hourly flow). The Zenon cassettes (Fig. 4.15 and Table 4.2) are readily placed trains in trains (Fig. 4.16) to attain the desired capacity.

A further innovative feature of the Zenon system is the use of cyclic aeration, which was commercially introduced in 2000 following the introduction of the ZeeWeed® 500 series. In this aeration mode, the blowers are operated continuously at a fixed speed and cyclic aeration is achieved by cycling the airflow from one air header to the other using pneumatically actuated valves. Since one half of each cassette is connected to one air header, this permits the air to be cycled. This was preferred over continuous aeration primarily for technical and commercial reasons. Technically, continuous aeration often resulted in air channelling through the module and could only thoroughly clean portions of the module. Introducing shear and

Table 4.2 ZeeWeed®500 and 1000 series modules

	500a	500b*	500c	500d	1000*
Alignment (horizonal/ vertical)	V	V	V	V	H
Reinforced	Y	Y	Y	Y	N
Internal diameter (mm)	0.8	0.8	0.8	0.8	0.5
External diameter (mm)	1.9	1.9	1.9	1.9	0.8
Length × depth × height (mm)	688 × 184 × 2017	688 × 184 × 2017	678 × 60 × 1940	844 × 56 × 2198	684 × 104 × 690
Membrane area (m^2)	46.5	60.4	23.2	31.6	37.2
Packing density (m^2/m^3)	182.1	236.6	294.0	304.2	757.9

All membrane materials hydrophilicised PVDF of nominal pore size 0.04 μm.
* For potable water applications.

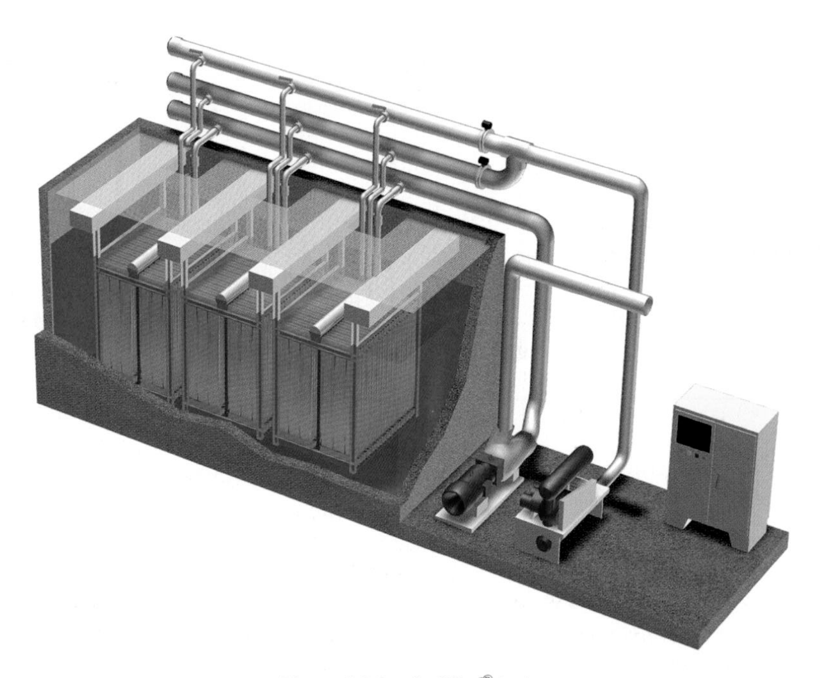

Figure 4.16 ZeeWeed® train

airflow instability into the process through cyclic aeration eliminated channelling. Continuously operating aerators also have a tendency to plug, but effectively flushing the aerators through the cyclic aeration action significantly reduces this problem. Membrane aeration contributes around 35% of the energy demand (Fig. 4.17a) and almost the same proportion of the MBR lifecycle costs (Fig. 4.17b). A filtration cycle of 10 s on/10 s off has been determined as being the most effective and is typically employed for medium- and large-size municipal MBR plants.

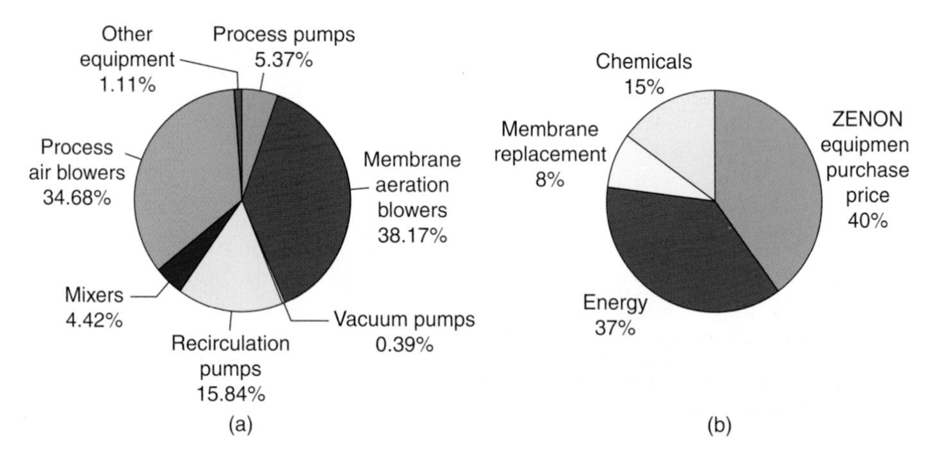

Figure 4.17 Representative costs for: (a) energy demand and (b) life cycle for the Zenon MBR system (5.7 MLD plant)

4.3.2 Mitsubishi Rayon Engineering

Mitsubishi Rayon represents the third largest MBR membrane supplier worldwide, with respect to installed capacity, after Zenon and Kubota. Mitsubishi Rayon Engineering (MRE) was spun out of the Mitsubishi Rayon Company Limited in 1975. The company, which has a turnover of around $330 million, operates in a number of areas relating to polymeric materials, and their product range includes membrane filtration as applied to both the industrial and municipal sectors. They first introduced an MBR membrane product in 1995 and, as of March 2005, the company had over 1500 of its systems installed worldwide, over 900 of these being in Japan. Its commercial profile is thus similar to that of Kubota.

There are currently two Mitsubishi Rayon elements. The first, based on the company's existing Sterapore™ PE HF membrane, comprises horizontally oriented filaments of 0.54 mm outer diameter (o.d.) and 60–70 μm wall thickness (Fig. 4.18a). For this membrane, slit-like pores (Fig. 4.18b) of nominally 0.4 μm are generated by stretching to produce a simple isotropic membrane material. The fibres are potted with polyurethane resin at either end within ABS plastic permeate collection pipes to form 1.5 m^2 SUR234L or 3 m^2 SUR334LA SteraporeSUN™ elements (Fig. 4.18c). The larger area element is wider (524 mm vs. 446 mm), the other dimensions being the same for the two elements: 1035 mm high by 13 mm thick.

Elements are mounted within a stainless steel frame to form units containing up to 70 elements, providing a membrane area of up to 210 m^2 for the larger element. Modules can be either single (Fig. 4.19a) or double deck (Fig. 4.19b), with permeate withdrawn from each deck. The largest (i.e. 210 m^2) single-deck module is 1538 mm wide × 1442 mm high × 725 mm deep and weighs 184 kg. A standard net flux of 10 LMH is recommended for this membrane based on a cycle of 9–13 min filtration/2–3 min relaxation and operating between TMP limits of 0.1 and 0.4 bar. The modules

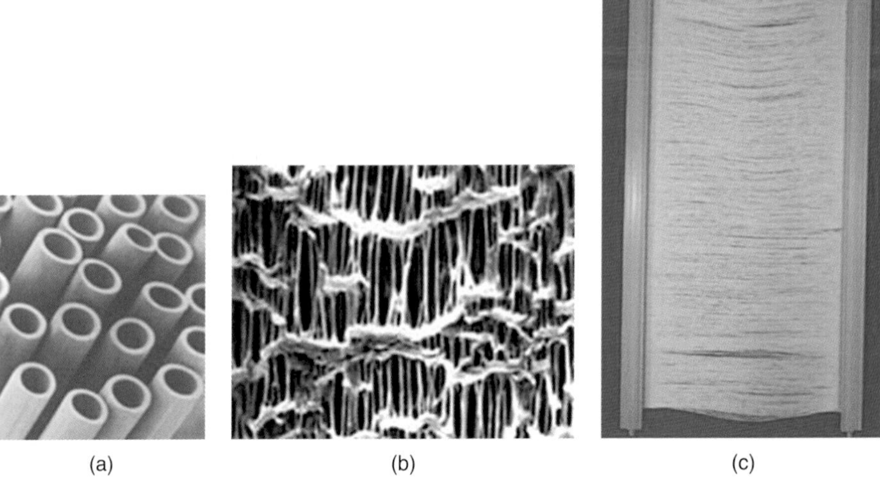

(a) (b) (c)

Figure 4.18 The Sterapore™ HF membrane: (a) fibres, (b) surface and (c) element

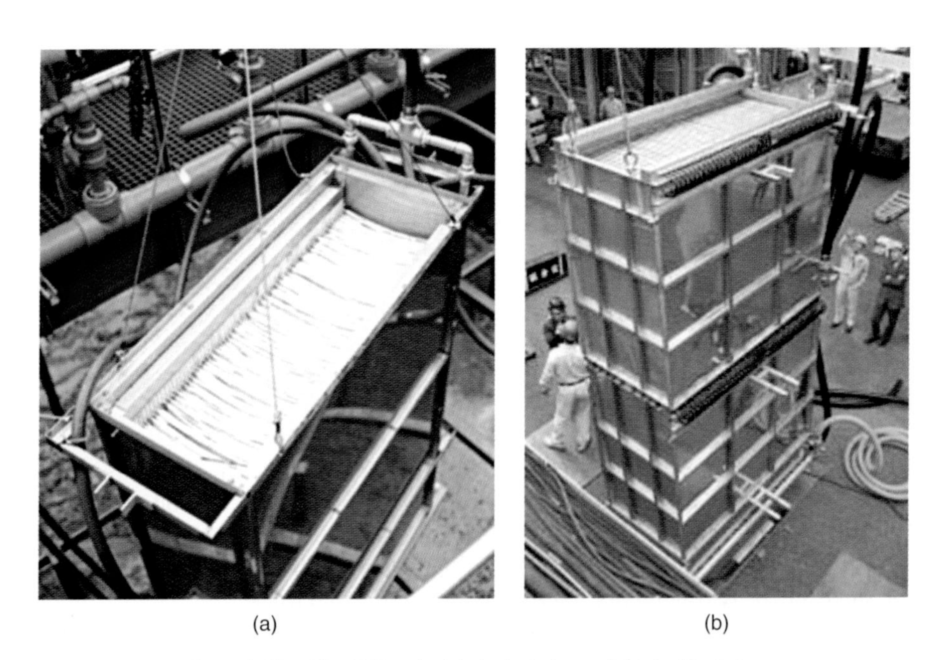

(a) (b)

Figure 4.19 The SUR unit: (a) single-decker and (b) double-decker

are positioned ~200 mm above the coarse bubble membrane aerator. Aeration rates for municipal sewage are normally in the region of 100–150 $Nm^3/(m^2 h)$, where in this case m^2 refers to the footprint area of the unit. This means that, for a single-decker SUR334LA unit, the larger of the two, SAD_m is 0.34 $Nm^3/(m^2 h)$.

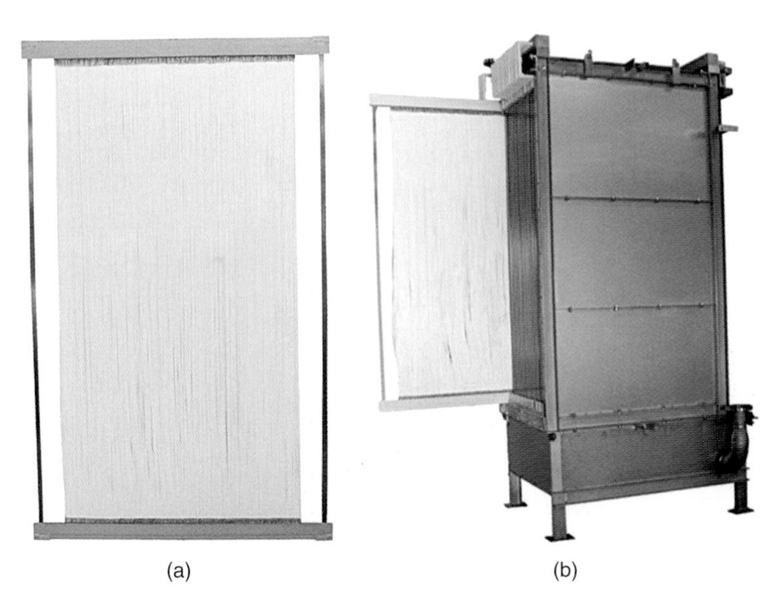

Figure 4.20 *The SteraporeSADF™: (a) element and (b) unit*

Thus far, almost all of the installed MR MBR plant have been based on the SUR module. More recently, the company has introduced the SteraporeSADF™ element, which is based on vertically oriented PVDF fibres. These fibres have a pore size of 0.4 μm and are 2.8 mm o.d. and are potted and framed using the same materials as for the SUR element. The PVDF elements (Fig. 4.20(a)) are much larger than the PE ones, being 2000 mm high × 1250 mm wide and 30 mm thick, and provide a smaller area of 25 m^2. The units contain 20 elements (and hence provide 500 m^2 total membrane area) and have dimensions 1553 mm wide × 1443 mm thick × 3124 mm high. They have an integrated coarse bubble aerator which contributes to the overall module dry weight of 700 kg (Fig. 4.20(b)).

4.3.3 Memcor

Memcor, now part of the Siemens group, is a well-established membrane product supplier dating back to 1982. The membrane product originates from Australia, but at various times the company has been acquired by USFilter in 1997 and Vivendi (now Veolia) in 1999 before being acquired by Siemens in July 2004 for $1.4 billion. At the time of the last acquisition, USFilter had a turnover of $1.2 billion and 5800 employees worldwide. In the potable water sector, their PP and PVDF products account for >2.5 gigalitres/day (GLD) of total installed membrane treatment capacity, about 50% of installed capacity in the USA, and almost 800 MLD, close to 30% of the total installed membrane capacity, in the UK (2005 figures).

The Memcor MemJet® MBR system was introduced in 2002. It comprises vertically oriented modules fitted with PVDF 0.04 μm-pore HF membranes. The air bubbles

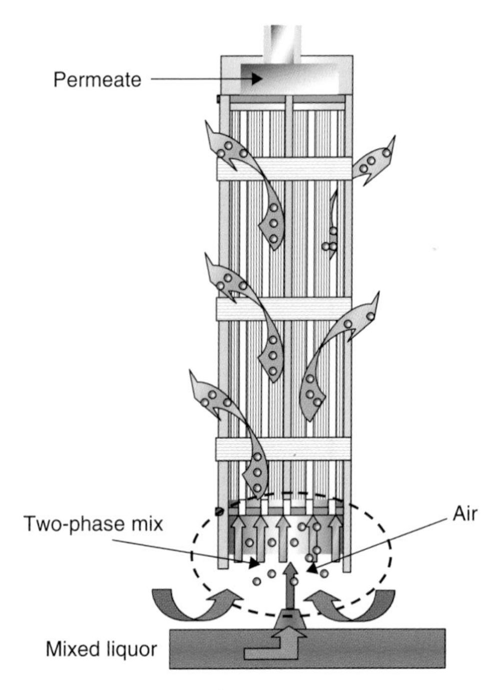

Figure 4.21 The MemJet® module with integral jet aerator

entrained in mixed liquor are introduced into the module using a patented two-phase mixing system (Fig. 4.21). This creates high-shear conditions, thereby minimizing solids concentration polarisation at the membrane surface near the jet. Higher membrane packing densities are permitted as a result, along with SAD_m values below 0.3 $Nm^3/h\,m^2$ membrane. Mean fluxes between 20 and 50 LMH are claimed, implying specific aeration demand values below 15 m^3air/m^3 permeate on average. The predicted TMP range is 0.05–0.5 bar, though the membranes can apparently tolerate pressures of up to 3 bar.

Modules are provided in two sizes: 10 m^2 (the B10R) and 38 m^2 (the B30R). A rack of modules (Fig. 4.22) may contain up to 40 B10R or 16 B30R modules, which can then be arranged in rows within a cell. This modularisation follows the same format as that used in potable water Memcor systems. The recommended mixed liquor suspended solids (MLSS) concentration range for the bioreactor is 5–15 g/L at an associated solids retention time (SRT) of 5–50 days and loading rates of 0.3–1.5 kg biochemical oxygen demand (BOD)/(m^3 day), generating sludge at rates of 0.1–0.7 kg SS/kg BOD removed. The expected cleaning in place (CIP) interval under the operating conditions stipulated is >3 months. The typical operation protocol recommended by Memcor is 12 min filtration followed by 1 min relaxation.

Memcor also produce a package plant for flows up to 0.5 MLD. As with the larger scale systems, the MemJet® Xpress (Fig. 4.23) has separate biotreatment and membrane tanks to assist operation and maintenance of the membrane and, in particular, membrane CIP.

Figure 4.22 A rack of Memcor B10R modules

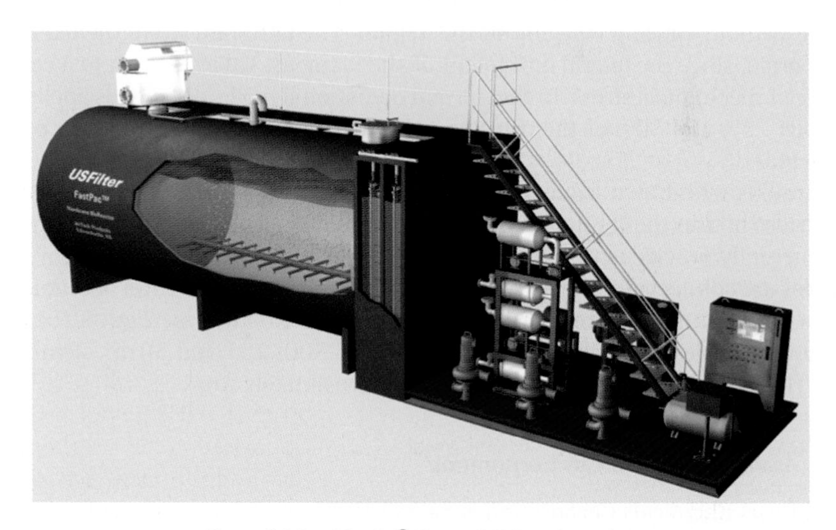

Figure 4.23 MemJet® Xpress MBR package plant

4.3.4 Koch Membrane Systems – PURON®

PURON was formed in late 2001 as a spin out company from the University of Aachen, and was subsequently acquired by Koch Membrane Systems in 2004. The membrane element is the submerged HF-type based on 2.6 mm o.d. and 1.2 mm internal diameter (i.d.) filaments. Unlike most other submerged HF MBR modules, the membrane material is PES. A unique feature of the PURON® system is the securing

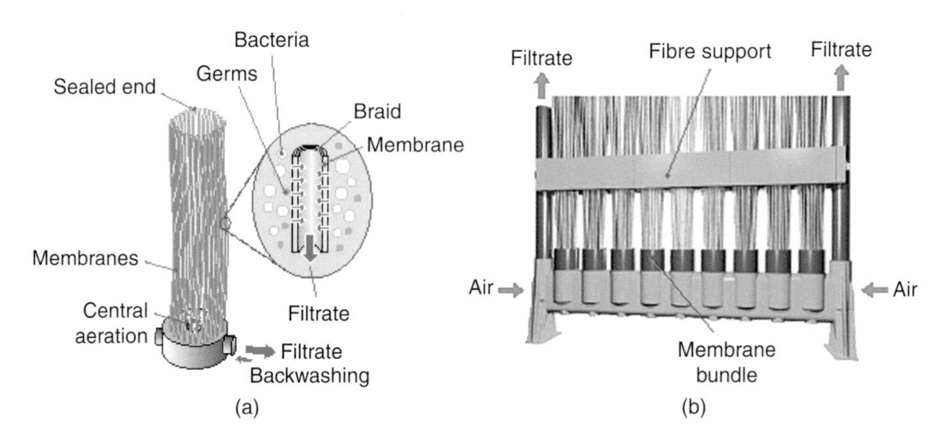

Figure 4.24 PURON® membrane (a) element and (b) row

of the fibres only at the base, with the membrane filaments individually sealed at the top (Fig. 4.24a). Scouring air is injected between the filaments intermittently by means of a central air nozzle at the module base so as to limit the degree of clogging at the module base. The free movement of filaments at the top of the module is designed to allow gross solids, such as hair and agglomerated cellulose fibres, to escape without causing clogging in this region. The fibres are strengthened by an inner braid, since the lateral movement of the filaments subjects them to a certain degree of mechanical stress. In the normal operational mode, aeration is applied for between 25% and 50% of the operation at a SAD_m of 0.15–0.35 Nm^3h air per m^3 permeate.

Filtrate is withdrawn from the manifold at the base of the cylindrical element which also houses the aerator (Fig. 4.24a). The individual fibre bundles are 2 m high and currently provide a membrane area of ~3.3 m^2. In a commercial system these bundles are connected in rows (see Fig. 4.24b). Several of these rows are mounted into a stainless steel frame forming a membrane module (see Fig. 4.25). The PURON® module is available in standard sizes of 500, 235 and 30 m^2 membrane area. The footprint of the 500 m^2 module is approximately 2 m^2.

4.3.5 Asahi Kasei Chemicals Corporation

Asahi Kasei Chemicals Corporation is part of the Asahi Kasei group in Japan. The company has been manufacturing HF UF/MF membranes and modules for various industrial applications since the 1970s. The Microza® PVDF MF membrane for water treatment applications was introduced in 1998, and Asahi Kasei Chemicals currently produces a million square metres of MF membrane each year, mainly for drinking water treatment. The Microza® membrane module has been used as a classic pressurised module by Pall Corporation in their industrial and municipal water treatment systems, providing about 500 MLD potable water capacity from over 50 membrane filtration plants in the USA alone, as of May 2005. Key features of the 1.3 mm o.d. Microza® membrane fibre are the narrow pore size distribution

Figure 4.25 A train of five PURON® modules

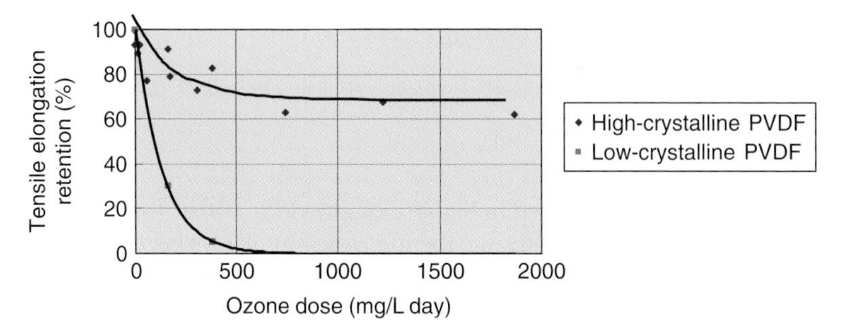

Figure 4.26 Tensile strength against ozone loading, Microza® fibre

(rated 0.1 μm), as determined by rejection of uniform latex (Fig. 2.4), and the high tolerance to oxidative chemicals (Fig. 4.26). Tests conducted against both 0.5 wt% sodium hypochlorite and 4 wt% sodium hydroxide have shown no significant decrease in tensile strength over a 60-day period. This has been attributed by the suppliers to the high crystallinity of the PVDF.

The MBR membrane module has been under development since 1999 and was launched in June 2004. The rack comprises a number of vertical elements mounted onto a steel frame (Fig. 4.27a), each module providing a membrane area of 25 m^2 and having dimensions of 150 mm diameter × 2 m. The fibres are potted at both ends and gathered into bundles at the top of the element (Fig. 4.27b) to encourage the escape of suspended matter. Air is introduced via a separate aerator and directed into the element by a skirt fitted to the element base (Fig. 4.27c). A four-element

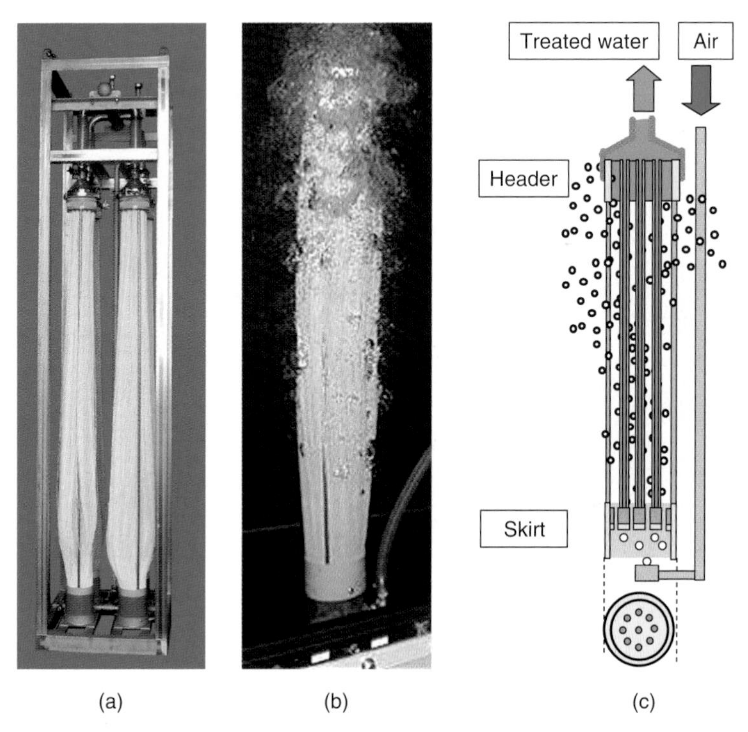

Figure 4.27 Microza® EFM HF membrane: (a) four-element test module (b) aerated module and (c) module schematic

module/rack, such as that shown in Fig. 4.27a, provides 100 m^2 membrane area in a 0.8 m × 0.7 m × 2.7 m high frame, having a weight of 185 kg.

Pilot trials have been conducted on the module in collaboration with the Japan Sewage Works Agency at Mooka (Fig. 4.28), based on a flow of 0.036 MLD of municipal wastewater entering a 9 m^3 tank divided equally into two compartments for denitrification/nitrification and providing a 6 h hydraulic retention time (HRT) overall. The maximum tank depth was 4.5 m, and the membrane was placed in the nitrification tank. According to these data (Fig. 4.29), the module can be operated at a net flux of 30 LMH with the TMP generally varying between 0.2 and 0.3 bar (and hence a mean permeability of 120 LMH/bar), with the MLSS level at 8–12 g/L. The total SAD_p demand reported during the trial was 13.3 m^3 air/m^3 permeate, with only half this amount relating to direct membrane aeration (though the membrane was placed directly in the aeration tank). The recommended filtration cycle for the MBR module is 540–580 s on/20–60 s off. A CEB (a chemically enhanced backflush, referred to by the company as EFM or enhanced flux maintenance) of sodium hypochlorite is recommended every 3 months supplemented by a yearly CIP soak. However, according to the pilot trial data, monthly CEBs were required for a period between January and March, and this was attributed to high organic loadings during this period. Complete recovery of permeability was achieved by soaking in highly alkaline (1 wt% NaOH) sodium hypochlorite (5 g/L) for 5 h, followed by a 1 h soak in 2 wt% nitric acid.

Figure 4.28 Pilot plant at Mooka

As of November 2005, only one MBR based on the Microza® membrane had been installed, this being a food effluent treatment application (Section 5.3.5). Several pilot trials on a number of different wastewater streams have been conducted since 1999, however, and given the commercial success of the Microza® membrane for other various water applications, it seems likely that the company will continue to persist with the MBR technology in the future.

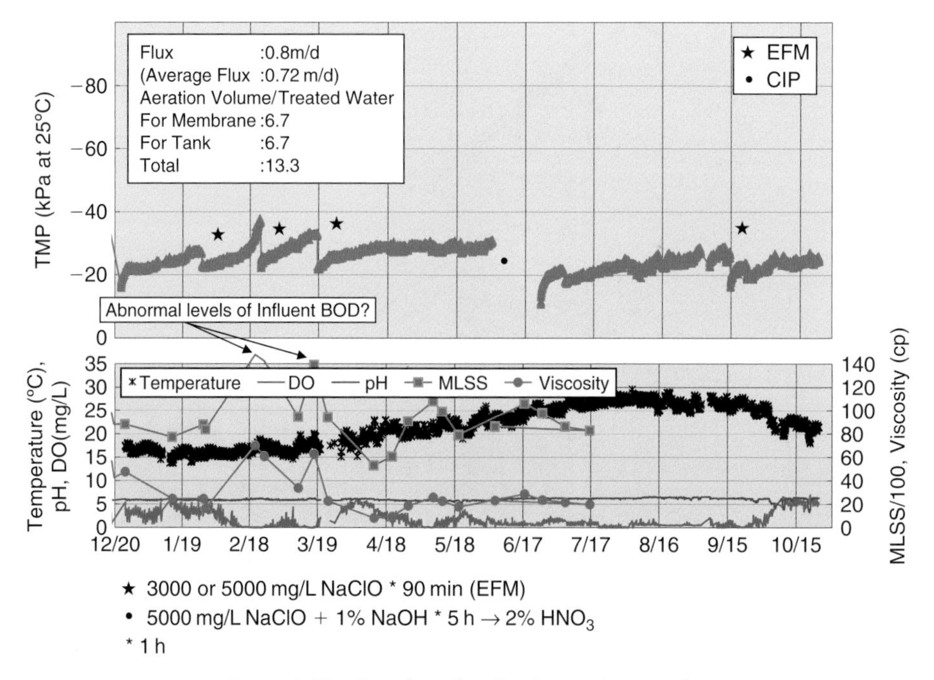

Figure 4.29 Data from the pilot plant trials at Mooka

4.3.6 ITT Industries

ITT Industries has a $8.0 billion turnover and offers a number of product lines relating to water and wastewater treatment. The company acquired PCI Membranes from Thames Water (now part of RWE) and WET, a supplier of RO systems in 2002 as well as Wedeco Disinfection in 2005. The new ITT Industries' Dual Stage MBR systems, launched in 2005, take advantage of synergies between two of the company's subsidiaries: Sanitaire and Aquious-PCI Membrane Systems. Sanitaire is the world's leading supplier of aeration systems for wastewater treatment and also supplies complete wastewater systems based on the Sanitaire ABJ sequencing batch reactor (SBR). Aquious, a unit of ITT Advanced Water Treatment, is a relatively new brand that envelopes ITT Industries' existing membrane filtration products and technology. As well as retaining the manufacturing capability for tubular membranes provided by PCI Membranes, Aquious supplies spiral-wound or HF membranes for clean feed streams, and ceramic membranes for applications requiring high chemical compatibility and thermal resistance.

The new Dual Stage MBR technology is targeted both at municipal and industrial wastewater applications, particularly in food and beverage production, chemical and pharmaceutical manufacturing, pulp and paper processing, metal finishing and steel production. As with many other HF MBRs, biological treatment is separated from the membrane filtration stage to allow independent control of retention time and aeration. The biological stage employs high-efficiency Sanitaire® diffused aeration technology based on Gold Series™ diffusers. MLSS levels can be maintained

Figure 4.30 The ITT MBR

at 10 000–12 000 mg/L. The membrane filtration stage employs immersed 0.05 μm reinforced PES membranes. Membranes are supplied in two module configurations, horizontally mounted and in single manifolds, which are open at one end and connected to a manifold for permeate extraction under vacuum. The modules have integral aerators, with air-scour nozzles placed within the fibre bundles (Fig. 4.30). The company is also incorporating the Dual Stage MBR in its new Integrated Reuse System, together with ITT RO, Ozone and UV technology.

4.4 Sidestream MBR technologies

There are a number of MT membrane module suppliers providing standard size cylindrical modules, generally 100 or 200 mm in diameter, which are then employed by sidestream MBR (sMBR) process suppliers in their proprietary processes. Details of two-membrane module product suppliers are given below. The proprietary MBR technologies of one established European MT MBR process supplier are described in Section 4.5.3.

4.4.1 Berghof Membrane Technology

Berghof is a German supplier who provide two MT UF 3 m-long membrane module products: HyPerm-AE and HyperFlux, the main difference being the module and membrane tube diameters. The HyPerm-AE is a 225 mm diameter module fitted with 11.5 mm diameter tubes, 112 in all, providing a total membrane area of 11.8 m^2, in PVC-C housing. The membrane material is either 250 kD PVDF or PES, offered with a range of selectivities between 20 and 150 kDa, both materials being on a PP substrate support. The HyperFlux is a standard 8″ (200 mm) diameter module fitted with 8 or 10 mm diameter tubes, thereby respectively providing a total membrane area of 27.2 or 21.2 m^2 from 365 or 224 membrane tubes, in glass–fibre reinforced plastic (GRP) housing. The membranes in this case are either of 250 kDa PVDF or 150 kDa PES. For both products, the membrane element is replaceable within the GRP housing.

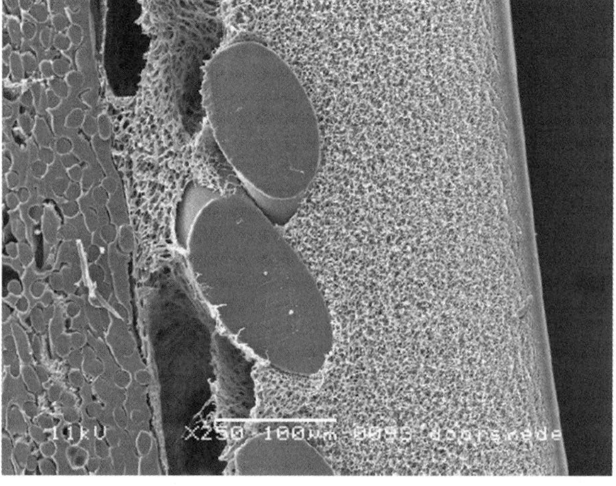

Figure 4.31 Norit X-Flow membrane, showing support on the left and active membrane surface on the right

Figure 4.32 The Norit X-Flow MT 38PRH F4385 module

4.4.2 Norit X-Flow

Norit X-Flow is a major supplier of MT and capillary tube (CT) membrane modules, the latter thus far being dedicated to potable water filtration. The tubular membrane is supported by a mechanically robust substrate (Fig. 4.31) by a two-fold polyester backing. The membrane material itself is either PES for the CT configuration, <3 mm external diameter, or PVDF for the MT type used for sMBR technologies. These membranes comprise type F4385, 5.2 mm i.d., and type F5385, 8 mm i.d., each having a pore size of 0.03 μm. The PVDF MT modules are supplied as both PVC (38PR) or glass fibre reinforced (38GR) shells. At 3 m in length, the respective total membrane area and number of lumens is 30 m^2 from 620 lumens (for the 5.2 mm i.d.) or ~27 m^2 from 365 lumen (8 mm i.d.).

Thus far, only the MT configuration (Fig. 4.32) has found use in sMBRs. The most established of these are based on pumped, horizontally mounted crossflow modules (Fig. 4.33a), with some of the more recent plants fitted with vertically mounted

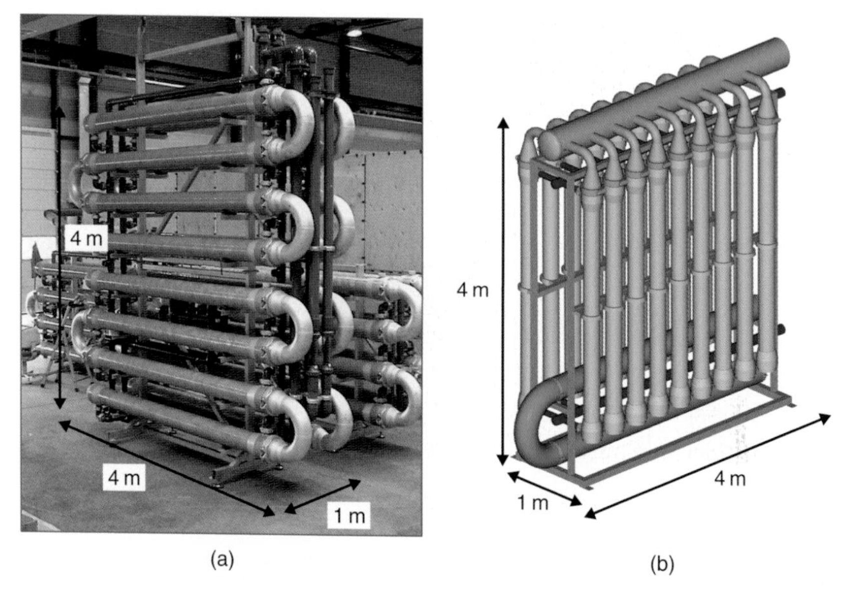

Figure 4.33 *Norit X-Flow modules: (a) pumped and (b) airlift, with accompanying dimensions*

Table 4.3 Norit X-Flow sMBR quoted operational parameters

Parameter	Pumped	Airlift
MLSS (g/L)	12–30	8–12
TMP (bar)	1–5	0.05–0.3
Flux (LMH)	80–200	30–60
Permeability (LMH/bar)	40–80	90–500
Footprint (m^3/h capacity m^2 projected area)[a]	10.8[c]	7.5[b]
Footprint (m^2 membrane/m^2 projected area)[a]	108	131
Specific energy demand (kWh/m^3)	1.5–4	0.5–0.7
Processing	More simple	More complex
Mode of operation	Continuous	Discontinuous

[a]Based on a single skid, 1 m × 4 m × 4 m high.
[b]Based on 55 LMH.
[c]Based on 100 LMH.

membranes operating with airlift (Fig. 4.33b) combined with liquid pumping at much lower flow rates (Table 4.3). This product has been used by a number of process designers for proprietary MBR technologies, including Wehrle (Germany), Aquabio (UK) and Dynatec (US). Whilst the majority of Norit X-Flow-based MBR plant are used for industrial effluent applications, some examples exist of small-scale municipal and domestic applications where the inherently higher specific energy demand is less of an issue than the system robustness, simplicity and control. The airlift system is recommended at relatively low chemical oxygen demand (COD) concentrations and high flows (<1 g/L COD and >250 m^3/h), whereas at high COD levels (>5 g/L) and low flows (<100 m^3/h), the pumped system is recommended. This leaves a range of flows and COD levels for which either system may be suited; the overall COD loading rate over the entire range is normally between 100 and 1000 kgCOD/h.

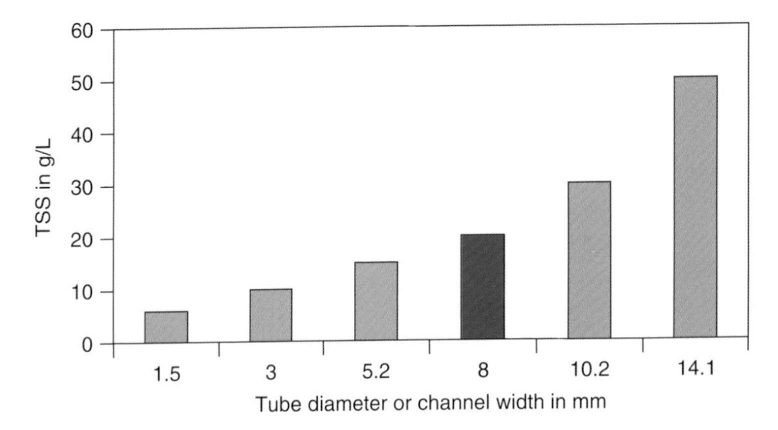

Figure 4.34 Suspended solids level vs. tube diameter (Robinson, 2005)

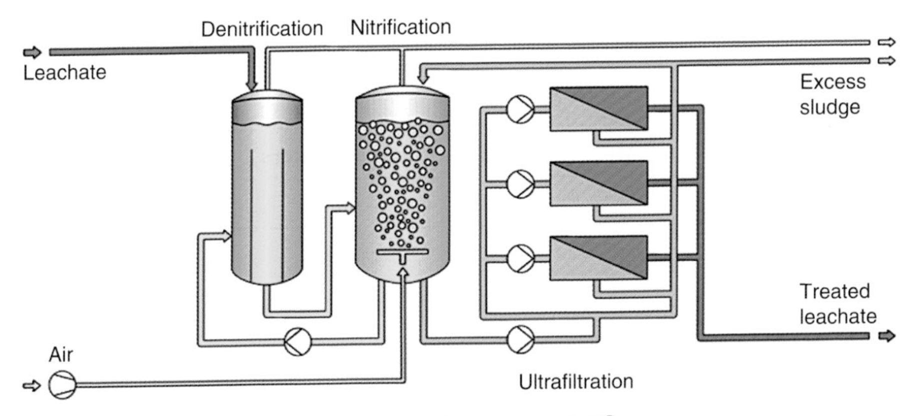

Figure 4.35 The Wehrle BIOMEMBRAT® process

4.4.3 Wehrle Environmental

Wehrle use off-the-shelf membranes for its MBR designs, including Norit X-Flow and Berghof. All the membranes used for sMBRs are MTs. Initially, 1″ (25 mm ID) tubes were used, and these were subsequently replaced by 11.5 and 10.5 mm units. Since 2000, all the crossflow MBR plants have been designed and built using the standard 8″ (200 mm diameter) × 3 m module incorporating 8 mm diameter tubes, since this tube diameter appears to be sufficient for moderate to high-MLSS bioreactor liquors (Fig. 4.34). The 8 mm tubular membrane design was originally conceived in co-operation between Stork Friesland (now part of Norit X-Flow) and Wehrle.

The Wehrle BIOMEMBRAT® process (Fig. 4.35) is an sMBR which can be combined with either an atmospheric or pressurised bioreactor depending on the process circumstances. The bioreactors can be operated at up to 25 g/L MLSS, which minimises

Figure 4.36 *Ejector aerator*

reactor volume for a given sludge loading. However, membrane flux performance deteriorates with an increase in MLSS, and the optimal sludge concentration is on average 17–20 g/L for the 8 mm tubes used (Fig. 4.36). Bioreactor pressurisation, up to a pressure of 3 bar, offers a number of advantages of:

- control of sludge foaming;
- enhanced oxygen dissolution, thus permitting higher organic loadings and/or reduced tank size;
- reduced risk of stripping of volatile organic matter, and so reducing the size of any air scrubbers which might be required for off-gas treatment.

The atmospheric bioreactor can be constructed of GRP or PP for capacities up to 20 m^3. At capacities greater than this, it is usual to use glass coated steel. The maximum tank height in either case is 10 m. The pressurised system tends to be used for more recalcitrant feedwaters (COD/BOD ratio $>\sim 4$). For these reactors, coated steel is used for capacities up to 200 m^3 and the maximum tank height is 15 m. Disc membrane aerators are used for low loads and ejectors (high-shear jet aerator, Fig. 4.36) for high loads.

The use of a smaller aeration tank tends to elevate the temperature to between 30°C to 35°C due to energy generated from the exothermic bio-process, aeration blowers and pumps. This process then operates more efficiently due to increased bioactivity and reduced permeate viscosity. On the other hand, the specific energy demand is high compared with an immersed process (Section 2.3.1).

Two other processes have been developed by Wehrle: a vertically mounted airlift sidestream BIOMEMBRAT® Airlift process, which provides a lower flux combined with a lower energy demand (Fig. 4.37), and a low-crossflow process called the BIOMEMBRAT®-LE (Fig. 4.38) where the modules are placed in series. The BIOMEMBRAT®-LE is designed to allow an adjustable crossflow velocity which permits a wide range of hydraulic loads. This means that peak loads are dealt with by

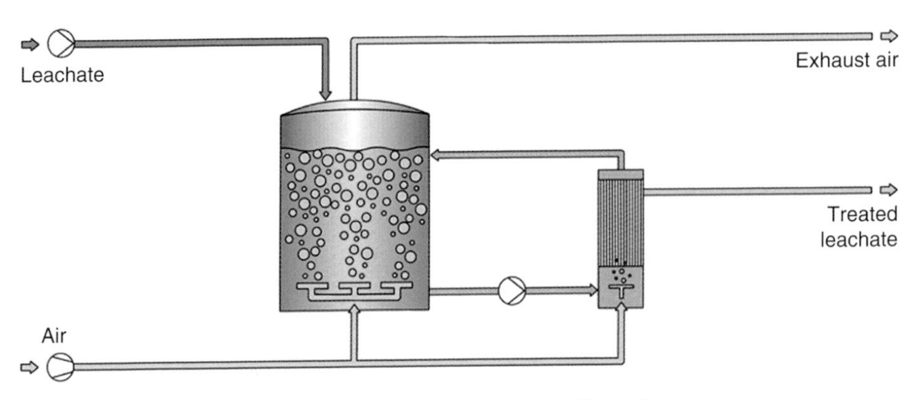

Figure 4.37 The Wehrle BIOMEMBRAT® Airlift MBR

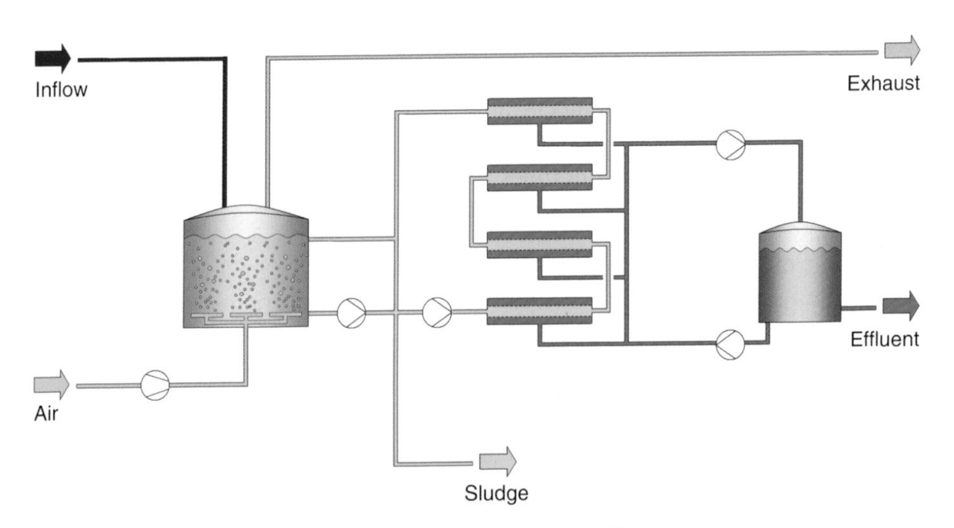

Figure 4.38 The Wehrle BIOMEMBRAT®-LE MBR

higher crossflows, and thus a higher energy demand, rather than having redundant membrane area at low-to-normal loads. So far, this process has been piloted at bioreactor process flows of ~0.15 m^3/h (up to ~0.4 m^3/h permeate flows). Net fluxes of ~55 LMH have been achieved at specific energy demands of around 1.5 kWh/m^3 arising at moderate crossflow velocities (1–2 m/s). It is thought that the BIOMEMBRAT®-LE will be more suited to industrial effluents with low to average COD than the airlift process, though it operates at a slightly higher energy demand than the latter. The airlift configuration will continue to be employed for municipal effluents.

Wehrle have also developed integrated post-treatment processes to provide additional purification. Unit operations have included both activated carbon (AC) and nanofiltration (NF) where the NF concentrate is fed to the AC for organics removal, with the AC effluent (Fig. 4.39).

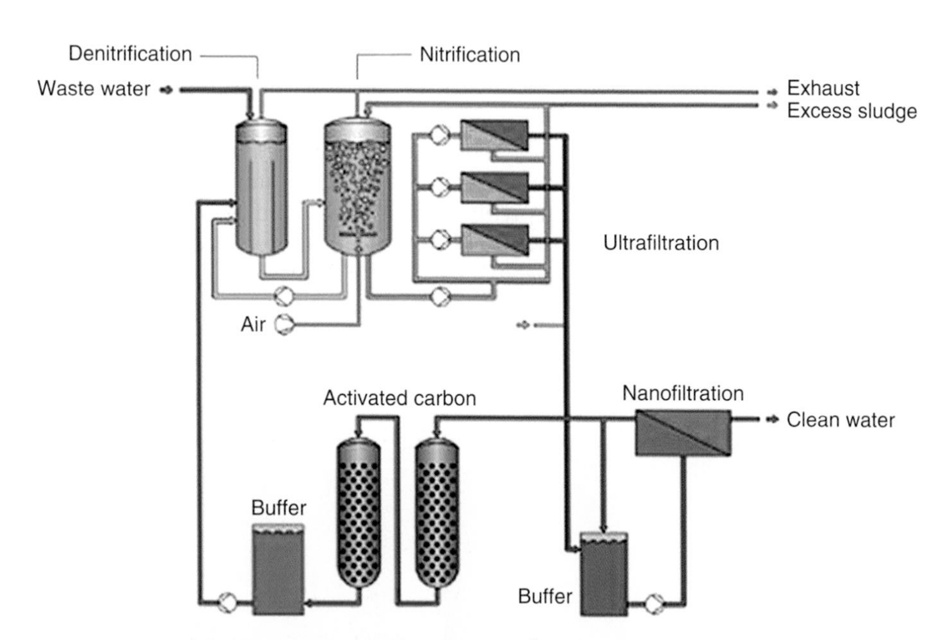

Figure 4.39 The Wehrle BIOMEMBRAT® with integrated post-treatment

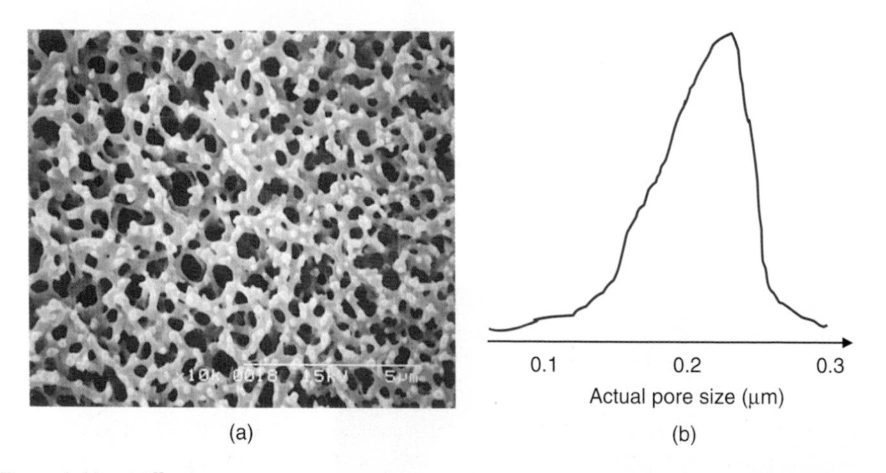

Figure 4.40 Millenniumpore 0.1 μm-rated MF membrane: (a) surface and (b) corresponding pore size distribution

4.4.4 Millenniumpore

Millenniumpore is a UK SME formed in 1997. The company manufactures who are technically CT membranes and modules, in that they are self-supporting cylindrical membranes, but they are of a wider bore – between 2.5 and 15 mm i.d. They are thus, in effect, tubular membranes but are not supported by substrate and, with a wall thickness of 1.5–2 mm, have sufficient mechanical integrity to be backflushable at moderate TMP levels. They are of PES (Fig. 4.40), which is hydrophilicised by immobilisation of an ethylene oxide–propylene oxide copolymer onto the membrane surface. The tubes are double skinned; they have a permselectivity on both the inner and outer surfaces (Fig. 4.41).

The tubes are potted in epoxy resin to form MT modules (Fig. 4.42a) which are used for both conventional membrane filtration and MBRs. Millenniumpore have developed both iMBRs (Fig. 4.42b) and sMBRs with vertically mounted MT modules which have found applications in niche, small-scale industrial effluent treatment applications. In the last few years, commercial applications have been limited to the sidestream configuration. These MBRs are fitted with 5 mm diameter MTs of 0.05 μm pore size. However, the company is able to offer bespoke membrane materials and

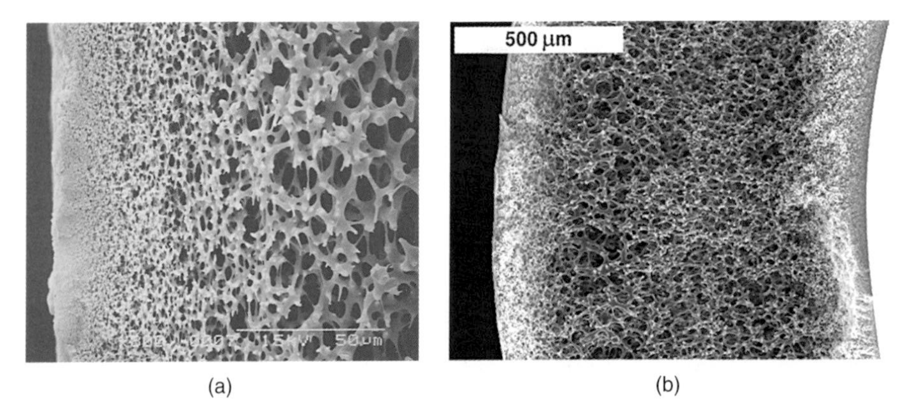

Figure 4.41 Cross-section of a Millenniumpore membrane: (a) single surface and (b) both surfaces

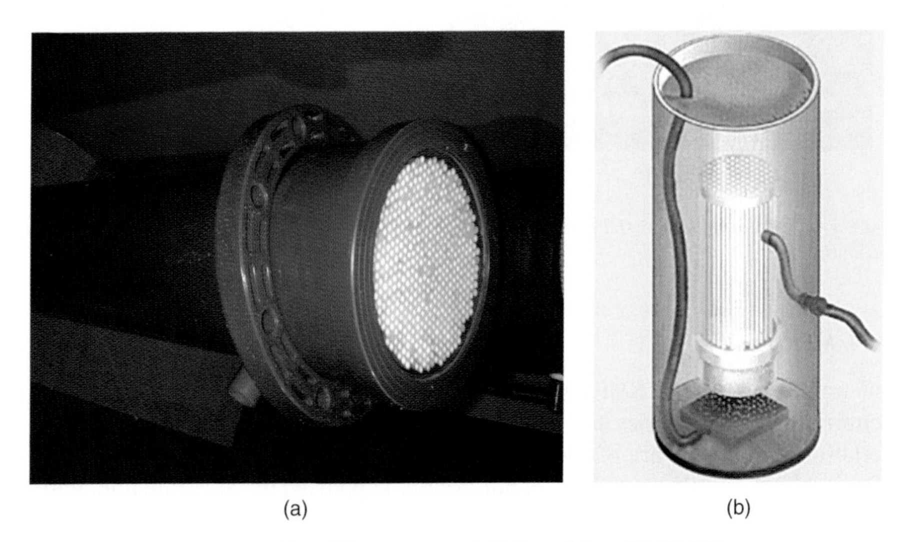

Figure 4.42 Millenniumpore: (a) MT module and (b) iMBR

modules, with pore sizes ranging from 20 kDa UF to 0.5 μm MF and i.d. between 2.5 and 15 mm. The membranes are robust to TMPs up to 2.5 bar.

4.5 Other sidestream membrane module suppliers

4.5.1 Novasep Orelis

Orelis, now part of the Novasep group and originally formed as part of the break-up of the Rhone Poulenc group, has a number of membrane products including ceramic monoliths (Carbosep® and Kerasep™), spiral-wound UF (Persep™) and FS UF (Pleiade®). It is the latter (Fig. 4.43) which has been used in sMBR applications, although the ceramic module is offered for MBR duties. The Pleiade® element comprises 2610 mm by 438 mm polyacrilonitrile-based FS rated at 40 kDa pore size. The membrane elements have inlet and outlet ports, which allow passage between each cell (i.e. FS channel), and the membrane channel width is 3 mm. This means that a module depth of 1710 mm provides a membrane area of 70 m^2 (Fig. 4.44).

4.5.2 Polymem

The Polymem technology is unusual in that it represents a rare example of an HF membrane module being employed in a sidestream configuration, the proprietary name of the module being Immem. The polysulphone (PS) membranes used are

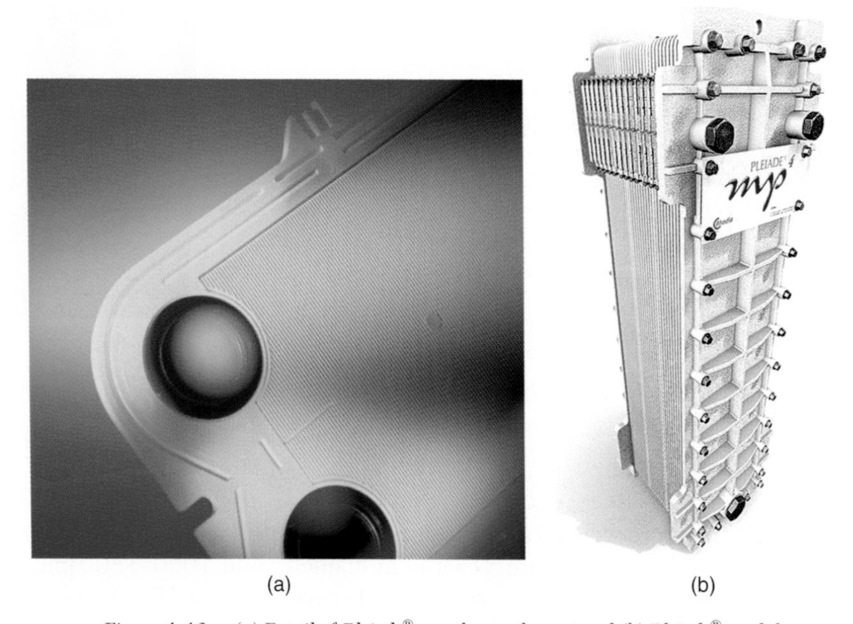

Figure 4.43 (a) Detail of Pleiade® membrane element and (b) Pleiade® module

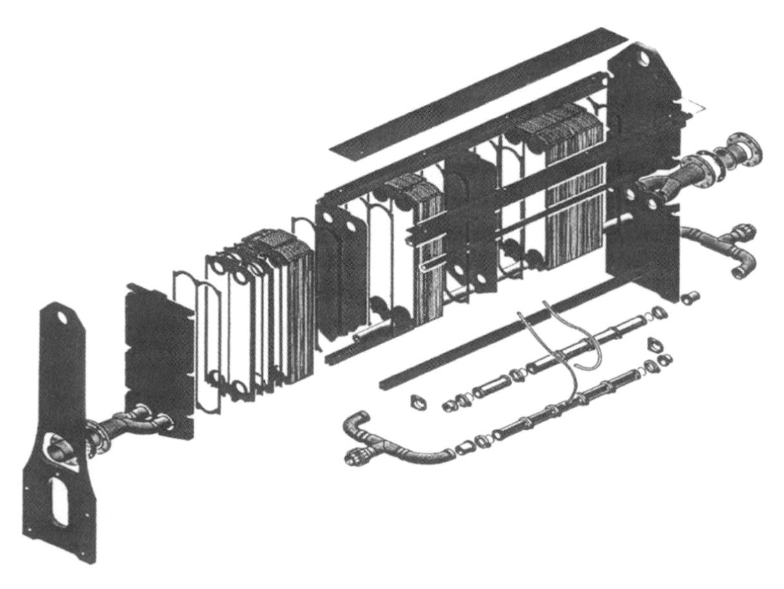

Figure 4.44 Pleiade® stack, exploded view

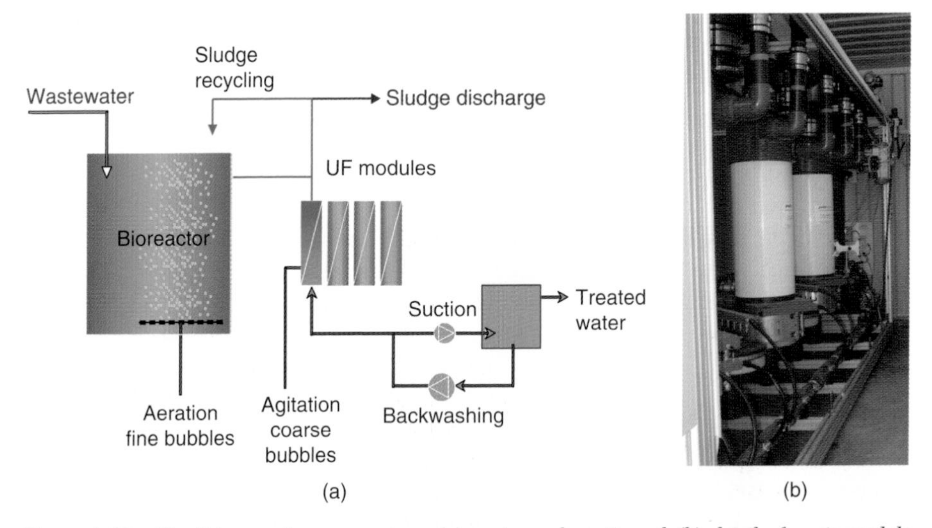

Figure 4.45 The Polymem Immem system: (a) system schematic and (b) detail of a six-module WW120 skid (in Livna, Israel)

0.08 μm (300 kDa) pore size and are provided with external diameters between 0.7 and 1.4 mm in modules which are 315 mm in diameter and 1000–1500 mm in length (Fig. 4.45b). The membrane area is thus between 60 and 100 m^2/module. The SAD_m value quoted by the manufacturer is 0.25 $Nm^3/h\,m^2$ membrane at fluxes

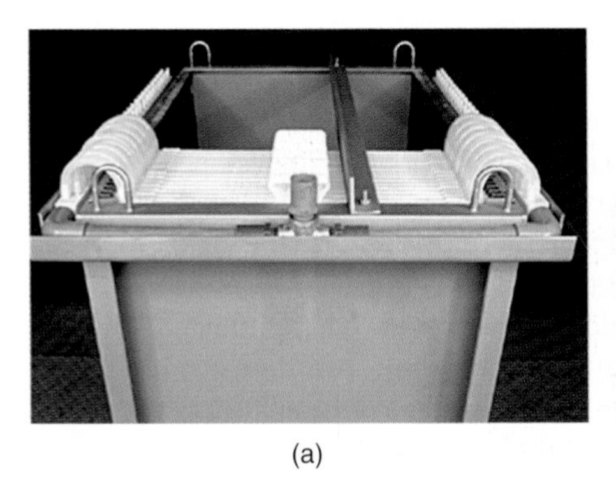

(a)

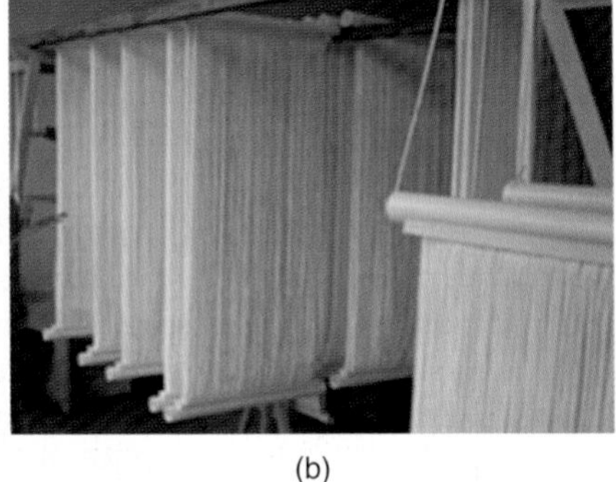

(b)

Figure 4.46 (a) Han-S Environmental FS and (b) Motimo HF membrane

between 10 and 20 LMH, yielding SAD_p values between 13 and 25 m^3 air/m^3 permeate.

4.6 Other MBR membrane products

A number of other companies have either recently introduced or are due to shortly introduce their own MBR systems. These all appear to be based on either rectangular FS or vertically oriented HF membrane configurations. A comprehensive listing of these companies is practically impossible, and the product details given below represent merely a few examples.

A significant proportion of the newer membrane and/or membrane module manufacturers are from South East Asia. Examples include KMS, Kolon and the Han-S Environmental Engineering Company in Korea, Ultraflo in Singapore and Puri Envi-Tech and Motimo in China. The Han-S product (Fig. 4.46a) is based on a 1000 × 560 mm (0.98 m^2) element, employing a 0.25–0.4 μm pore size MF membrane. Fluxes of 13–21 LMH are reported for TMP values up to 0.5 bar. Motimo MBR membranes, employed by Lam Environmental for their iMBR, are (somewhat confusingly) referred to as "Flat Plat" but are actually 1 mm o.d. rectangular (1010–1510 mm × 530 mm) PVDF HF membrane modules (Fig 4.46b). Ultraflo are a Singaporean company producing double-skinned thin-walled polyacetonitrile (PAN) HF membranes ranging from 0.5 to 2.7 mm o.d. A 0.19 MLD MBR using the Ultraflo membrane has been trialled at the Bedok Sewage Treatment Works in Singapore, although not as part of the trials sponsored by the Public Utilities Board (Section 3.2.4).

Microdyn-Nadir (Germany) have recently launched their BIO-CEL® membrane module (Fig. 4.47), based on self-supporting FS PES-based UF membranes rated at 150 kDa MWCO (0.08 μm). The membrane sheets are 2 mm thick and self-supporting; they are mounted to provide a channel width of 8 mm. Standard BIO-CEL® modules (pre-fixed "BC") of 10, 100 and 400 m^2 are offered with membrane panel dimensions of 1200 × 650 mm for the BC-100 and 1340 × 1140 mm for the BC-400. The modules provide a specific surface area of 70 and 90 m^2/m^3 for the BC-100 and

BC-400 respectively. The company's literature indicates a very wide range of airflow rates (0.15–1 m^3/h airflow m^2 membrane, depending on module size) and TMP values (0.03–0.3 bar) without providing a target flux rate.

There are also a number of other German MBR products developed for niche applications (ZEK, 2004). These include that developed by A3 (Abfall-Abwasser-Anlagentechnik GmbH). A3 provide an FS membrane and have had a textile wastewater treatment reference site since 2000. A3 also appear to have interests in the ship industry, a market sector for which MBRs provide a key candidate technology for wastewater treatment and, ultimately, recycling. Another company, ATB Umwelttechnologien GmbH – also in the North Rhine-Westphalian region – have produced a single house-scale MBR, sized for 4–16 population equivalent (p.e.). This product also appears to be based on FS membranes, and would appear to be similar in concept to the more established BusseMF system. This technology, originally commercialised in 1997, comprises a two-stage sedimentation + FS MBR process, apparently employing Kubota membranes, and is available for 4 and 8 p.e. As of December 2005 there are over 200 such units installed in Germany alone, with perhaps another 30 installed across Europe.

Finally, worthy of a mention is the earliest commercially established anaerobic MBR. The anaerobic digester ultrafiltration (ADUF) process is based on sidestream treatment and was commercialised in the early 1990s in South Africa following successful demonstration based on 0.05–3.0 m^3 reactors. The process is used to treat industrial wastewaters with high COD concentrations (3.5–37 kg/m^3), achieving removals generally >90% depending on the biodegradability of the organic matter. A 3 m long, 12 mm diameter MT PES UF module, the membrane pore size typically being 0.10 μm, is employed with flow paths of over 60 m (i.e. >20 modules). MLSS levels of 10–50 g/L are used, coupled with high crossflow velocities to maintain fluxes above 50 LMH. As with aerobic processes, the permeate produced is of a consistently high quality with regard to particulate material, bacteria and viruses. Whilst this process achieved some technical success and was licensed to Bioscan in Denmark in the mid-1990s, it has not achieved the commercial success of the

Figure 4.47 The Microdyn-Nadir module

aerobic counterpart. This may be partly attributable to improvements in more conventional anaerobic treatment processes such as the upflow anaerobic sludge blanket (UASB) reactor (Section 2.2.7). It is also the case that complete clarification from anaerobic treatment is rarely necessary. Having said this, anaerobic MBR systems continue to be explored and developed.

4.7 Membrane products: summary

A review of the membrane products available for MBR technologies (Table 4.4) reveals that almost all iMBRs are either rectangular FS or HF, and that most sMBR technologies are MT. The exceptions to these general observations appear to be:

(a) the Orelis Pleiade® PAN FS membrane used for sidestream treatment;
(b) the Polymem (PS) and Ultraflo (PAN) sidestream HF systems;
(c) the hexagonal/octagonal rotating immersed Huber FS membrane;
(d) the Millenniumpore MT membrane, which has been used as an immersed module as well as for airlift sidestream.

Airlift sidestream MT systems are offered by some process suppliers. These provide higher permeabilities than the classical pumped systems. No examples of pumped

Table 4.4 Membrane product categories

		Process configuration	
		Immersed	Sidestream
Membrane configuration	FS	A3 Colloide Brightwater Huber[a] ITRI *NWF* Kubota Microdyn-Nadir Toray	Novasep-Orelis
	HF	Asahi Kasei Han-S Environmental ITT Koch-Puron Kolon Mitsubishi Rayon Motimo Siemens-Memcor Zenon	Polymem Ultraflo
	MT	Millenniumpore	Berghof[b] Millenniumpore Norit X-Flow[b]

[a] Rotating membrane.
[b] MT membrane products used by process suppliers such as Aquabio, Dynatec, Triqua, Wehrle.

Table 4.5 Membrane product specifications (see also Appendix D)

Supplier	Membrane (configuration, material)	Pore size, μm	Diameter *d* or separation (δ), mm	Specific surface area* ϕ, m^{-1}	Proprietary name, membrane or module
Berghof	MT, PES	0.08	9	110	Hy*Perm-AE*
	or PVDF	0.12			Hyper*Flux*
Brightwater	FS, PES	0.08	9	110, *47*	MEMBRIGHT®
Toray	FS, PVDF	0.08	7	135	Toray
Kubota	FS, PE	0.4	8	115	Kubota
Colloide	FS, PES	0.04	10	133	Sub Snake
Huber	FS, PES	0.038	6	160, *90*	VRM
Millenniumpore	MT, PES	0.1	5.3	180	Millenniumpore
Koch-Puron	HF, PES	0.05	2.6 (3.5)	314, *125*	Puron
Zenon	HF, PVDF	0.04	1.9 (3.0)	300	ZW500C-D
Norit X-Flow	MT, PVDF	0.038	5.2	320, *30*	F4385
			8	290, *27*	F5385
Siemens-Memcor	HF, PVDF	0.04	1.3 (2.5)	334	B10R, B30R
Mitsubishi Rayon	HF, PE	0.4	0.54 (1.7)	485, *131*	SUR
	HF, PVDF	0.4	2.8 (2.9)	333, *71*	SADF™
Asahi Kasei	HF, PVDF	0.1	1.3 (1.3)	710, *66*	Microza
Polymem	HF, PS	0.08	0.7 (1.1)–1.4	800	WW120
Ultraflo	HF, PAN	0.01–0.1	2.1 (0.7)	1020	SS60
Motimo	HF, PVDF	0.1–0.2	1.0 (0.9)	1100, *735*	Flat Plat

* Refers to elements; italicised figures refer to modules.

sidestream HF systems exist for full-scale operating plant, although such systems have been commercialized (by Ultraflo and Polymem). Unlike most FS systems, almost all HF products are immersed in a separate tank from the main bioreactor. Moreover, the majority of HF MBR membrane products currently on the market are vertically mounted and PVDF based (Table 4.5), the exceptions being the Koch-Puron membrane which is PES, the Polymem PS membrane, the Ultraflo PAN membrane and the Mitsubishi Rayon SUR module, which is PE material and also horizontally oriented. All HF products are in the coarse UF/tight MF region of selectivity, having pore sizes predominantly between 0.03 and 0.4 μm. All vertically mounted HF systems are between 0.7 and 2.5 mm in external diameter. Distinctions in HF MBR systems can be found mainly in the use of reinforcement of the membrane (essential for those HF elements designed to provide significant lateral movement) and, perhaps most crucially, the air:membrane contacting.

The impact of aeration is influenced by the membrane spatial distribution. An examination of the specific surface area (or packing density ϕ: the membrane area per unit module volume) reveals that, as expected, the packing density tends to increase with decreasing filament or tube diameter. The reinforced braided HF membranes, namely the Zenon and Puron products, thus tend to provide lower specific surface areas. The MTs, although larger in i.d. than the HFs are in outside diameter, can nonetheless provide a high packing density because of the degree of tessellation possible (Fig. 4.32): tubes can be packed together without any requirement for separation, whereas HFs must provide flow channels for the sludge retentate. With the exception

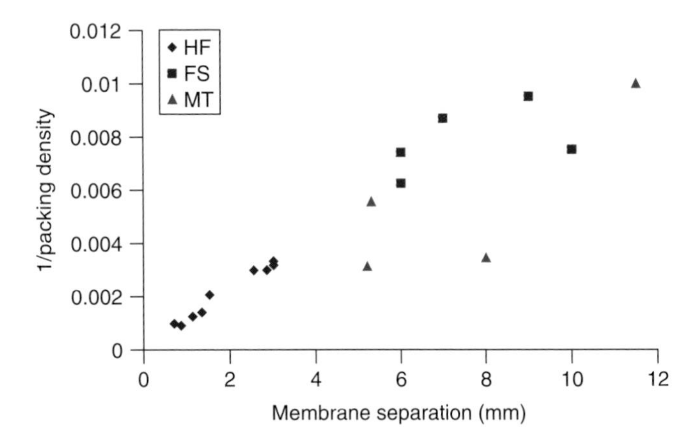

Figure 4.48 Inverse packing density (1/ϕ) vs. membrane separation for the three module configurations, data from Table 4.5

of the particularly large-diameter Berghof HyPerm-AE module, all of the commercial MBR membrane elements identified which are of specific surface area (ϕ) below 170/m are FS in geometry, and only one FS technology is above 130/m.

Perhaps the most pertinent parameter for comparison of the various systems is the membrane separation itself. For an FS system, this is simply given by the thickness of the flow channel. For HF elements, the mean minimum separation can be determined from trigonometry (Appendix C) as being related to filament ID and ϕ by:

$$\delta = 1.95\sqrt{\frac{d}{\phi}} - d \tag{4.1}$$

Values for δ are included in parenthesis in Table 4.5. As might be expected, the largest separations arise with the largest diameter filaments, but even the largest HF δ values are half those of the FS modules, as is inevitable from the packing densities. There appears to be a roughly inverse relationship between δ and ϕ (Fig. 4.48), with the coefficient or proportionality being around 10^{-3}, regardless of the configuration. Outliers arise mainly from the high packing densities attainable with some of the MT products. On the other hand, HF configurations generally provide a higher minimum separation distance, relative to the filament size (hence d/δ), than the FS and MT counterparts, where $d = \delta$ (Fig. 4.49).

The deliberation over the optimum membrane configuration may proceed firstly through a consideration of shear, which is given by the ratio of the velocity to channel thickness (Equation (2.31)). If shear can be assumed to relate to the aeration velocity, it follows that shear rate can be represented by:

$$\gamma \propto \frac{U_G}{\delta} = \frac{Q_{A,m}}{A_x \delta} \tag{4.2}$$

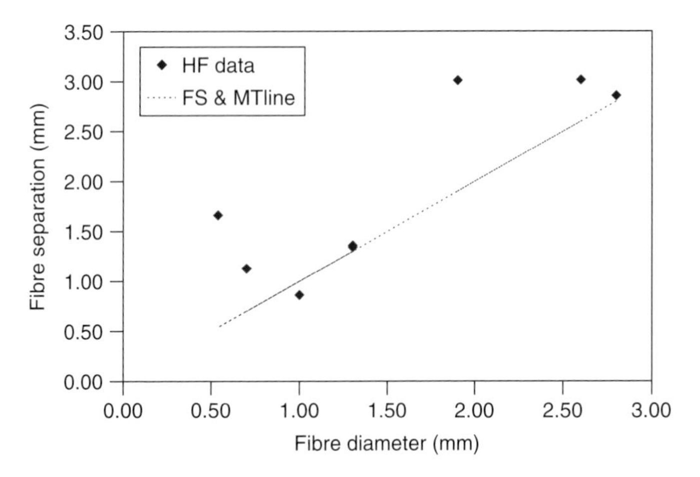

Figure 4.49 Membrane separation vs. filament diameter for HF membrane modules

where A_x is the open cross-sectional area and Q_A is the aeration rate. For an HF module, A_x can be determined from the same trigonometric approach as that yielding Equation (4.1) as being:

$$A_x = \frac{A_f}{L}\left(\frac{1}{\phi} - \frac{d}{4}\right) \tag{4.3}$$

where A_f is the module or element membrane area. The area associated with each filament is given by (Appendix C):

$$A_{x,\text{filament}} = \frac{\pi}{4}\left(\frac{4d}{\phi} - d^2\right) \tag{4.4}$$

This compares with the more straightforward case of the FS module, where A_x is simply the total cross-sectional area provided by the channels and each panel cross-sectional area is given by:

$$A_{x,\text{panel}} = dw \tag{4.5}$$

where w is the panel width.

According to Equations (4.2) and (4.4), for a fixed aeration rate shear increases both with increasing packing density and with decreasing filament diameter in the case of an HF module. For an FS module it increases with decreasing channel width (Equation (4.5)). However, a number of important parameters mitigate against small filaments and high packing densities:

(a) the bubble rise velocity decreases with bubble size, and thus decreasing δ, with correspondingly increasing downward drag force;
(b) clogging, the risk of which increases with decreasing δ;

(c) air channelling, the risk of which is likely to be a function of δ;
(d) hydraulic losses on the permeate side, leading to differential fouling (Section 2.3.8.3) and which increase with decreasing i.d.

It can be envisaged that there exists a range of optimum values of d (and corresponding i.d. values) and δ at which, when coupled with the appropriate aeration rate $Q_{A,m}$, clogging, air-channelling and permeate-side hydraulic losses are minimised. However, little research into the first two of these specific areas of MBR module design appears to have made its way into the open literature. Moreover, and as already stated (Section 2.3.10), research into clogging is difficult given its generally insidious nature. Given the challenges imposed, it is perhaps more apposite to review information provided from case studies and practical experience (Chapter 5) to shed light on the practical aspects of MBR technology and membrane module design.

References

Alavi Moghaddam, M.R., Satoh, H. and Mino, T. (2002) Performance of coarse pore filtration activated sludge system. *Water Sci. Techol.*, **46** (11–12), 71–76.

Chang, I.S., Gander, M., Jefferson, B. and Judd, S. (2001) Low-cost membranes for use in a submerged MBR. *Trans. IchemE.*, **79**, Part B, May, 183–188.

Chang, M.C., Tzou, W.Y., Chuang, S.H. and Chang, W.K. (2003) Application of non-woven fabric material in membrane bioreactor processes for industrial wastewater treatment, *Proceedings of 5th International Membrane Science and Technology Conference (IMSTEC)*, November 10–14, Sydney, Australia.

Fuchs, W., Resch, C., Kernstock, M., Mayer, M., Schoeberl, P. and Braun, R. (2005) Influence of operational conditions on the performance of a mesh filter activated sludge process. *Water Res.*, **39**, 803–810.

Kennedy, S. and Churchouse, S.J. (2005) Progress in membrane bioreactors: new advances. *Proceedings of Water and Wastewater Europe Conference*, Milan, June 2005.

Meng, Z.G., Yang, F.L. and Zhang, X.W. (2005) Do non-wovens offer a cheaper option? *Filtr. Separat.*, June, 28–30.

Robinson, A. (2005) Landfill leachate treatment. *MBR5*, Cranfield University, January 2005.

Seo, G.T., Moon, B.H., Lee, T.S., Lim, T.J. and Kim, I.S. (2002) Non-woven fabric filter separation activated sludge reactor for domestic wastewater reclamation. *Water Sci. Technol.*, **47**, 133–138.

ZEK (2004) Project references: technologies, products, companies; North Rhine-Westphalian water management industry, Report by Zentrum für Entsorgungstechnik und Kreislaufwirtschaft (Centre for Waste Disposal Technologies and Recycling), VVA, Dusseldorf.

Chapter 5

Case Studies

With acknowledgements to:

Sections 5.2.1.1 and 5.2.1.2	Silas Warren Steve Churchouse	Wessex Water, UK –
Section 5.2.1.4	Dennis Livingstone Berinda Ross Running Springs plant data and photographs	Enviroquip, Inc., USA Water Environment Federation, USA Adapted with permission from "Membrane Systems for Wastewater Treatment", Copyright © 2006 Water Environment Federation: Alexandria, Virginia
Section 5.2.2.1	Paul Zuber	Brightwater Engineering, UK
Section 5.2.2.2	Paul Zuber Halfway plant photograph	Brightwater Engineering, UK Reproduced by permission of Cork County Council
Section 5.2.3	Paddy McGuinness	Colloide Engineering Systems, Northern Ireland
Section 5.2.4	Torsten Hackner	Hans Huber AG, Germany
Section 5.2.5	Shanshan Chou, Wang-Kuan Chang	Industrial Technology Research Institute, Taiwan
Section 5.2.6	Marc Fayaerts Derek Rodman Jan Willem Mulder	Keppel Seghers, Belgium Naston, UK Water Authority Hollandse Delta Photographs reproduced by permission of Water Authority Hollandse Delta
Section 5.2.6.1	Marc Fayaerts	Keppel Seghers, Belgium
Section 5.3.1	Nicholas David, – Jean Christophe Schrotter	Anjou Recherche, Generale des Eaux, France

Section 5.3.1.2	Luca Belli, Sandro Monti TullioMontagnoli	Zenon Environmental Inc, Italy ASM, Italy
Section 5.3.1.3	Christoph Brepols	Erftverband, Germany
Section 5.3.1.4	Enrico Vonghia, Jeff Peters	Zenon Environmental Inc, Canada
Section 5.3.1.5	Shanshan Chou, Wang-Kuan Chang Chen-Hung Ni	Industrial Technology Research Institute, Taiwan Green Environmental Technology Co, Ltd, Taiwan
Section 5.3.2.2	John Minnery	GE Water and Process Technologies, USA
Sections 5.3.3.1 and 5.3.3.2	Nathan Haralson, Ed Jordan, Scott Pallwitz	Siemens Water Technologies – Memcor Products, USA
Sections 5.3.4.1 and 5.3.4.2	Klaus Vossenkaul	Koch Membrane Systems GmbH, Germany
Section 5.3.5	Atsuo Kubota	Microza Division, Asahi Kasei Chemicals Corporation, Japan
Section 5.4.1	Ronald van't Oever	Norit X-Flow BV, Netherlands
Sections 5.4.2.1, to 5.4.2.3	Steve Goodwin	Aquabio Ltd, UK
Sections 5.4.3 to 5.4.3.4	Gunter Gehlert, Tony Robinson	Wehrle Werk AG, Germany
Section 5.4.4	Ingrid Werdler	Triqua BV, The Netherlands
Sections 5.4.5.1 and 5.4.5.2	Steve Wilkes	Millenniumpore, UK
Sections 5.4.6.1 and 5.4.7	Marine Bence, Sylvie Fraval	Novasep Process, Orelis, France

5.1 Introduction

Membrane Bioreactor (MBR) plant operation is largely characterised by hydraulic and purification performance. Purification is normally with respect to biochemical oxygen demand (BOD) and/or chemical oxygen demand (COD), total suspended solids (TSS), ammonia (NH_4^+-N), total nitrogen (TN) and total phosphorous (TP), and micro-organisms, though the discharge consents may not necessarily specify all of these. Hydraulic characteristics centre mainly on the flux, physical and chemical cleaning cycle times, downtime associated with cleaning, conversion and, in the case of immersed systems, aeration demand. Cleaning cycle times are normally dictated by the requirement to sustain a reasonable mean permeability for the system, and the absolute permeability value appropriate to an MBR treatment process is dependent on the technology, and more specifically the membrane configuration (Chapter 4). Similarly, the membrane aeration demand also varies between technologies, as well as with feedwater characteristics.

In the following sections, around 24 case studies are detailed, based on information provided from 14 different membrane process suppliers, contractors and end users. Contributor are listed in the Acknowledgements section at the front of this book. As with Chapter 4, the subject matter is divided according to membrane configuration and specific process technology. Summary data are provided in the Design chapter (Section 3.3.1).

5.2 Immersed flat sheet technologies

5.2.1 Kubota

5.2.1.1 Porlock

Background Porlock is the site of the first full-scale MBR plant to be installed in the UK. Interest in MBRs in the UK arose almost directly as a result of EU legislation on the quality of treated sewage effluent being discharged to recreational waters and specifically the Bathing Water Directive. This directive, originally promulgated in 1976 and revised in 2002, stipulated that such waters should meet stringent microbiological guide values of 500/100 mL total coliforms and 100/100 mL faecal coliforms and faecal streptococci. The first membrane plant to be installed for sewage treatment was actually the groundbreaking abiotic plant at Aberporth, a Welsh Water site, in 1994. This plant employs enhanced upward flow clarification with lamella plates (the "Densadeg" process), followed by polishing with Memcor hollow fibre (HF) microfiltration (MF) membranes, with upstream and intermediate screening by a 0.5 mm perforated plate (Fig. 5.1). This process has since been superseded by the MBR process whose widespread installation in the southwest of England followed successful pilot trials of the Kubota MBR process in Kingston Seymour in the mid-1990s. The trials subsequently led to the installation of plants at Porlock in February 1998, and the larger plant at Swanage in 2000.

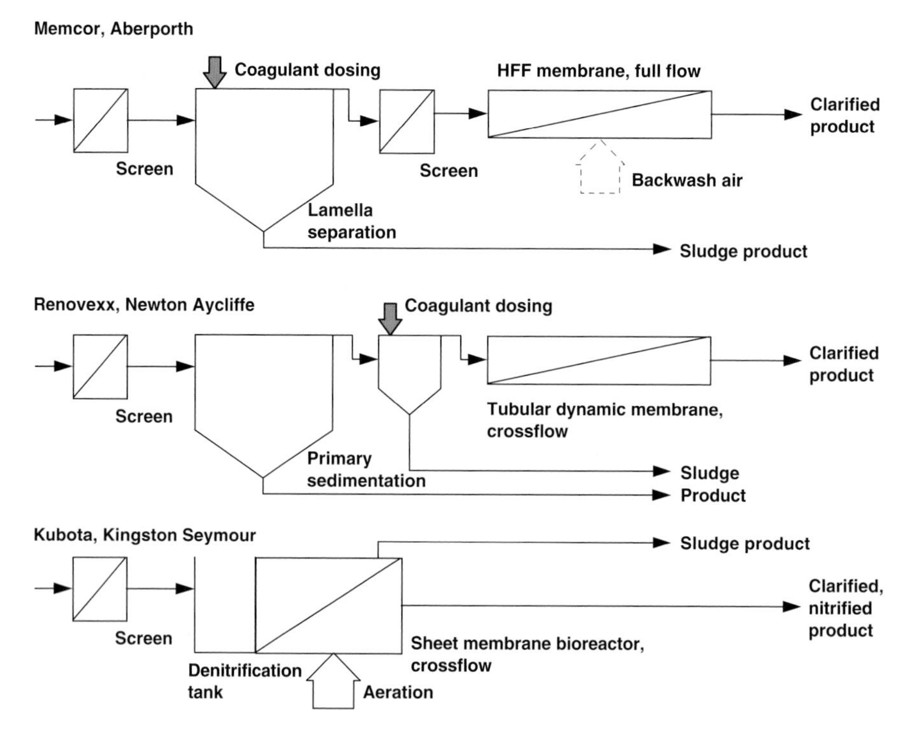

Figure 5.1 Schematics of the plants at Aberporth, Newton Aycliffe and Kingston Seymour

The trials conducted Kingston Seymour, a Wessex Water site, arose through a collaboration between Welsh Water, South West Water and Wessex Water. It was largely through these trials that Wessex Water was able to gain considerable know-how in this process, eventually leading to the formation of the MBR Technology group which was then subsequently spun out through a management buy-out. The pilot plant was operated for 2 years prior to the installation of the full-scale plant at Porlock and continued for several years thereafter.

At the time at which the pilot trials were taking place, a second pilot trial based at Newton Aycliffe in Northumberland was being conducted on a membrane system developed by Renovexx. This latter system was based on a dynamic membrane of aluminium flocculant material formed on an inert woven-cloth porous substrate from washout from a primary tank dosed at 30 ppm as $Al_2(SO_4)_3$. There was thus downtime associated with the formation and removal of the dynamic membrane, though the actual cost of the woven-cloth substrate was relatively low. However, because this was a crossflow-filtration process, the membrane path length was very long – about 22.5 m – in order to obtain a reasonable conversion of 50%. The substrate module (a "curtain", of which there were 22 in total) comprised a 30 × 25 mm diameter multitube (MT) of a single piece of material, around 1 m high. Whilst low in cost with respect to the membrane itself, the engineering needed to successfully operate the plant was complicated and made the technology over-expensive for most practical purposes.

Table 5.1 Comparison of performance of the plants at Aberporth, Newton Aycliffe and Kingston Seymour

Parameter	Kubota	Renovexx	Memcor
Location	Kingston Seymour	Newton Aycliffe	Aberporth
Function	TSS, COD/BOD removal, and nitrification/ denitrification	TSS/BOD, precipitated and coagulated solids removal	TSS/BOD removal and disinfection
Membrane material	Hydrophilicised PE	Dynamic Al floc on substrate	PP
Membrane configuration	Immersed FS	Crossflow MF MT	Full flow HF
Membrane pore size, μm:			
nominal	0.4	10 (for substrate)	0.2–0.3
operating	<0.1	nominal <1 μm	0.2–0.3
Feedwater specification	Screened raw Sewage	Screened, settled sewage	Coagulated, settled sewage
TMP (bar)	0.1	2 inlet, 1 bar outlet	0.2
Flux (LMH)	22	140 capacity	87 capacity
Production rate (MLD)	0.13	4 capacity	3.75 capacity
Membrane module	0.8 m^2 panel	25 mm dia., 22.5 m $\times$ 30 multi-tube, 53 m^2	0.55 mm diameter 10 m^2
Total number of modules	300	22	180
Opernational/ regenerational cycle (min)	Continuous	180/25[a]	45/2
Cleaning			
Mechanical	Water jet, 1/year	Brush/water jet	Compressed air
Chemical	2/year		every 3–4 days
Total membrane area (m^2) (duty/standby)[a]	240	1166	2700
Membrane life, years	~7	1.5–2 (w. brush cleaning)	~5
Membrane cost £/m^2	80[b]	30–40	35
Capital cost, £m/MLD plant	0.6–0.8/1	0.5/4	na/3.75

[a]2:1 duty:standby, hence 1800 m^2 duty.
[b]Projected cost: the cost of the Kubota membrane at the start of the 1995 Kingston Seymour trial was >£150/m^2 including housing. The cost in 2005 for a large installation would be likely to be <£40/m^2.
PP: Polypropylene; PE: Polyethylene; na: not available.
Adapted from Judd (1997).

Data provided at the time (Judd, 1997) for the three plants, the Kubota and Renovexx pilot plants and the full-scale Memcor plant at Aberporth, are given in Table 5.1 and the schematics in Fig. 5.1. It was evident from the trials that, whilst operating at a much lower flux, the MBR technology offered a number of advantages over the other two membrane technologies:

- It achieved denitrification.
- It achieved significant dissolved organic carbon removal.
- It operated on screened sewage, with no further pretreatment demanded.
- It operated at low pressure.
- It demanded minimal mechanical cleaning.

Whilst the projected capital costs of the MBR plant were higher because of the lower flux, this was more than offset by the additional capability offered by the MBR. It is noteworthy that whilst Aberporth remains the only UK site where microfiltration of sewage is employed abiotically, there were at least 44 MBRs for sewage treatment operating in the UK by June 2005, 75–80% of these being Kubota and all but two based on the single-deck membrane (Section 4.2.1).

Plant design and operation The stipulations regarding the Porlock plant were that it should be a small-footprint plant, and able to disinfect substantially the sewage whilst imposing little visible impact. The plant therefore had to be housed in a stone-faced building identical in appearance to an adjacent farmhouse (Fig. 5.2). It comprises four aeration tanks (Fig. 5.3), 89 m^3 in liquid volume size and having dimension 3.3 m wide × 7.4 m long × 4.5 m deep. The actual liquid depth is ~3.8 m, the dimension allowing for future expansion to seven modules per tank. The tank volume allows for the volume displaced by the membrane units, about

Figure 5.2 *A view of the MBR plant at Porlock*

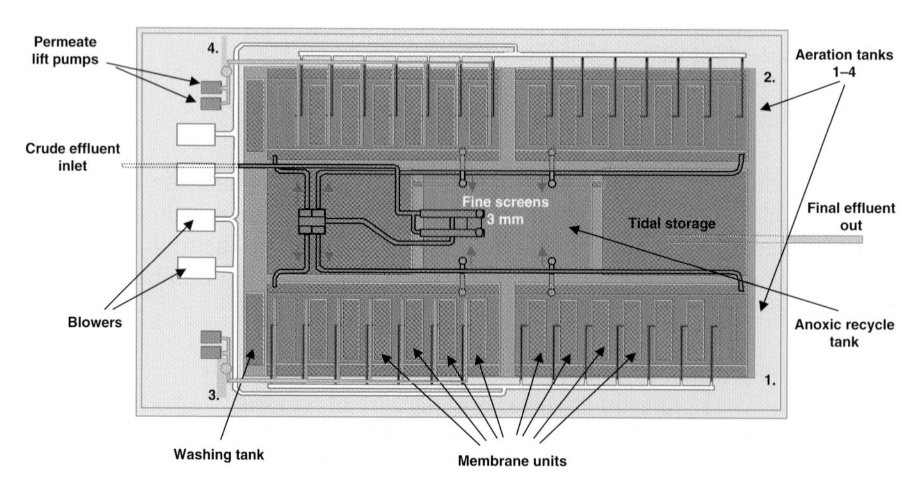

Figure 5.3 *A schematic of the MBR plant at Porlock*

$0.6\,m^3$ per unit. Each tank contains six membrane units with 150 panels per unit, giving a total membrane panel area of $2880\,m^2$ provided by 3600 membrane panels at $0.8\,m^2$ membrane area per panel. Pretreatment comprises a 3 mm perforated screen, this having replaced the original 2 mm wedgewire screen during the first year of operation.

At a peak flow of 1.9 megalitres per day (MLD), the plant operates at a flux of 27 litres per m^2 per hour (LMH) and generally operates at a transmembrane pressure (TMP) between 0.02 and 0.11 bar under gravity-feed conditions (governed by the tank depth and pipework losses) and can sustain a flux of ~20 LMH. This means that the permeability is normally between 200 and 500 LMH per bar, but can rise higher when the fouling and clogging propensity is low. The membrane module is aerated at a rate of $0.75\,Nm^3/h$ per m^2 membrane area (SAD_m) the standard coarse bubble aeration rate for the Kubota system, which means that each m^3 of permeate product demands around $32\,Nm^3$ air ($SAD_p = 32$). The membrane is relaxed for 30–60 min daily as a result of low overnight flows, and cleaning in place with 0.5 wt% hypochlorite is undertaken every 8–9 months. Diffusers are flushed manually without air scour. The plant operates at solids retention times (SRTs) of 30–60 days, producing 0.38–0.5 kg sludge per kg of BOD, $3–6.5\,m^3$/day of 2% dry solids (DS) sludge. The recommended mixed liquor suspended solids (MLSS) concentration range is 12–18 g/L, but it has been operated as high as 30 g/L and is regularly over 20 g/L.

The plant is of a simple and robust, possibly conservative, design and construction and has generally proved very reliable. Membrane replacement due to actual membrane failure for this plant during the first 8 years of operation has been minimal. Indeed, a consideration of all UK and US membrane plants based on this technology reveals membrane replacement per annum to be very low (Table 5.2) – generally ~1% and always <3%. Permeate water quality has always been well within consent, and the plant produces very low odour levels ($<2\,mg\,m^{-3}\,H_2S$). Porlock was originally costed on the basis of a membrane life of 7 years, a milestone that has already been passed. Periodic fouling events have been encountered at this plant, and some of these have been linked with seawater intrusion. Shock loads of salinity

Table 5.2 Total number of membrane failures by age of original panel

Within	No installed panels of age in year	Panel failures	% p.a.	Cumulative (%)
Year 1 (i.e. 0–1)	141 061	192	0.14	0.14
Year 2 (i.e. 1–2)	115 985	598	0.52	0.65
Year 3 (i.e. 2–3)	89 714	690	0.77	1.42
Year 4 (i.e. 3–4)	74 606	1 388	1.86	3.28
Year 5 (i.e. 4–5)	36 247	33	0.09	3.37
Year 6 (i.e. 5–6)	32 772	95	0.29	3.66
Year 7 (i.e. 6–7)	14 938	20	0.13	3.80
Year 8 (i.e. 7–8)	4 200	0*	0	3.80
Year 9/10 (i.e. 8–10)	386			

Note: Data as at June 2005, 45 operational plants (including *all* 40 UK and Ireland, both municipal and industrial). Data excludes ~300 panels repaired by client following accidental damage.
*No failures reported in this year as at June 2005.
Kennedy and Churchouse (2005).

are known to cause increases in soluble COD levels in the mixed liquor, and this has been shown to increase fouling propensity (Section 2.3.7.3) (Reid, 2006).

5.2.1.2 Other Wessex Water regional Kubota plant

Subsequent Wessex Water MBR plant at Swanage (12.7 MLD) and the small plant at South Wraxall (0.26 MLD) were designed based on slightly higher peak fluxes than that of Porlock: 33 LMH rather than 27 and operates at a lower MLSS (8–12 g/L). Other plants in the region, specifically Westbury (4.8 MLD) and Longbridge (1.6 MLD), have similar design fluxes to that of Porlock. Westbury has a significant industrial effluent input, and chemical phosphorous removal has been successfully retrofitted at this plant.

The plant at Swanage is of some significance, it being the largest MBR installation in the UK at the time of installation, though Daldowie (Section 5.1.3) is significantly larger in membrane area (25 600 panels) and larger in average flow but similar on design full flow at 12.9 MLD. The Swanage plant is also one of the least visible large-scale sewage treatment works. The plant has been completely landscaped into the Dorset coastline (Fig. 5.4), a considerable feat of civil engineering incurring a correspondingly considerable cost. The plant has six aeration tanks of $3.3 \times 22.5 \times 5$ m dimension with average liquid depth 3.5 m, giving a volume of 250 m^3. It contains a total of 132 units (22 per tank) with 150 panels per unit providing a total membrane

(a)

(b)

Figure 5.4 View of Swanage sewage treatment works: (a) from the sea and (b) from the air

area of 15 840 m^2. This plant operates under similar aeration conditions as that of Porlock and also operates with manual diffuser flushing with no diffuser air-scour.

The most recent Kubota MBR plants installed for sewage treatment in the UK have an automated diffuser maintenance programme, whereby they are periodically flushed with water and air-scoured. It is generally recognised by the operator that these plants require careful maintenance to suppress clogging (or sludging), since the filling of the channels with sludge represents a very significant constraint to the viable operation of these plant. To this end, more conservative peak fluxes of 27 LMH appear to be appropriate, though mean operating fluxes are much lower, coupled with rigorously cleaned aerators to maintain the aeration rate, and thus air-scour, in the membrane flow channels.

5.2.1.3 Daldowie

Sludge liquor and treatment The plant at Daldowie is a sludge liquor treatment plant owned and operated by Scottish Water. Sludge liquor is the aqueous fraction of sewage sludge which has undergone dewatering by processes such as belt pressing, rotary drum vacuum filtration and centrifugation following conditioning with coagulant and/or polymeric flocculant reagents. It may contain up to 25% of the TN load in the original sludge and contribute as little as 2% of the total influent flow. It is thus highly concentrated in ammonia (Table 5.3), as well as in dissolved organic matter. Composition and flow are extremely variable and dependent on the upstream sludge handling and treatment processes. Whilst the most economical option is to return the liquor to the head of works, this is not always possible either due to logistical limitations or due to the excessive load it would place on the existing sewage treatment process. Under such circumstances, a number of options for its treatment can be considered (Table 5.4). Of key importance in choosing the most suitable option are confidence in a robust solution, cost and environmental factors such as the impact of the wastes generated. Speed of start-up may also be of importance if the plant will only run seasonally.

Plant design and operation As is often the case for this duty, the choice of technology at Daldowie was seen as being between a sequencing batch reactor (SBR) and an MBR. However, the large fluctuations in organic loading arising from periodical releases of poor-quality effluent from the centrifuges and the belt presses meant that operation of an SBR was seen as being more problematic than the MBR. This was

Table 5.3 Sludge liquor properties

Parameter	Average concentration (mg/L)	Range (mg/L)
BOD	1221	162–3004
COD	941	470–1411
NH_4-N	750	162–1218
SS	1442	50–5000
Alkalinity	2200	1900–2500
pH	7.2	7–7.5

Table 5.4 Summary of treatment options (modified from Jeavons *et al.*, 1998)

Method	Advantages	Disadvantages
Magnesium Ammonium phosphate (struvite) precipitation	Low capital costs TN removal Instantaneous start-up	High operating costs Difficult to control Centrafugation required High sludge production No large-scale experience
Ammonia stripping	TN removal Limited experience of similar technology Instantaneous start-up	Disposal of high ammonia liquid waste No large-scale experience in UK water industry Nitrates returned to ASP
Biological	Tried and tested technology Simple to control No problem wastes Surplus sludge boosts nitrifier population in ASP	Start-up not instantaneous

(a)

(b)

Figure 5.5 The Daldowie sludge liquor treatment plant: (a) under construction and (b) operational. The photograph shows the plant with the membrane units and manifolding to the right

principally due to the buffering capacity offered by the MBR, which can operate at an MLSS of 12–18 g/L rather than the 5 g/L of a conventional activated sludge process (ASP) or SBR. This means that the MBR is relatively tolerant to significant increases in organic and hydraulic loading, provided these do not exceed periods of 2–3 days.

The sludge treatment facility incorporates 6 dryers and 12 centrifuges and was built to process most of the sludge produced by the city of Glasgow (>1.5 million people). As a result, the combined effluent liquors produced from this processing require a separate treatment facility as the liquors produced are very high in ammoniacal nitrogen, BOD and COD. The MBR plant was built to treat the liquors to a high enough standard to allow direct separate discharge to the river Clyde (Fig. 5.5), the discharge consent being 20:30:8 (BOD:SS:NH_3). It has been operating since December 2001, and the most recent report is from 2003 (Churchouse and Brindle, 2003).

The main effluent streams are centrifuge centrate and dryer condensate, although other site liquors are also treated. The feedwater typically has a mean composition of

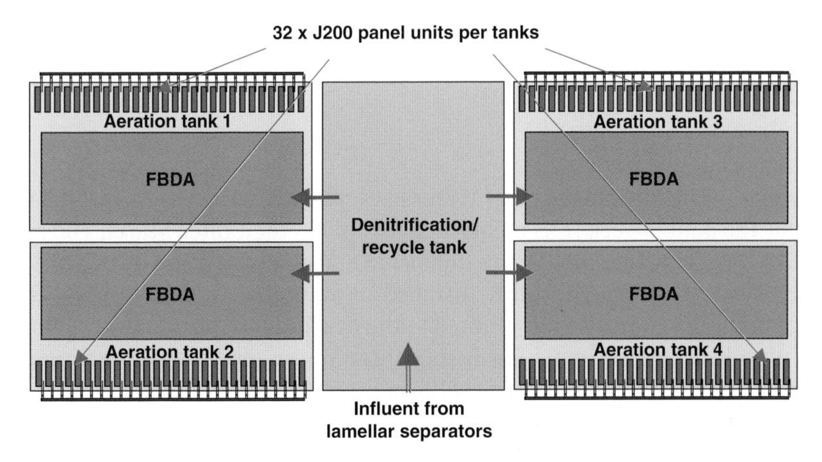

Figure 5.6 Schematic layout of the Daldowie sludge liquor treatment plant

280 mgN/L as NH_3 and 1500 mg/L as BOD, and mean COD levels are in the region of 2600 mg/L; corresponding permeate levels are around 100–400 mg/L COD (hence >90% removal) with ammonia levels of 1–2 mg/L (>95% denitrification). The maximum design flow is 12.8 MLD (yielding a minimum hydraulic retention time (HRT) of 15 h), obtained with six dryers running without a standby, with an average of ~9 MLD being sustainable according to information from site operations. Flows pass through lamellar separators and a 3 mm screen into a central denitrification/ recycle tank. From there, flows are distributed into four combined membrane and aeration tanks, each of 2000 m^3 volume (Fig. 5.6), and a sludge recycle stream returned from the tanks to the denitrification section. In each of the aeration tanks there are 1378 diffusers and 32 × 200-panel membrane units, giving a combined 25 600 panels in the plant (20 480 m^2). The flows thus equate to a mean design flux of 18 LMH and a maximum of 26 LMH.

A great deal of time, effort and money has been spent on identifying the most appropriate polymer and its dose for chemical conditioning to be effective in the belt press and centrifugation dewatering operations without detriment to membrane permeability in the downstream MBR. Despite the problems associated with optimising the polymer dosing, the plant has maintained reasonable performance in terms of COD and ammonia removal. Moreover, the membranes have proven very robust; no more than 40 of the 25 600 panels were replaced in the first 3 years of operation, despite there being no *ex situ* chemical clean instigated over that period for permeability recovery. Routine cleaning in place (CIP) with hypochlorite takes place every 6 weeks or so.

One unusual operational problem that was encountered during an early period of operation was an abrupt increase in differential pressures accompanied by sudden foaming. The foam on the top of the tanks was green-grey in colour, whereas the sludge remained brown. Samples of the foam revealed it to consist almost entirely of non-settling discrete chlorella algae of about 15 μm diameter. The algae were neutrally buoyant and non-flocculating and appeared to pass through the centrifuge as a result. The incident appeared to be due to the sludge plant processing waterworks sludge from a site where an algal bloom had been experienced. Whilst differential pressures had initially doubled, the foam gradually subsided over a number of days and differential pressure gradually recovered without intervention over a 1–2 week period.

5.2.1.4 Running Springs

Background The Running Springs Water Recycling Plant represents the first installation of the double-deck (EK) membrane module in the United States. The plant is located at Big Bear National Park in California and is operated and owned by the CA County Water District. The MBR installed at Running Springs resulted from a change in the discharge permit granted to the existing ASP plant, designed for BOD removal only, in which limits for nitrogen and phosphorous were tightened. The operators originally attempted to meet the new consents by extending the sludge age to increase the MLSS concentration in the aeration tanks. However, since the existing rectangular secondary clarifiers could not handle the increased loading, other options had to be explored. Between April and September 2003, the existing plant was converted to an MBR system by retrofitting with Kubota membranes (Fig. 5.7) coupled with the SymBio® process, a proprietary technology designed to achieve simultaneous nitrification/denitrification (SNdN) and partial enhanced biological phosphorus removal in the same zone.

Process design The plant has a design flow of 2.3 MLD with a peak flow of 4.5 MLD. It is an example of a treatment strategy referred to as the UNR™ (Ultimate Nutrient Removal) by Enviroquip, the suppliers, essentially referring to the combined SNdN process and the MBR. It is thought that such MBR systems reject more coagulated phosphorous than conventional clarification technologies and are better suited to handle seasonal and diurnal load variations through MLSS control.

Figure 5.7 Retrofitting of membranes to the aeration tanks at Running Springs

In this UNR™ configuration, the MBRs are placed downstream of sequential anoxic (AX) and anoxic/anaerobic (AX/AN) basins, converted from existing clarifiers, which promote denitrification and enhanced biological phosphate removal (EBPR) as well as acting as equalisation chambers. The anoxic treatment is preceded by screening, originally a 3 mm-step screen which was subsequently replaced with a 3.2 mm centre flow band screen, followed by degritting in an aerated grit chamber. The aerobic part of the UNR™ process is operated at a low dissolved oxygen (DO) concentration using the SymBio® process to promote SNdN. The SymBio® zone is fitted with Sanitaire fine bubble aeration equipment and is fed air at a controlled rate based on the measured DO and biological potential activity (BPA) – a parameter based on the level of the energy transfer co-enzyme nictotinamide adenine dinucleotide hydrogenase(NADH). Submersible pumps continuously recycle mixed liquor to the aerated reactors, which are partitioned into MBRs and SymBio® zones, at roughly four times design flow, that is 9.1 MLD. The retained sludge is returned to the anoxic basin as thickened sludge to complete an internal recycle loop for denitrification. To control sludge age, mixed liquor is periodically wasted from the system and pressed before disposal and the filtrate is returned to the head of the plant for treatment. The process schematic is shown in Fig. 5.8.

Process operation and performance The MBR plant controls include relaxation. The supervisory control and data acquisitions (SCADA) system allows the operator to adjust the relax frequency and duration. Typically, the system is set to relax for 1–2 min out of each 10–20 min filtration cycle. Membrane performance is optimised by automatically matching permeate flow to hydraulic demand while minimising changes in flux. Hydraulic loading rate is estimated from the liquid level in the AN/AX basin, which then determines the mode of operation of the MBR, which can be anything from zero to peak flow. Using the real-time NADH and DO signals, the SymBio® process controller automatically adjusts the speed of a dedicated blower to keep the MLSS (predominantly between 10 and 22 g/L) DO concentration between

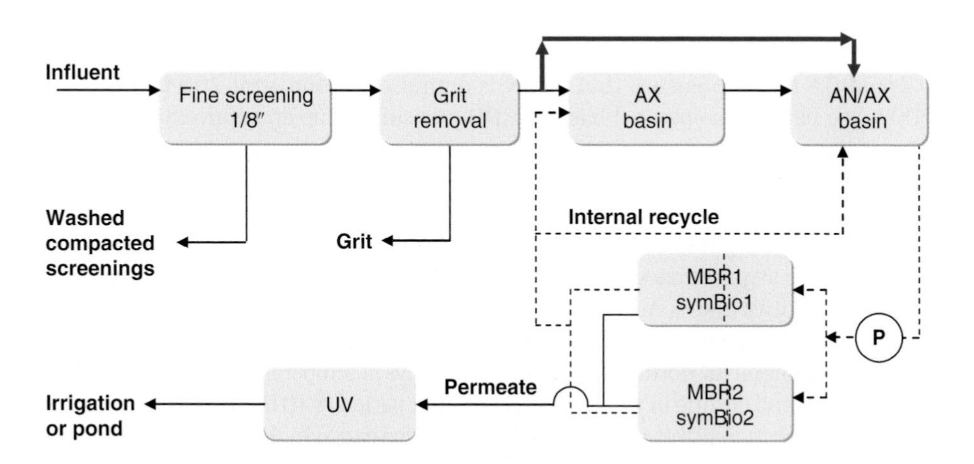

Figure 5.8 Running Springs plant, schematic

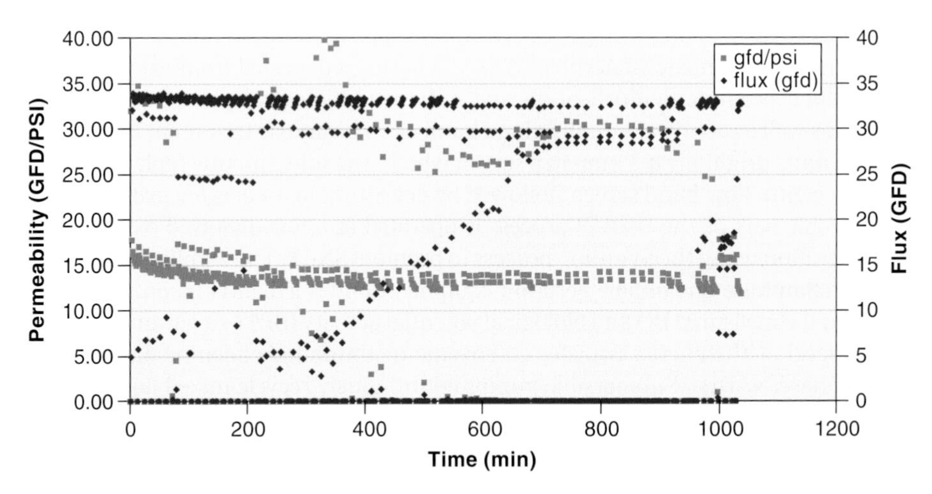

Figure 5.9 Flux and permeability with 400 ppm MPE50™ at 13.3°C

Table 5.5 Water quality data and permitted limits for discharge, Running Springs

Parameter	Post 2004 limits	Ave. influent	Ave. effluent*
$cBOD_5$ (mg/L)	<5	223 mg/L	<2.0 mg/L
TSS (mg/L)	<5 mg/L	165 mg/L	<1.0 mg/L
TKN (mg/L)	Not permitted	41 mg/L	<1.0 mg/L
TN (mg/L)	Not permitted	Not available	<4.0 mg/L
N-NH_4^+ (mg/L)	<5 mg/L	Not available	Not available
P-PO_4 (mg/L)	<2 mg/L	15 mg/L	<5.0 mg/L
Turbidity (NTU)	<0.2 NTU	–	Not available

*Effluent values based on samples taken during September 2003.

0.2 and 0.8 mg/L to ensure that SNdN is optimised. From April 2003 to August 2003, the plant was operated based on DO measurements and conventional treatment strategy. Following optimisation based on the SymBio® process during September 2003, effluent TN levels decreased from 14.4 mg/L to 3.9 mg/L and effluent TP levels from 7.4 mg/L to 4.9 mg/L.

The peak loading rate coupled with limited equalisation demands that the membranes operate at gross fluxes >50 LMH for up to several hours a day, and in some cases for 24 consecutive hours. According to operator feedback, during November 2004, a net flux of 30.5 LMH was sustained, some 8 LMH above the membrane manufacturer's recommended limit for normal operation, for 19 days at temperatures of 11.4–12.7°C with only a small decline in permeability over that period (840 to 500 LMH/bar, arising from a TMP increase of 0.025 bar). To handle extended higher peak loads during extreme cold weather conditions (<10°C), a flux-enhancing polymer [PermaCare® MPE50™] is occasionally added to increase the critical flux. A 20 h trial conducted

Table 5.6 Design criteria, Running Springs

Parameters	Units	Current	Future
Maximum month average daily flow	MLD	2.3	3.8
Peak day flow	MLD	4.6	7.6
Minimum day average process temperature	°C	15	15
Maximum day average process temperature	°C	20	20
Annual average flux	LMH	25[a]	25
Peak hour average flux	LMH	50	50
Membrane cassette type	–	EK300	EK200
Number of cartridges per cassette	–	150	100
Surface area per cassette	m^2	120	120
Number of cassettes installed per train	–	8	16
Number of MBRs (membrane tanks)	–	2	2
Coarse bubble aeration rate (20°C), per cartridge	l/min	7.5	7.5
Typical cleaning cycle time	days	180	180
Typical cleaning reagent	–	hypo	hypo
Typical cleaning reagent strength[b]	mg/L	8000	8000
Typical Cleaning (soak) time/protocol	h	2	2

[a]Assuming an operating temperature of 15°C.
[b]An existing sodium hypochlorite system is used that generates a 0.8% solution that is stronger than the 0.5% solution recommended by the system supplier.

in December 2003 demonstrated that a stable net flux of around 54 LMH (32 GFD, Fig. 5.9) was achieved following the addition of 400 mg/L of the polymer. Performance with respect to purification and system hydraulics is given in Tables 5.5 and 5.6, respectively, with the latter including a summary of the process design.

5.2.2 Brightwater Engineering

5.2.2.1 Coill Dubh, Ireland

Background Coill Dubh is a small village about a 1 h drive west of Dublin. The existing works consisted of basic concrete primary/septic tanks followed by peat filter beds. Effluent was directly discharged from these peat beds to a small local river. The works was in a poor condition and, on completion of the MBR scheme, was demolished.

The new works (Fig. 5.10) was provided by a developer as part of his planning consent for a housing estate close to the site. This works was to treat the sewage from this new estate plus the existing village. At the tender and design stage in 2001–2003, little information was available on the influent, other than it was to be mainly domestic in nature, contain storm flows, have a maximum flow to treatment of 50 m^3/h and contain loads of 2000 p.e. (population equivalent). The treated effluent consent was: BOD 10 mg/L, SS 10 mg/L, NH_3 5 mg/L and total phosphorus 1 mg/L. Though this standard could arguably be achieved with a conventional treatment process, the small footprint and relatively shallow tanks afforded by the MBR process led to this technology being selected; the new works was to be built on a peat bog with a very high water

Figure 5.10 The Coill Dubh site. The four MBR tanks are adjacent to the lifting davits

table, making deep excavations difficult and expensive. The simplicity of the MEMBRIGHT® design made this option competitive and attractive.

No piloting was undertaken as this influent was considered "normal" domestic sewerage based on an inland catchment with an observable existing treatment process. The contract was led by the house developer MDY Construction. MDY also supplied the civil element to the contract based on supplied designs. The process, mechanical and electrical design element of the contract was delivered by Bord na Mona Environmental and supplied by Brightwater Engineering. M&E installation was supplied by BnM subcontractors, while process commissioning, verification and pre-handover operational support was supplied by Brightwater. The plant was commissioned in June 2004.

Design The works (Fig. 5.11) consists of an inlet flume, followed by an overflow sump from where flows in excess of 50 m^3/h are diverted to a storm tank with 2 h retention at flows between 50 and 100 m^3/h. Excess flows from the storm tank discharge from it direct to the stream. Storm water is pumped back to the inlet sump once works inlet flows have reduced.

Flows to treatment are dosed with ferric for phosphorous control and pumped to a 6mm free-standing wedgewire rotary screen. Flows drop from the screen into a stilling zone located in a settling tank that also acts as the works sludge and scum storage. From the stilling zone, flows pass under a fat-retaining submerged weir and onto a 3 mm drum screen. Flows are then diverted via a common manifold to four membrane tanks, each containing six modules. Each module provides a total effective surface area of 92 m^2 from 50 panels. Aeration is supplied to the modules only by duty/duty blowers. Treated effluent/permeate gravitates by siphoning from each reactor to a common collection sump via flow measurement on each reactor. Permeate

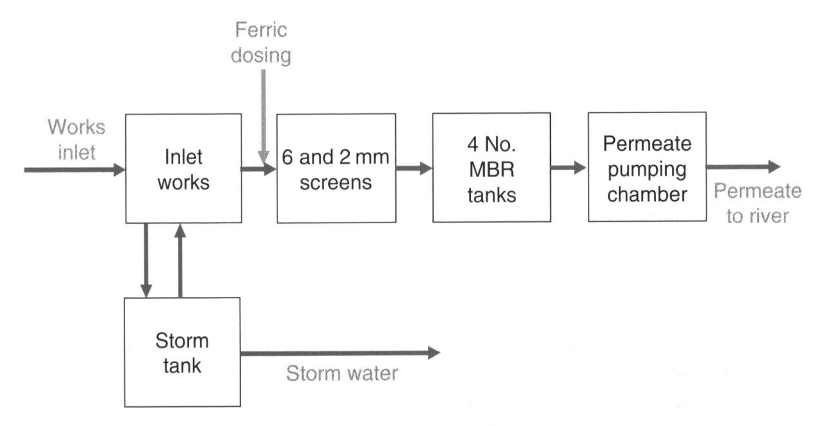

Figure 5.11 Schematic of Coill Dubh wastewater treatment works

is then, on level, pumped to outfall, or can be recycled back to the inlet of the works via a divert valve, conducted to assist in commissioning.

Excess sludge is removed manually from each reactor via a single progressive cavity pump and discharged into the inlet sludge storage tank. Fat, scum and grit are also collected in this tank, and the accumulated waste periodically removed from site. Also supplied on-site are a control/mess building, ferric storage and a standby generator. A remote monitoring system provides Brightwater with site feedback.

Operation and maintenance The MBR operates at a maximum TMP of 300 mbar and a gross flux of 27 LMH (hence >90 LMH/bar permeability) with relaxation for 5 min every hour. Aeration is supplied by an aeration lateral system located at the base of each plate providing coarse-bubble aeration for both membrane scour and process aeration at a rate of up to 118 Nm^3/h per module, or a SAD_m of 1.28 $Nm^3/(m^2\,h)$. No additional fine bubble aeration system is required for the combined sewer loads supplied to Coill Dubh. Although chemical cleaning is suggested at 12-monthly intervals, none has been applied over the first 18 months of operation; the performance of the membranes is maintained with the air scour rate supplied. Cleaning is to be conducted *ex situ* by sluicing, followed by a 4–12 h soak in either hypochlorite or citric acid, depending on the assessed fouling. Flushing of the aeration system can be conducted at the same time. The design sludge age is 21 days, though this often increases to >30days such that the design MLSS concentration of 12 g/L increases to >15 g/L. The HRT is ~11 h at full flow.

The process at Coill Dubh was built as a low-tech approach to membrane technology and has performed better than anticipated despite periods of operational stress. The permeability remains acceptable at flux rates between 22 and 27 LMH and the permeate water contaminant levels have consistently been below most of the parameter limits of detection (i.e. <5 mg/L BOD, <5 mg/L SS, <0.5 mg/L $N\text{-}NH_4^+$ and <0.05 mg/L TP). While the effluent quality exceeds that required for discharge, the simplicity of the system and the robustness of the membranes have made this plant very simple to operate, despite fat and rag levels far in excess of those anticipated. The continued operation without the requirement for chemical cleaning makes the system even more attractive. A preventative inspection/clean is currently being discussed with the client, who is now Cork County Council.

5.2.2.2 Other Brightwater plant

Another of Brightwater's MBR plants (450 p.e. or 0.276 MLD flow design capacity), operated by Bord na Mona on behalf of the local council, is located at Halfway WWTW near Cork in Ireland (Fig. 5.12). This plant is fitted with a 3 mm screw wedgewire screen and has an anoxic reactor (~45 min HRT) for TN removal (the consented concentration being 5 mg/L TN) and ferric dosing for phosphorous removal. The plant is fitted with six membrane modules in a single aeration reactor, providing a total area of 552 m^2. All process aeration is supplied by the scour air system described for Coill Dubh. Mixed liquors are recycled from the aeration reactor to the baffled and mixed anoxic reactor to enable nitrate removal. Screened and ferric-dosed influent is also fed into the anoxic reactor at this point.

5.2.3 Colloide Engineering Systems

5.2.3.1 Fisherman cottages, Lough Erne, Northern Ireland

At this site a small wastewater treatment plant was required for a group of cottages on the side of Lough Erne, Co. Fermanagh in Northern Ireland. The wastewater had

Figure 5.12 The plant at Halfway. The screen is located mid/top picture, with the inlet pump chamber behind it. The anoxic tank is below the screen, and MBR is the covered tank with the swan-neck air feed lines. The sludge tank is located behind the MBR and to the right of the anoxic tank. The permeate pump chamber is behind the sludge tank

to be treated to a 10/10 (BOD/TSS) standard or better. The flow rate to be treated was a maximum of 0.012 MLD. Given the very low flow and relatively tight discharge consent, an MBR plant was offered as the most attractive process option based on the comparatively low capital and whole life costs of the MBR over that of conventional activated sludge treatment with tertiary treatment polishing. Colloide Engineering Systems was commissioned to provide the plant by the developer on a design and build contract. The plant was installed and commissioned in August 2004.

The reactor comprises a pre-existing elliptically shaped concrete tank divided into two sections, each of $2\,m^3$ volume (Fig. 5.13a). The first compartment is used as an inlet section for settlement of gross solids; settlement comprises the only pretreatment used at this site. The wastewater is allowed to flow to the second chamber via an interconnecting pipe fitted near the top of the tank. The tank is installed below ground, with its top at ground level. The membrane assembly comprises two $10\,m^2$ Sub Snake modules, providing a total membrane area of $20\,m^2$. The modules are mounted in a PVC frame which also holds a tubular fine bubble air diffuser, positioned directly below each membrane module. This complete membrane assembly is immersed in the second chamber of the tank. A float switch is also mounted in this chamber to control the water levels. A control kiosk is positioned above ground adjacent to the wastewater treatment tank with a duct running between the tank and the kiosk. The kiosk houses a 0.2 kW air blower delivering air to the diffusers, the permeate suction pump discharging treated water and all process control equipment.

The system is operated at a TMP of 0.4 bar (4 m head), which provides a flux of 20–30 LMH. A filtration cycle of 6 min on/2 min relaxation is employed, with a cleaning cycle time estimated as being 6–12 months, using 200 ppm sodium hypochlorite. No coarse bubble aeration is used, fine bubble aeration being provided at a rate of $10\,m^3/h$ to the total membrane assembly, yielding a SAD_p of $20\,m^3$ air per m^3.

The plant has generally operated very well. The hydraulics of the system have never been an issue with this plant, and the membranes remove practically all

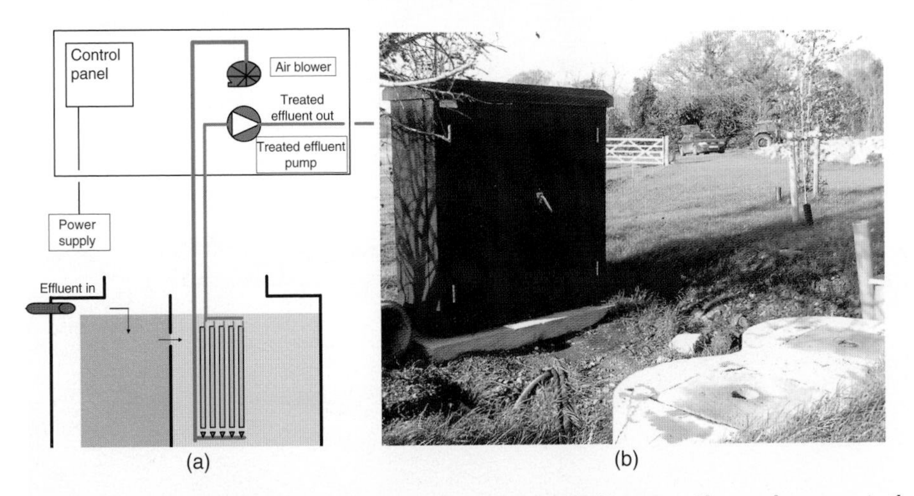

Figure 5.13 The Fisherman Cottages MBR: (a) schematic and (b) in situ. *The membranes are in the second of the two underground concrete tanks in the picture, only the kiosk housing the control equipment is mounted above ground*

suspended solids. The reactor biomass took some time to develop and acclimatise due to a combination of the plant not being properly seeded and disinfectant chemicals being discharged into the plant. However, the plant has been operating to the design specification over a year.

5.2.4 Huber Technology

5.2.4.1 Knautnaundorf

The water and wastewater company in Leipzig (KWL) is responsible for the operation of 30 sewage treatment plants in and around Leipzig. Many of these plants were found to be no longer able to meet the tighter effluent quality standards imposed by the government and so required upgrading. Prior to upgrading, the sewage treatment plant at Knautnaundorf was equipped with a balancing tank, two primary tanks and four small activated sludge plants for biological clarification of the wastewater. Oxygen supply to the plant had been controlled using simple timers, with no feedback control based on DO levels in the mixed liquor. The quality of the treated water, which is discharged into the Weiße Elster river, was therefore poor. After a tendering process, the vaccum rotating time (VRM)® system was selected to provide water meeting the new effluent standards, which specify microbiological parameters.

The new treatment plant (Fig. 5.14), was designed for a maximum flow rate of 18 m^3/h and a daily flow of 0.113 MLD. It contains a stack of trapezoidal plate

Figure 5.14 Municipal WWTP at Knautnaundorf

membranes, assembled so as to form a complete hexagonal plate (Fig. 5.15), with the module providing a total membrane area of 756 m^2. The total investment costs amounted to approximately €400 000. The plant was commissioned in August 2002 following a test period lasting most of July 2002 which successfully demonstrated the system robustness. Fluxes of up to 28 LMH were maintained at a constant permeability of 300 LMH/bar (and hence a TMP below 0.1 bar). The membrane is aerated at a rate of 265 Nm^3/h, corresponding to a SAD_p of 14.4 at the maximum throughput and 22 at the nominal flow rate, which equates to a flux of 16 LMH and contributes to an energy demand for membrane filtration (MF) of 0.54–0.59 kWh/m^3. The total energy (including aeration and pretreatment) demand of the plant is ~1.3 kWh/m^3. The membrane units are operated on a 9 min filtration/1 min relaxation cycle, the efficacy of the physical cleaning being affected by rotation of the stack at 2 rpm (0.033 Hz). The plant operates at an SRT of 25 days and an HRT of 14 h, providing an MLSS of 8–14 g/L.

In addition to the municipality's domestic wastewater, the influent contains a small amount of industrial wastewater from an industrial kitchen. Pollutants and nutrients in the wastewater are reliably reduced by the process. The system achieves a COD reduction of 96%, from a mean feed concentration of 750 mg/L, nitrification down to 1 mg/L $N\text{-}NH^+{}_4$, and around 30% phosphorous removal. No measurable amounts of solids or *Escherichia* coli have been detected in the outlet.

5.2.4.2 Other Huber plant

A number of other Huber plant have subsequently been installed in Europe, including a starch products production facility at Veurne in Belgium (formerly Raisio Chemicals

Figure 5.15 The VRM® 20/252 membrane stack at Knautnaundorf

Table 5.7 Summary of municipal wastewater treatment data, Huber MBRs

Parameter	Knautnaundorf	Schwägalp
Commissioned	March 2002	April 2002
Dry weather flow (based on nominal flux) (MLD)	0.113	0.1
Peak flow (m^3/h)	18.4	6.5
Number of membrane elements	1	1
Total membrane area (m^2)	756	270
Feed COD concentration (mg/L)	750	233
COD removal (%)	96	97
Nominal flux (LMH)	16	13
Peak flux (LMH)	25	25
Permeability (LMH/bar)	<300	<300
SAD_m (m^3 air/m^2h)	0.35	0.35
SAD_p (m^3 air/m^3)normal (peak)	22 (14)	27(14)
Energy demand (kWhr/m^3)	1.3	1.4

and now Ciba Speciality Chemicals) and another small municipal wastewater treatment works at an alpine hotel at Schwägalp railway station in Switzerland, which has been in operation since April 2002. The plant at Schwägalp receives effluent from a dairy as well as an hotel and a guest house. The dry weather and peak flows are 0.1 MLD and 6.5 m^3/h, respectively and, as such, the plant is similar to that at Knautnaundorf (Table 5.7).

5.2.5 The Industrial Technology Research Institute non-woven fabric MBR

The NWF MBR developed by the Industrial Technology Research Institute (ITRI) of Taiwan (Section 4.2.5) has been trialled for food processing effluent at a company manufacturing vinegar from grains and vegetables. The wastewater has a COD of around 2600 mg/L, mainly soluble, and BOD, TSS and TKN concentrations of 1400, 600 and 60 mg/L respectively. Additionally, the effluent has an average fats, oils and grease (FOG) concentration of 34 g/L. The MBR was retrofitted into the existing 75 m^3 square-sided SBR tank, since the latter process was found to be unable to cope with fluctuations in loading and space for expansion of the plant was limited. The membrane module has dimensions of 2.4 × 1.1 × 1.0 m (L × W × D), offering a total filtration area of 270 m^2. The module was found to consistently achieve over 95% COD removal efficiency, with <90 mg/L of COD in permeate when the influent COD was <2935 mg/L, SS was less than 1010 mg/L and volumetric loading maintained at 1.6 kg COD/m^3-day. The operational flux attained was around 4.5 LMH. Two other industrial effluent treatment plants in Taiwan are also based on an NWF MBR.

5.2.6 Toray

Toray membranes have been employed by Keppel Seghers (Singapore–Belgium) for their UNIBRANE® MBR and by Naston (UK) for their package plant (Fig. 5.16). There are about 25 operating plant worldwide, all but three being below 0.5 MLD. The installation at Heenvliet in Belgium (due early 2006) appears to be the first reference

Figure 5.16 A Naston package plant at Inversnaid in Scotland

Figure 5.17 The plant at Heenvliet in the Netherlands

site for large-scale municipal wastewater treatment and the largest Toray plant to date. The capacity of the plant (Fig. 5.17) is 2.4 MLD and will contain 3000 Toray flat sheet (FS) membranes. Naston had installed six UK plants, ranging between 0.13 and 0.56 MLD in capacity, by the end of 2005.

Table 5.8 Effluent characteristics, photographic production plant

Parameter	Unit	Design value
Mean flow (peak)	MLD	0.84 (1.080[St1]) check
Temperature	°C	37
pH	–	7–8
COD concentration (load)	mg/L (kg/day)	1800 (1512)
BOD concentration (load)	mg/L (kg/day)	850 (714)
BOD/COD	–	0.7
TSS concentration (load)	mg/L (kg/day)	48 (40)
Total solids concentration (load)	mg/L (kg/day)	79 (94)
TKN	mg/L	35
NO_3^--N (NO_2^--N)	mg/L	30 (25)
Total P concentration (load)	mg/L (kg/day)	5 (4.2)
BOD/P ratio	–	170
Ag concentration (load)	mg/L (kg/day)	47 (40)

5.2.6.1 Fuji photographic production plant

The plant at Tilburg in the Netherlands is a photographic paper production plant, owned by Fuji, discharging waste high in dissolved organic matter (Table 5.8). This project was initiated in 1998 when a number of drivers combined to make the existing management scheme untenable. Firstly, the province where the factory was based was advocating reduced ground water usage. Trace chemicals discharged in the wastewater, although not blacklisted at the time, were viewed as being onerous. Costs levied by the municipality for wastewater treatment were increasing. Also, the temperature of the water discharged in the summer months was close to the permitted limit of 30°C. There was also a supplementary cost associated with silver recovery from one of the sludge byproducts, this process being outsourced to a specialist company.

A cost benefit analysis revealed that for an estimated capital investment of €2.3 million for the recovery plant, savings in sludge treatment, energy and chemicals (amounting to €600000/a), would provide a payback of <4 years. Additional savings arising from reuse of reverse osmosis (RO) permeate from the recycling process were estimated to reduce the payback time further to 2.5–3 years. It was on this basis that piloting was initiated in 1998 using a 50 L/h sMBR pilot plant followed by a 500 L/h plant in 1999, coupled with an RO pilot plant with a 400 L/h capacity. A 3-month pilot trial based on an iMBR pilot plant, estimated to further improve the cost benefit by ~100 k€/a in energy costs, was successfully conducted in 2002 by the consultants for the project (Keppel Seghers) and led to the installation of the full-scale plant.

Plant design The plant comprises an MBR-RO plant with a 0.75 mm pre-screen. The biological treatment in the MBR consists of an aeration basin with separate denitrification from which the sludge flows under gravity to the aeration basin (Fig. 5.18). Aeration is achieved using disc aerators powered by one or two air compressors which are frequency-controlled to attain a target reactor DO level. A third basin contains the membrane modules, comprising 12 stacks of 135 m^2 area, each fitted

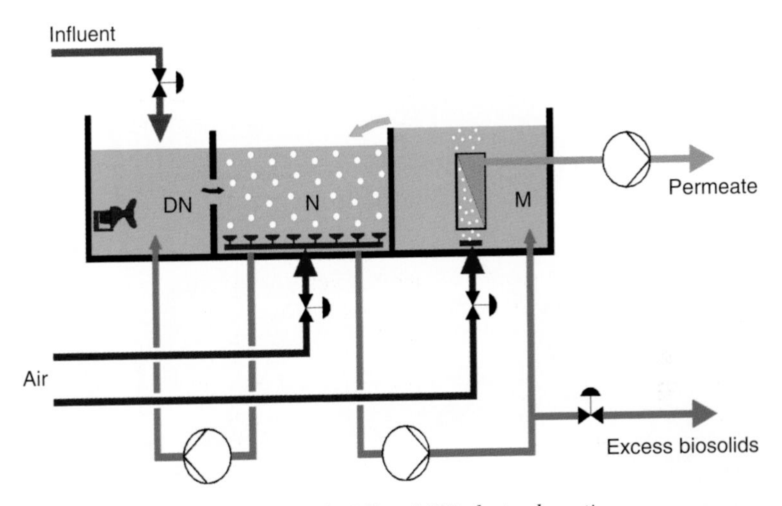

Figure 5.18 The Tilburg MBR plant, schematic

with 100 panels and providing 1620 m^2 membrane area in total, which are continuously aerated. Excess sludge from the membrane basin is pumped to a peristaltic decanter where the sludge and silver are separated from the water, the supernatant being returned to the denitrification basin. Residual sludge at ~18 g/L is fed into the decanter to produce a dewatered product of 20–24% from which the silver is ultimately recovered (300–400 ton annually).

The ultra filtration (UF) unit (Fig. 5.19) is designed for a maximum throughput of 45 m^3/h and a nominal flow of 35 m^3/h (0.840 MLD), equating to a flux of 21.6 LMH (27.8 LMH peak). Membranes are projected to operate at a TMP of 15 mbar and thus a mean permeability of around 1500 LMH/bar at 20°C. At the SRT and HRT of 24 days and $<$14 h respectively, the MLSS concentration is 15 g/L, of which ~12 g/L is estimated to be biomass. Membranes are operated on a cycle of 8 min on/2 min relaxation with chemical cleaning for 4 h with up to 6 g/L NaOCl planned on a yearly cycle. Membranes are aerated at 0.4 Nm^3/h per m^2 membrane, providing a SAD_p of 18.5 Nm^3 air per m^3. This compares with fine bubble aeration rates of 11–22 Nm^3 air per m^3 permeate for the biotreatment.

The permeate from the UF membranes is pumped to the RO system, housed in an adjacent building (Fig. 5.20), via two heat exchangers to recover heat for part of the process water to the boilers. The inlet design temperature of the RO is set between 20°C and 25°C. The RO permeate is delivered to the cooling towers as make-up water. Studies revealed that the water cannot be used for duties in which it comes into contact with the product due to the variable ion content. The three-stage (4/5-2-1, six elements per module) RO unit is fitted with Toray SU-720 modules and normally operates between 11 and 14 bar with a recovery of up to 85%. The first stage has a spare line in case an additional pressure vessel is needed. If maximum flow conditions prevail over a long period, this vessel may be fitted with six extra elements to reduce the hydraulic load. The RO permeate is stored in a buffer tank prior to transporting to the ion-exchange filter feed tank. The brine is discharged to sewer since it contains only inert salts.

Figure 5.19 *Membrane modules at the plant at Tilburg*

Figure 5.20 *Overview of plant at Tilburg. Front right: MBR plant building; front left: building housing RO, dewatering unit, storage of chemicals and permeate and control room*

Table 5.9 Toray membrane-based MBR plant, design parameters

Parameter	Fuji	Heenvliet	Naston
Effluent	Industrial	Municipal	Municipal
Capacity (MLD)	0.84 (1.08 peak)	2.4 (5.6 peak)*	0.13–0.56
Flux (LMH)	21.6 (27.8 peak)	24.3 (56.3 peak)*	17–25
Permeability (LMH/bar)	1500	<1000*	210
SAD_m ($m^3/m^2.h$)	0.4	0.3	0.25–0.82
SAD_p (m^3 air/m^3)	19 (14 peak)	12 (5.3 peak)	~15

*design data from Mulder *et al.* (Mulder and Kramer, 2005).

The plant is guaranteed to meet effluent concentration limits of <51 mg/L COD (>97% removal), undetectable TSS and <5 mg/L TN. Thus far, some minor teething problems have been experienced since commissioning. These have included foaming, linked to wider than expected variation in feedwater quality, and concrete erosion. Further improvements, and specifically automated control, are in progress.

5.2.6.2 Comparison of plant O&M data

Design parameters for the Heenvliet and Fuji, both of which were designed by Keppel Seghers (who are also responsible for the Unibrane MBR systems in Singapore), and the Naston package plants are provided in Table 5.9. All of the six UK examples of Naston package plants are operated at 6–18 (average 12) g/L MLSS. The data collated for all three plant or plant types are characterised by high projected permeabilities and low aeration demands. It remains to be seen whether these are sustainable over extended periods of time.

Thus far, the largest operational Toray plant in Europe, Reynoldston in Wales with a capacity of 0.69 MLD at peak flow, operates at a permeability of around 210 LMH/bar. As with all Naston package plants, this plant has not been chemically cleaned. Instead, the membranes are subjected to a mild backwash, every 6 months or so, which causes sufficient membrane displacement to dislodge accumulated solids. The solids are then subsequently flushed away when the air scour is applied. Clogging which has been noted has been attributed to blocked diffusers, which is overcome by their physical cleaning.

5.3 Immersed HF technologies

5.3.1 Zenon Environmental

5.3.1.1 Perthes en Gatinais sewage treatment works

The plant at Perthes en Gatinais is one of the earliest examples of a Zenon MBR sewage treatment plant and the earliest in Europe. It was commissioned in 1999 and was built and is operated by Veolia Water (formerly Vivendi). The plant treats the effluent from four small communities in the Paris region, and there are currently 24

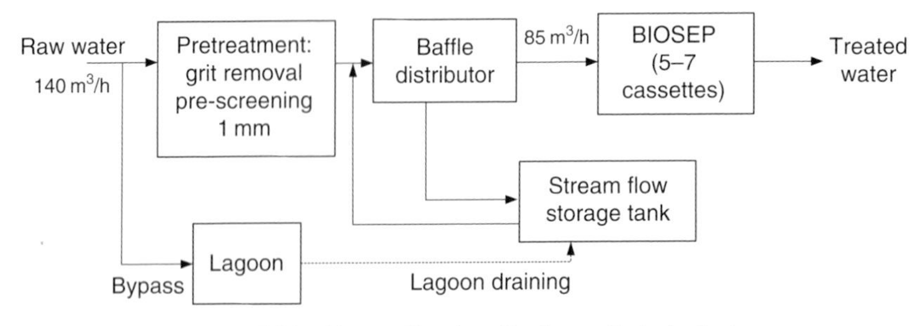

Figure 5.21 Process flowsheet, Perthes en Gatinais plant

Figure 5.22 Dewatered sludge, Perthes en Gatinais plant

industrial references in France for this process, which Veolia commercialised as the Biosep® process in the early 1990s.

The main characteristics of the plant, with reference to discharge consents (Table 5.10) and loading rates (Table 5.11), are given below. The complete process treatment scheme (Fig. 5.21) includes pre-screening to 1 mm and post-dewatering of the sludge (Fig. 5.22). The MBR is based around a cylindrical aeration tank (Fig. 5.23) of around 500 m^3 in volume, with the water depth of between 5 and 6 m: the level variation is due to buffering during rainfall. This equates to a HRT of around 13 h, and the plant operates at a sludge age of 25 days. The MLSS concentration in the aeration tank is ~15 g/L. The membrane plant comprises seven cassettes, each fitted with eight ZW500 modules, providing a total area 2604 m^2 (372 m^2 per cassette). At the hydraulic loading given, the flux through the membrane is around 18 LMH at average flow and 29 LMH at peak flow.

Figure 5.23 Aeration tank and covers, Perthes en Gatinais plant

Table 5.10 Discharge consents, Perthes en Gatinais plant

Parameter	Units	Value
COD	mg/L	40
BOD	mg/L	5
TSS	mg/L	5
TN	mg/L	10
Total P	mg/L	2.5
Total coliforms	Log removal	5

Table 5.11 Characteristics of plant and effluent, Perthes en Gatinais plant

Parameter	Units	Value
Population equivalent	p.e.	4500
Dry weather daily flow rate	m^3/day	900
Rain weather flow	m^3/day	1440
Peak hourly flow	m^3/h	140
Maximum flow rate at inlet Biosep	m^3/h	85
COD load	kg/day	675
TSS load	kg/day	315
TN-load	kg/day	67.5
Total P-load	kg/day	18
Temperature range	°C	10–20

Table 5.12 Membrane characteristics and maintenance, Perthes en Gatinais plant

Parameter	Value
TMP range	0.2–0.5 bar
Backwash frequency and duration	Every 10 min for 45 s
CIP	Once a week with hypochlorite
External chemical cleaning	Hypochlorite/acid soak 1–4 times a year

Coarse bubble aeration is applied at a rate of 1 m^3/h per m^2 membrane area. The permeability is routinely between 70 and 120 LMH/bar over the range of TMP indicated in Table 5.12. Regular flushing, around once every 10 min, together with the CIP procedures outlined (Table 5.12), enable the required steady-state flux rates to be maintained. External recovery cleaning with sodium hypoclorite is conducted by taking the membrane out of the aeration tank and transferring it to an adjacent cleaning tank where both cleaning chemical reagents (acid and sodium hypochlorite) and cleaning durations can be employed more flexibly. Cassettes in the aeration tank can be backflushed separately, and the backflushing process demands around 25% of the product permeate.

Full nutrient removal is achieved by the plant. Phosphate removal is achieved by dosing ferric chloride into the effluent, the phosphate being removed with the excess sludge. The resulting excess sludge production, including the solids created by chemical dosing, is around 0.5 kg TSS/kg COD.

The dewatering unit consists of a sludge thickener combined with a belt press. The total sludge production is ~270 kg/day. To achieve a satisfactory thickness of 16% after the belt press, a polymer dose of 6 kg/ton DS is required. After dewatering, the solids content of the sludge product is increased and the sludge is further stabilised with lime at a rate of 0.52 kg CaO/kg. The end product (Fig. 5.22) has a DS concentration 25 wt% and is distributed to local farmers and disposed to land.

5.3.1.2 Brescia sewage treatment works

Background At the existing conventional plant at Brescia, the Verziano WWTP had three trains, each consisting of primary clarification, biological oxidation, secondary clarification and final chlorination (Fig. 5.24). Pretreatment of the incoming untreated wastewater prior to the three trains was undertaken by coarse and fine screening and with sand for oil and grease removal. Trains A and C were identical in design capacity, treating 24 MLD each, while Train B had an original capacity of 12 MLD. Shortly after assuming operations at the Verziano WWTP in July 1995, the water company ASM Brescia undertook a series of evaluations and identified problems relating to additional discharges. The most significant problem was plant capacity, which was about two-thirds of that required. In response, several important projects were undertaken over the subsequent 5-year period (1995–2000) to remedy these issues.

The principal remedial step was the upgrading of Train B, the oldest of the three trains, from 12 MLD to 38 MLD annual average flow capacity by converting the

Figure 5.24 The original plant at Brescia

Table 5.13 Discharge design criteria, Brescia plant (from Table 1, Annex 5 of D.Lgs 152/99)

Parameter	Units	Regulation limit	Design Value
COD	mg/L	125	125
BOD	mg/L	25	25
TSS	mg/L	35	5
TN	mg/L	–	10
Total P	mg/L	–	10
Escherichia	UFC/100 ml	–	10 on 80% of the samples 100 on 100% of the samples

activated sludge treatment system to an MBR. High effluent quality and limited available footprint were the key driving forces leading to the decision to employ MBR technology for the plant upgrade. The need for improved effluent quality resulted from new Italian legislation (Community Directive 91/271/CEE and D.Lgs. 152/99), requiring mandatory nitrification/denitrification and effluent TSS limits lowered from 80 mg/L to 35 mg/L (Table 5.13). In addition, ASM had extremely limited available land to extend the plant footprint. Estimates for conventional processes treating the same capacity and incoming loads showed that the footprint required would be twice that of an MBR for the biological reactor and roughly six times greater for the final sedimentation. The decision to pursue the MBR solution was also made easier by prior experience with the same membrane technology at full scale from the summer of 1999 onwards for the plant's leachate treatment system, with positive results. The upgrade was initiated in 2001 and completed in October 2002. The MBR system was expanded

Figure 5.25 Membrane tank at the commissioning phase, Brescia

to 42 MLD capacity through the installation of additional membrane cassettes after about 1 year of operation.

Plant design and operation The MBR system, which is protected by a fine screen upstream, consists of a denitrification tank followed by one nitrification tank which precedes four independent filtration trains. Each train is capable of producing 438 m^3/h of permeate, yielding an overall installed capacity of 42 MLD from a total membrane area of 73 000 m^2 (hence 24 LMH mean flux at peak load). In each train permeate is withdrawn from two membrane tanks (Fig. 5.25), each having a capacity of about 115 m^3. Each tank thus holds 20 ZW500c cassettes (Fig. 5.26), giving 160 cassettes in total. Each cassette contains 22 membrane elements providing an area of 456 m^2 per cassette (20.7 m^2 per element). Fine bubble aeration takes place at between 6000 and 11 000 Nm^3/h depending on the organic loading. An MLSS concentration of 8–10 g/L is maintained by an SRT of 17 days and an HRT of 7–8 h.

The plant is equipped with fully automated operation and maintenance (O&M) cleaning. It is operated with a filtration cycle of 10.5 min production and 90 s relaxation: no backflushing is employed at this plant. Since this is one of the more recent Zenon plant, intermittent aeration is used, cycling 10 s on/10 s off by switching the aeration between two banks of membranes. Coarse bubble aeration of the membranes is conducted at a rate of 6 Nm^3/h per membrane element overall (i.e. 12 Nm^3/h during

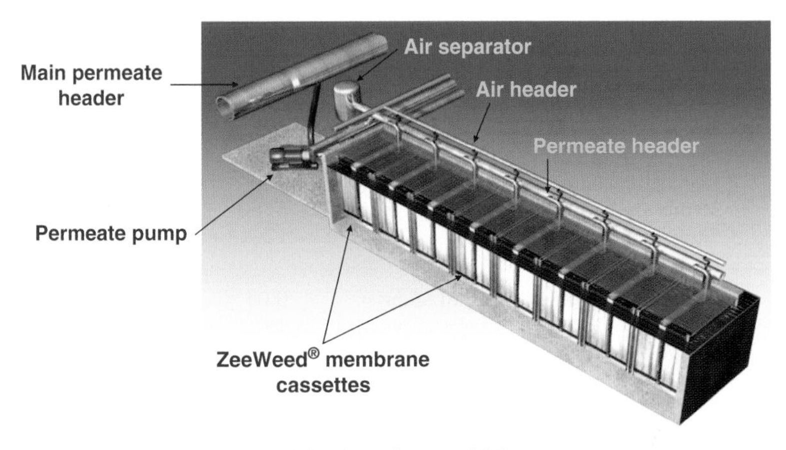

Figure 5.26 A single train of 16 cassettes

the on cycle), equating to a SAD_m of 0.29 Nm^3/h per m^2. This is sufficient to maintain a mean net flux of 17–19 LMH and a TMP of 0.1–0.2 bar, yielding a permeability of ~144 LMH/bar. The SAD_p is thus 15–17 m^3 air per m^3. The typical totalised flow through the membrane per month is ~1 Mm^3, and the typical maximum totalised daily flow of ~40 800 m^3.

Maintenance cleaning *in situ* is conducted weekly using 500 mg/L NaOCl at pH 8–8.5. The chemical is back-pulsed slowly through the membrane, whilst still immersed in the mixed liquor, for a period of 45 min. Only two recovery cleans, that is *ex situ* chemical cleaning by soaking the membranes in 1000 mg/L NaOCl for 6 h, were performed over the first 30 months of operation, with none conducted over the first year.

To date, there have been no operational problems with this plant, other than one foaming incident which demanded dosing with anti-foaming agent during the first 2 months during start-up of the system (Côté *et al.*, 2004). Effluent quality has met or exceeded expectations for SS, BOD, Ammonia-N and TN. The high-quality MBR effluent is blended with effluent from the conventional trains A and C, allowing the overall plant to meet present discharge standards. The MBR effluent quality is also of high-enough quality to be considered for reuse as irrigation water.

5.3.1.3 Nordkanal wastewater treatment works at Kaarst

Background and plant description The plant at Kaarst in Germany, owned and operated by the Erftverband (Erft Association), treats wastewater from the nearby towns of Kaarst, Korschenbroich and Neuss. It was installed and commissioned in January 2004 following the success of the group's first MBR plant at Rödingen, a smaller plant which was commissioned in 1999 (Table 5.14). As of November 2005, the Kaarst plant was the largest MBR plant in the world, with a population equivalent of 80 000 and a capacity of 48 MLD.

The site has five buildings which, respectively, house the sludge mechanical dewatering process, the fine screens, the coarse screen, the MBR and the process controls. Additional installations include lidded sludge and sludge liquor holding tanks, a grit

Table 5.14 Comparison of MBR plant at Rödingen and Kaarst

Parameter	Rödingen	Kaarst
Design		
Capacity (p.e.)	3000	80 000
Bioreactor volume (m^3)	400	9200
Membrane area (m^2)	5280	84 480
Sludge concentration (g/L)	12–18	12
Operation		
Sludge loading (kg BOD/(kg MLSS day))	0.04	$<$0.05
HRT, h	3.6	4.6
Sludge retention time (day)	25	25
Flux (LMH)	28	25

Figure 5.27 Aerial view of the Kaarst site

chamber and the denitrification tanks, the two latter operations being open to the atmosphere (Fig. 5.27). Water is pumped from the original WWTP 2.5 km east of the site to 5 mm step screens, followed by an aerated grit chamber. It is then fed to two Huber rotary drum 1 mm mesh-grid fine screens, changed from 0.5 mm fine screens originally installed, each providing a capacity of 24 MLD. There is a standby 1 mm fine screen which comes on-line in case of a mechanical breakdown of the former two. Screenings are discharged into a skip and subsequently disposed of by incineration off-site. The screened water is transferred to the MBR.

Biotreatment comprises four tanks each fitted with two membrane trains with an upstream denitrification zone of 3500 m^3 total capacity; the latter receives sludge from the subsequent membrane aeration tank at a recycle ratio of 4:1. The membrane trains are each fitted with 24 ZW500c membrane cassettes of 440 m^2 membrane area, such that the total membrane area is 84 480 m^2 (20 m^2/elements; 22 elements/cassette), and the aeration tanks receive supplementary mechanical agitation from impeller blade stirrers to maintain biomass suspension. The total volume of the tanks is 5800 m^3, of which about one-third comprise the membrane aeration, which together with the denitrification tanks provide a total HRT of $\sim$5 h at 48 MLD flow. Sludge wasted from the tank is dewatered by centrifuge to around 25 wt%. There is also a simultaneous precipitation for phosphorous removal.

Plant operation The plant is operated at an SRT of 25 days, which maintains the mixed liquor at between 10 and 15 g/L. The membranes are operated at a net flux of 25 LMH and a mean permeability of 150–200 LMH/bar. Intermittent coarse bubble aeration is provided at 34 000 m^3/h on a 10 s on/10 s off basis, giving a SAD_m of 0.40 Nm^3/h per m^2 and a SAD_p of 17 m^3 air per m^3. Physical cleaning comprises a backflush every 7 min for 60 s at 1.5 times the operating flux. A maintenance chemical clean is conducted every two weeks by draining the membrane tank and backflushing for 1 h alternately with different cleaning agents, including 500 mg/L hypochlorite. Cleaning is conducted on individual tanks, that is 12.5% of the installed capacity. The number of membranes on-line is adjusted according to the flow, but this is controlled in such a way as to ensure that no membrane train is off-line for more than a total of 70 min. The total specific power demand for all operations is 0.9 kWh/m^3, which compares to an average of 0.5 kWh/m^3 for all conventional sewage treatment plants operated by the Erftverband.

The plant, which incurred a total capital cost of €25 m, has been successfully operating for 23 months without major incidents.

5.3.1.4 Unifine Richardson Foods effluent treatment plant

Background The plant at Unifine Richardson Foods in St Mary's, Ontario (Figs 5.28 and 5.29) is an industrial pretreatment facility that discharges to a municipal wastewater treatment works and must meet 300 mg/L BOD and 350 mg/L TSS. Any organic load in excess of this is levied an additional surcharge by the municipality. Reliance on the municipality to receive and treat any high-strength waste leaves the industrial client in a vulnerable position should the works not be able to receive the industrial wastewater in the future. Richardson Foods moved to its new production facility in 1999 and needed a new wastewater pretreatment facility to replace the existing SBR, which was over 20 years old. The latter was labour-intensive, generated nuisance odours and was subject to upsets due to the proliferation of filamentous bacteria. Richardson Foods hired CH2MHill to select bidders for a new wastewater pretreatment system and, as a result of the selection process, Zenon Environmental Inc. was selected to design, build and install the system on a complete turn-key basis. The total contract value was $1 102 000, and installation and commissioning were completed in November 2000.

Plant design and operation Unifine Richardson Foods Ltd. manufactures condiments for the food service industry. The plant makes over 200 products, including drink syrups, sundae toppings, sauces, mayonnaise and salad dressings. The wastewater strength and composition is highly variable and changes with product changes, tank flushes, etc. The mean influent COD and BOD concentrations are 7400 and 5500 mg/L, respectively, and the average daily flow is 0.15 MLD. There is a significant concentration of oil and grease in the wastewater. Wastewater from the

Figure 5.28 The Unifine Richardson Foods effluent treatment plant

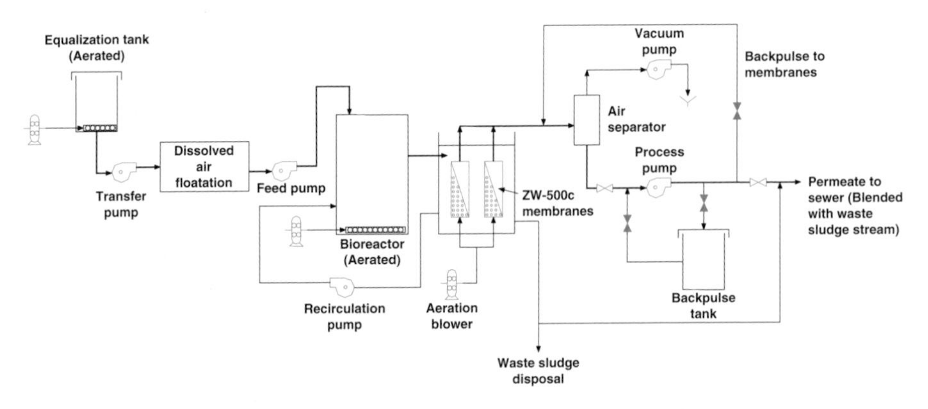

Figure 5.29 Schematic of Unifine Richardson Foods

industrial plant is pumped into an aerated equalisation tank, which is of bolted carbon steel construction, and is then pumped continuously into a dissolved air flotation (DAF) process for FOG and solids removal. The DAF float is stored and tankered to a rendering facility.

The DAF effluent is pumped to the bioreactor tank, also of bolted carbon steel construction. A coarse bubble diffuser grid distributes air into the bioreactor from positive displacement blowers. Caustic soda and phosphorous can be added to the biological process by chemical metering stations. Mixed liquor from the bioreactor

Table 5.15 Water quality data, Unifine Richardson Foods plant

	Influent	DAF effluent	MBR effluent
BOD (mg/L)	5500	3300	<20
COD (mg/L)	7400	5400	<145
TSS (mg/L)	–	–	<1

flows by gravity into an epoxy-lined carbon steel tank that houses the ZeeWeed® membranes. The ZeeWeed® tank contains two ZeeWeed® 500c cassettes with 12 modules in each cassette, providing a total membrane area of 557 m^2. The membrane cassettes are scoured with air by a positive displacement blower. A centrifugal pump extracts the permeate from the membrane system by applying a slight vacuum on the permeate side. Mixed liquor in the membrane tank is returned to the bioreactor at five times the forward flow rate, to ensure a uniform concentration between the two tanks (~10–15 g/L). Sodium hypochlorite and citric acid chemical metering stations are included for membrane cleaning. Waste activated sludge is discharged to a filter press for dewatering. The design capacity of the system is 0.15 MLD, and the treated effluent quality is COD $<$ 145 mg/L, BOD $<$ 20 mg/L and TSS $<$ 1 mg/L (Table 5.15).

Aeration of the membranes at a rate of 360 m^3/h at 0.33 bar is via a 7.5 kW blower, and the bioreactor blowers are sized for 940 m^3/h at 0.88 bar (37 kW). Aeration of both tanks is via coarse bubble aerators. The membranes operate at a net flux of approximately 12 LMH, back-pulsing for 30 s every 10 min. The TMP is maintained between 0.14 and 0.44 bar with an average of 0.17 bar. A maintenance clean is applied twice weekly, this comprising injection of a 200 mg/L solution of sodium hypochlorite during a backpulse for a total clean duration of 1 h. No recovery cleaning (i.e. an *ex situ* soak) has been required at the plant since commissioning. Waste sludge is dewatered in a filter press and then disposed of at a landfill.

The plant has been successfully operating for 5 years without undue problems and to the satisfaction of the client, who have cited the elimination of odour problems, ease of operation and system reliability as key benefits of the plant. The treated effluent is of such good quality that the client is able to blend wasted sludge back into the permeate and still remain under the limits set out by the local treatment facility. This dramatically reduces the costs associated with sludge disposal.

5.3.1.5 Industrial effluent Zenon MBR-RO, Taiwan

The manufacture of TFT–LCD (thin film transistor–liquid crystal display) visual display units (VDUs) generates wastewater which is high in organic solvents such as dimethyl sulphoxide (DMSO, $(CH_3)_2SO$) and isopropyl alcohol (IPA, CH_3CHOCH_3), as well as organic nitrogen compounds, such as ethanolamine (MEA, $C_2H_5ONH_2$) and tetra-methyl ammonium hydroxide (TMAH, $(CH_3)_4NOH$) (Table 5.16). The total organic carbon to total nitrogen ratio (TOC-TN) ratio is over 95%, significantly higher than the figure of 80% typical of municipal wastewaters. Development of an enhanced nitrification/denitrification process to contend with the higher organic nitrogen load was therefore required.

Table 5.16 TFT–LCD organic wastewater characteristics

Source	Stripper	Developer	Rinse	Average value
Solvent	DMSO/MEA	TMAH	IPA	–
pH	9–11	10–13	4–10	10–11
SS (mg/L)	<10	<10	<10	<10
COD (mg/L)	800–2000	100–600	500–3700	800–2000
TKN (mg/L)	70–200	60–90	90–240	100–200
NH_3-N (mg/L)	0–15	2–15	0.1–10	2
NO_x-N (mg/L)	0.1–0.4	0.0–0.3	0.1–1.3	0.2 3

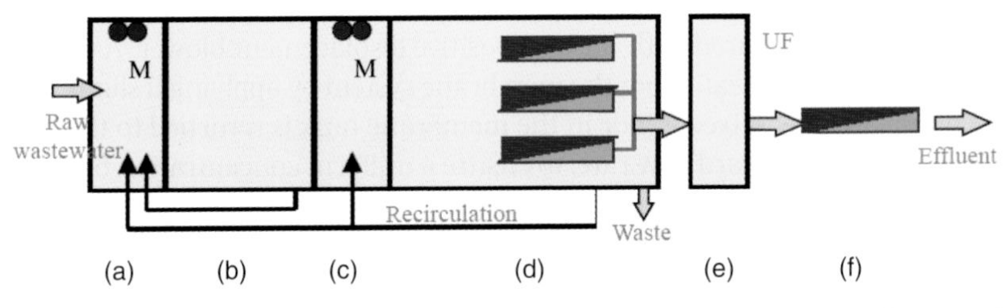

Figure 5.30 Schematic, Bah-Teh plant: (a) anoxic tank, (b) aerobic tank, (c) anoxic tank, (d) membrane tank, (e) UF water tank and (f) RO

Although MBRs are often identified as an effective pretreatment for reverse osmosis, the installation of MBR-RO is a fairly recent phenomenon within the municipal sector. Paranjape *et al.* (2003) reported only two such full-scale municipal facilities in North America – City of Colony Key (1.5 MLD) and City of Laguna, (1.9 MLD) – by 2003. The high-profile NEWater project in Singapore, in operation since May 2000, is based on MF of secondary sewage. It is only within the last 2–3 years that an MBR-RO system has been piloted for the duty of reclamation and reuse of treated sewage effluent for industrial process water in Singapore (Qin *et al.*, 2006). Pilot demonstration MBR-RO projects for municipal wastewater reuse duty in the USA are also comparatively recent (Lozier and Fernandez, 2001).

Over the last 5 years, a number of relatively large MBR-RO effluent treatment plants specifically for TFT–LCD wastewater treatment have been installed in Taiwan in response to the island's increasing stringent regulations concerning industrial water use and reuse. Trials were originally conducted on a 13 m^3 anoxic/aerobic tank pilot plant (3 m^3 anoxic tank, a 10 m^3 aerobic tank) using a separate ZeeWeed® 500a module followed by a four-stage RO plant, and achieved 80% recovery of purified water.

The first full-scale plant, 1.27 MLD capacity, was installed at Bah-Teh, Taoyuan County, Taiwan in December 2000, designed and installed by the Green Environmental Technology company. The treatment system involves an anoxic–aerobic–anoxic biological process followed by an aerobic MBR with a separate membrane tank followed by an RO unit (Fig. 5.30). Six membrane cassettes, each fitted with eight ZW500a modules, are used to provide a total membrane surface area of 2160 m^2. The RO array comprises 90 Dow Filmtec™ fouling-resistant membranes, providing a 1.5 MLD capacity and recovery rate of 80%.

The replacement of the conventional clarifier with an MBR obviated the occasional severe bulking problems associated with the former. The MLSS concentration

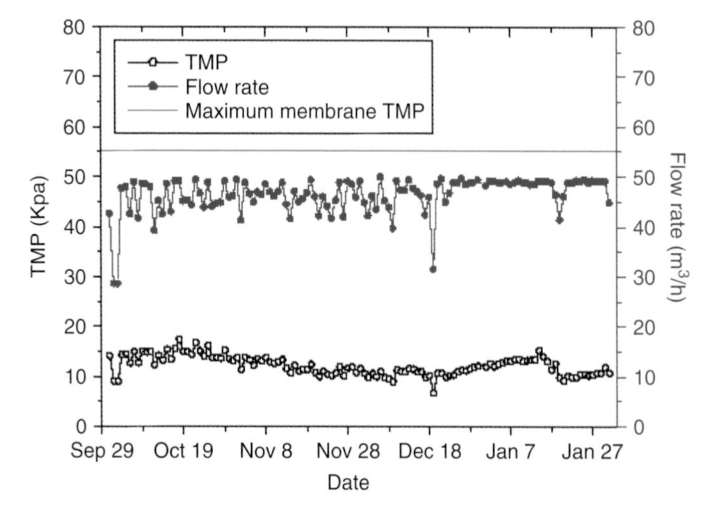

Figure 5.31 TMP and pressure transients over winter season, Bah-Teh MBR plant

has been maintained between 8–12 g/L, at an SRT of ~30 days, regardless of feedwater quality and organic loading (0.46–1.5 kg COD/m^3/day). The sludge yield is around 0.27 kg SS/kg COD. Purification performance by the plant is very satisfactory. Removals achieved by the MBR are 96% COD, 97% TOC, 92% TKN, 71% TN and 97% BOD5, combined with very low silt density index (SDI) of below 2. These values equate to average values of 35.2 mg/L, 20.2 mg/L and 6.5 mg/L for permeate COD, TOC and BOD, respectively. The RO permeate had a TOC of <2 mg/L, satisfying the stringent requirement for water reclamation.

At capacity, the plant achieves an MBR flux of 24.5 LMH; the mean net flux is 21.5 LMH, maintained by backwashing for 30 s every 15 min and applying a recovery clean every 3 months with sodium hypochlorite. The average TMP is 0.08–0.12 bar, considerably below the maximum allowable value of 0.55 bar, or 55 kPa (Fig. 5.31).

The success of the Bah-Teh plant has led to other similar effluent treatment plant being installed or planned. The TFT–LCD plant at Longtan, Taoyuan County was commissioned in May 2003 and is designed for a peak flow of 4.2 MLD from 20 cassettes of 20 ZeeWeed® 500c modules in four trains, providing a total membrane area of 8800 m^2. The planned installed capacity for VDU fabrication effluent was 50 MLD by 2005, with a 22.5 MLD plant due for commissioning in May 2006. Installations include two MBR-RO plants, of 5 and 1.5 MLD capacity commissioned in 2004 and 2005, respectively, which are fitted with anaerobic pretreatment to contend with the high COD loadings.

5.3.2 Mitsubishi Rayon Engineering

5.3.2.1 Early Japanese installations

Mitsubishi Rayon has almost as many installed plant in Japan as Kubota. The plants tend to operate at slightly lower fluxes than those of many of the other MBR technologies, but this is countered to a large extent by the lower cost of the membrane.

Early examples (from the mid-1990s) of installations for industrial effluent treatment include the JA (Japan Agricultural Cooperative) facility for processing milk and ume plums and a livestock wastewater recycling plant (Mitsubishi Rayon, 2000). The effluents involved in these early examples are particularly challenging, leading to very low membrane permeabilities.

The food processing effluent had BOD and SS levels of 810 and 710 mg/L, respectively, from a 0.02 MLD flow combined with a 0.007 m^3/day flow from ume plum processing wastewater with BOD and SS levels of 4600 and 32 mg/L, respectively. The company had voluntary discharge concentration limits of 25 mg/L BOD and 5 mg/L SS, with discharge to a nearby stream. The treatment scheme consisted of a holding tank followed by aerated equalisation upstream of the MBR unit, with the waste sludge being concentrated using the same immersed membrane technology. The MBR itself was fitted with two tiers of two 64 m^2 units (256 m^2 total), thus operating at a mean flux of ~4 LMH. The sludge concentration unit comprised 4 × 4 m^2 units. For both the MBR and the sludge processes, permeate was removed by suction. The plant operated at an MLSS of 8–12 g/L and 1.5 h HRT with chemical cleaning carried out twice yearly by a 16 h soak in 0.1 wt% NaOCl/4 wt% NaOH. No membrane replacement was required in the first 4 years of operation.

The 0.06 MLD livestock wastewater recycling plant was designed for zero liquid discharge and comprised two holding tanks, screening, equalisation, sedimentation (with the sludge fed to a drying unit, the condensate from which was returned to the equalisation tank) and an anoxic tank (with sludge recycled from MBR tank), all upstream of the MBR. The MBR permeate was treated by reverse osmosis, operating at 80% conversion, with the concentrate being used for composting. This plant used two tiers of five units (640 m^2 total) and operated at a mean flux of ~4 LMH with a TMP between 0.25 and 0.4 bar and hence a mean permeability of ~12 LMH/bar. The MLSS was maintained at 8–12 g/L and chemical cleaning applied every 8 months according to the same protocol as the dairy effluent treatment plant. No membrane replacement was needed in the first 3 years of operation.

5.3.2.2 US installations

The USA licence for the Mitsubishi Rayon technology is held by Ionics, now part of GE Water and Process Technologies. Between 2000 and end of 2005, 18 contracts were secured by the company for MBR installations, ranging in capacity from 0.1 to 1.9 MLD and 15 of these being in the USA or Caribbean regions. For all plants, a standard aeration rate of 68 Nm^3 per module is employed, regardless of the number of tiers (between one and three). Thus, the SAD_m can vary from 0.65 Nm^3/h air per m^2 for a single-tier, 70-element module (105 m^2) to 0.22 for a triple-tier module. Moreover, the newer elements provide double the membrane area of the old ones (3 m^2 vs. 1.5 m^2), reducing the aeration demand to as low as 0.11 Nm^3/m^2h for a triple-tier module.

The oldest of these installations is the Governor Dummer Academy plant in Byfield, Massachusetts. This plant, 0.38 MLD capacity, is fed with moderate strength sewage (200–300 mg/L BOD, 100–150 mg/L TSS, 40 mg/L TN) and produces high-quality effluent (<2 mg/L BOD, <2 mg/L TSS, <1 mg/L ammonia, <10 FCU/100ml sample). The plant operates at a mean MLSS of around 12 g/L. The design flux of the plant is

10 LMH; the plant operated at fluxes between 2 and 15 LMH and permeabilities between 10 and 34 LMH/bar in the period 11–16 March 2004. Since the plant comprises single-tier modules based on the older 1.5 m^2 elements, the SAD_m is around 0.65 Nm^3/h air per m^2 and thus the SAD_p is 65 Nm^3 air per m^3.

5.3.3 Memcor

5.3.3.1 Park Place, Lake Oconee, Georgia

The plant (0.61 MLD average., 0.89 peak) at Park Place in Greene County began operating in June 2003 and produces irrigation water for the local golf course from municipal effluent. The process was selected because of the quick turnaround from installation to start-up as well as the small land footprint required, since the land has a high value in this area. The facility (Figs. 5.32–5.34) is totally enclosed in a rectangular building 23 m by 16 m (Fig. 5.32). The feed and treated effluent quality is given in Table 5.17.

The process (Fig. 5.34) comprises 2 mm fine screens from which the wastewater is fed to two separate lines of sequenced anoxic and aerobic tanks which provide a mean HRT of 3 and 8 h, respectively, the recycle ratio to the anoxic tank being 5.7. The mixed liquor is recirculated between the membrane and aerobic tanks. The SRT is around 35 days and provides an MLSS concentration of 12 g/L.

The membrane tank is 3.6 m in depth and is fitted with 160 B10R modules in eight racks of 20 modules, providing a total membrane area of 1600 m^2. The membrane operates at a net flux of 16 LMH, with regular backflushing, and the maximum TMP is 0.64 bar. The membrane aeration rate is 1.8 m^3/h per module, yeilding a SAD_m of 0.18 Nm^3/h per m^2, which at the stated net flux equates to a

Figure 5.32 The facility at Park Place, outside

Figure 5.33 The facility at Park Place, inside

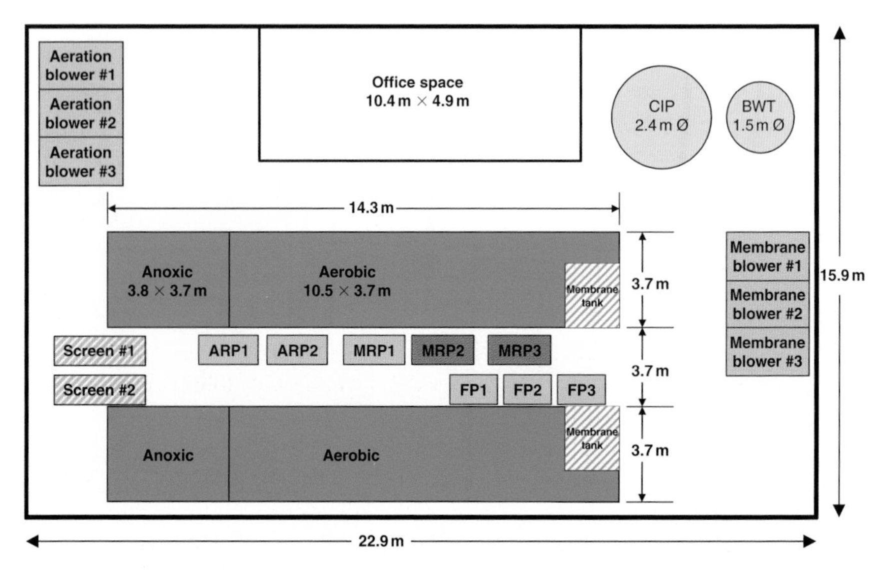

Figure 5.34 Schematic and layout, Park Place treatment scheme. Pumps denoted ARP (anoxic recycle), MRP (membrane recycle) and FP (feed); CIP and BWT denote clean-in-place and backwash tanks

Table 5.17 Feed and treated water quality, Park Place

Parameter	Feed	Effluent
BOD (mg/L)	250	<5
TSS (mg/L)	220	<5
NH_4^+-N (mg/L)	30	<1
TKN (mg/L)	45	<10 (TN)
Loading rate (kg COD/m^3/day)	0.84	
Sludge production (kg VSS/kg BOD) removed	0.45	
F:M ratio	0.058	

SAD_p of 11.3 m^3 air per m^3. Aeration power demand is estimated at 6.4 kW for membrane aeration (i.e. 0.25 kWh/m^3), 15.4 kW for biological treatment and 5.6 kW for pumping; total power demand is ~29 kW (1.1 kWh/m^3), about half of which is accounted for by the membrane aeration and permeation. A hypochlorite CIP is conducted by transferring the mixed liquor to bioreactor tanks, filling the membrane tank with the hypochlorite, soaking and recirculating the chemical via the aerator, and discharging the chemical to sewer prior to refilling the membrane tank ready for operation.

5.3.3.2 Other MemJet® plants

Other MemJet® plants have recently (2004) been installed at Calls Creek, GA (2.5 MLD) and Hawks Prairie (7.6 MLD, increasing to 18.9 MLD in the future). The Calls Creek plant is a retrofit to an existing activated sludge plant (the Orbal® system), and the Hawks Prairie plant is a new-build water reclamation plant for irrigation instigated by the Lacey, Olympia, Tumwater and Thurston County (LOTT) Wastewater Management Partnership as part of their Wastewater Resource Management Plan.

5.3.4 Koch Membrane Systems – PURON®

5.3.4.1 Simmerath (Germany)

The municipal MBR at Simmerath (Germany) was commissioned in March 2003. It is designed for a population equivalent of 750, and a mean flow equating to a mean flow of 0.63 MLD and is operated as a bypass to the existing WWTP. It is fitted with two modules providing a total membrane area of 1000 m^2. The situation at Simmerath is representative of many municipal WWTPs sited in the North Eifel region where the effluent is often discharged to surface waters that are subsequently used as a source for drinking water production. Low wastewater temperatures and high amounts of infiltration water represent an additional challenge for the technology.

The MBR (Fig. 5.35) is fed with the WWTP wastewater following screening to 3 mm. The membrane plant has its own aeration tank with a denitrification step followed by nitrification. The activated sludge is pumped from the aeration tank into the membrane tanks and then overflows back to the aeration tank. The MLSS is maintained at 10–12 g/L in the aeration tanks and 14–16 g/L in the membrane tanks.

The intermittent air scouring of the PURON® module is conducted in repeated cycles of about 300 s combined with the filtration interval. To support the flushing of the module, there is an additional backflush supplied after each filtration interval. The aeration

Figure 5.35 MBR demonstration plant at Simmerath showing baffled bioreactor and building housing the membrane tank

frequency varies between 25% and 50% of the filtration time, depending on the flux level of operation. The specific air consumption for the module aeration has been between 0.15 and 0.35 Nm^3/m^2h. Even during winter time, flow rates have been maintained at a high and stable level (between 20 and 30 LMH, and 120–200 LMH/bar). Maintenance CIP is applied every 2–4 weeks using 500 mg/L of sodium hypochlorite adjusted to pH 12, followed by citric acid at pH 3. The CIP takes around 2 h.

Since commissioning, the MBR effluent has always met the stringent requirements of the European Bathing Water Directive. The PURON® system has thus far been free from clogging and sludging problems, despite the relatively coarse screening (3 mm slot screens) being employed compared with that normally used for HF modules.

5.3.4.2 Sobelgra

The largest industrial MBR plant in Belgium (as of 2005) was commissioned in November 2004 and is equipped with 16 PURON® modules. The site at Sobelgra is located in Antwerp harbour. The MBR plant has a capacity of 2 MLD and treats effluent from a malting operation, where barley is converted to malt for beer by a natural enzymic process. Malting effluent is particularly challenging, being very high in organic content and with a relatively low BOD/COD ratio (Table 5.18). The MBR plant was selected as a result of the plant capacity being extended from 110 to 250 ktonne per annum, making it the largest independent malting company in

Table 5.18 Water quality, Sobelgra plant

Parameter (mg/L)	Feed concentration (mg/L)	Load (kg/day)	Effluent concentration (mg/L)
COD	1880–2100	4000	100–200
BOD	700–930	2000	2–5
TSS	330–460	800	0
TN	35–50	100	1–2
TP	13–15	30	<1

Figure 5.36 Koch membrane systems PURON® modules at Sobelgra

Belgium. This meant that the capacity of the existing conventional WWTP had to be doubled and, due to lack of space on the factory site, conventional technology could not be used. The MBR plant (Fig. 5.36) was piloted in the Spring of 2003 and installed by the Belgian turnkey constructor ENPROTECH with the Belgian Engineering company SEE:WATER supplying and assembling the MF system.

The MBR was retrofitted into the existing aeration tank, which was divided into two parts to provide an aeration tank and a separate membrane tank. The tank of the former clarifier has been converted to an additional bioreactor, such that, in the new process, 25% of the total treatment volume is occupied by the membranes. The membrane tank is equipped with 16 500 m^2 PURON® membrane modules, providing a total membrane area of 8000 m^2.

The plant consists of a screening at 0.25 mm prior to biotreatment using two tanks in series. The sludge is then transferred to two membrane compartments, with a third chamber being available for future extensions of the plant. The membrane compartments are fed from the bottom so that the sludge flows upwards through the membrane modules. Permeate is extracted from the membrane modules under a slight vacuum. The activated sludge is concentrated during this process and is led back into the biology tanks by means of spillways.

5.3.5 Asahi Kasei

The Asahi Kasei Microza HF membrane module was installed at a 0.6 MLD food effluent treatment plant in Japan in August 2004 and subsequently expanded to 0.9 MLD in early 2005. The respective mean BOD, COD and TSS values of the feed are 1080, 630 and 377 mg/L, respectively, and all are removed by at least 99.5% by the process. Originally, 16 racks of 100 m^2 modules (i.e. a total membrane area of 1600 m^2), each containing four elements (Fig. 4.27a), were inserted in the tank to achieve the total flow capacity required, the modules being retrofitted into an existing MBR membrane tank plant. The plant is operated at ~8 g/L MLSS. A stable TMP of around 0.2 bar was maintained over the first 3 months of operation at a flux close to 16 LMH at an aeration rate of 384 m^3/h, equating to a SAD_m of 0.24 m^3/h per m^2 and a SAD_p of 15 m^3 air per m^3.

5.4 Sidestream membrane plants

5.4.1 Norit X-Flow airlift process

The earliest example of the use of an airlift sidestream membrane bioreactor (sMBR) for municipal wastewater is probably that of the pilot plant at Vienna (Fig. 5.37), installed in 1999. This plant has a capacity of 0.24 MLD and the feed strength is 300–700 mg/L COD. The plant provides an effluent of <30 mg/L COD at all times.

The plant comprises pretreatment using a coarse grid followed by a sand and FOG trap. A submersible pump, which also provides a degree of comminution, pumps the raw effluent into a sedimentation tank with an HRT of 30–60 min. The supernatant is then fed through a 0.8 mm screen and into the bioreactor, into which ferric coagulant is dosed directly. Membrane separation is achieved by four vertically-mounted 38PR-type MT modules fitted with 5.2 mm F4385 membranes. The modules are operated at a TMP of 0.05–0.15 bar with a combined pumped air and liquid crossflow rate of <1 m/s. This achieves a flux of 40 LMH for dry weather flow, rising to 55 LMH at full capacity when the TMP is at its highest value of 0.3 bar. The permeability is generally maintained between 300 and 400 LMH/bar, only falling below this range when fluxes of >50 LMH are sustained for several days. Membrane cleaning is achieved by chemical soaking every 15–30 days using a combined chemical clean of hypochlorite and citric acid (Fig 5.37).

Figure 5.37 Airlift municipal WWTP, Vienna

5.4.2 Food wastewater recycling plant, Aquabio, UK

5.4.2.1 Background and drivers

The drivers and issues surrounding industrial water recycling are well known and have been debated extensively (Judd and Jefferson, 2003). However, in most cases, the principal issue is cost. Users with an incoming mains water supply are charged for both potable water into the factory and trade effluent out of it. Recycling becomes an attractive option if the cost savings made by obviating water supply and discharge are more than the capital cost of the plant over some critical payback time period (in many cases, >2 years). However, there is usually one or more other motivating factor which strengthens the case for recycling. This can include pressure on abstraction, business growth and/or factory expansion, or an unexpected dramatic increase in water supply or discharge costs. Abstraction is affected by more global concerns such as decreasing water levels caused by a higher general demand for water and/or climate changes. In such cases, companies who already have water treatment infrastructure in place for potable water polishing may find the recycling option more attractive.

The plants at Kanes Food and Bourne Salads in the UK both represent examples of food effluent recycling with an sMBR, based on a pumped UF MT membrane, employed upstream of a reverse osmosis plant and, finally, a UV polishing unit. The Kanes Food plant in Worcestershire was commissioned in February 2001 following pilot plant trials conducted between January and June 1999 by the contractor Aquabio, who refer to their process as an advanced membrane bioreactor (AMBR). The plant has a 0.815 MLD capacity (Table 5.19), with 80% of this flow being recycled.

5.4.2.2 Plant configuration

The process treatment scheme (Fig. 5.38 and Table 5.20) comprises upstream screening, flow balancing, the AMBR and downstream treatment by reverse osmosis followed by UV disinfection to yield potable quality water suitable for blending with mains water for use within the factory. Screening down to 0.75 mm is achieved by a 150 m^3/h rotary drum screen. Flow balancing is carried out in a 900 m^3 balance tank, thus providing 18 h HRT, where aeration and mixing is carried out using two self-entraining JETOX aerators. A proportion of the wastewater collected at the balance tank is discharged directly to the public sewer, having undergone screening, balancing and pH correction. The remaining wastewater (up to 0.815 MLD) is passed forward to the bioreactors for biological treatment in the AMBR process, comprising two 250 m^3 bioreactors with four banks of crossflow membrane modules. The two bioreactors provide process flexibility, allowing them to operate in series or in parallel. Currently, series operation is utilised successfully to provide “roughing” treatment in the first tank, followed by “polishing” in the second.

The *JETOX aeration systems* provide a total of 1980 kgO_2/day with reference to clean water and standard conditions. The maximum MLSS concentration employed

Table 5.19 Design basis for plant at Kanes Foods

Parameter	Value
Volume to screening/balancing	1.2 MLD
COD concentration (average)	1000 mg/L
Volume to MBR process	0.815 MLD
Volume to RO system	0.815 MLD
Volume of potable quality water for reuse	0.65 MLD

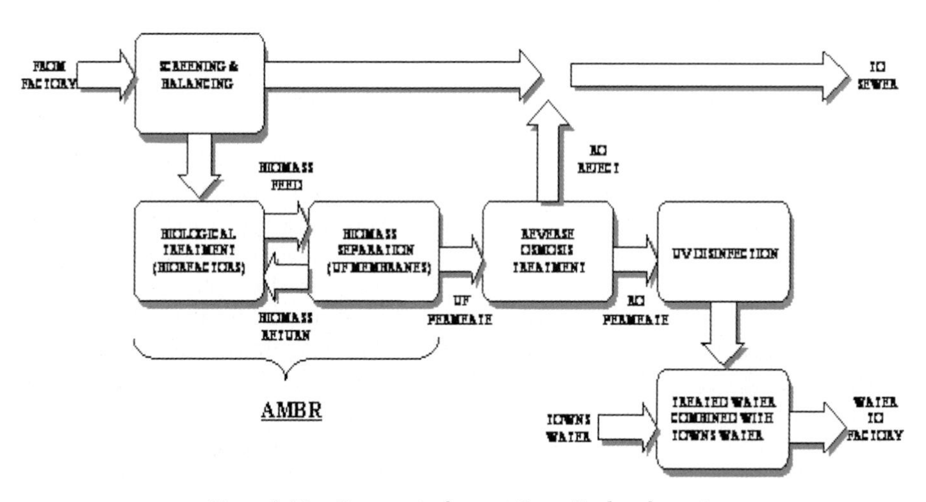

Figure 5.38 Treatment scheme at Kanes Foods, schematic

has been 20 g/L, but the bioreactor is generally operated at around 10 g/L resulting in food-to-micro-organism (F:M) ratios of around 0.13 kg COD/kg MLSS day. Sludge production is calculated as being 0.14 kg DS/kg COD removed from a sludge age of over 100 days. Each membrane bank (Fig. 5.39) comprises three 200 mm diameter modules, fitted with MT UF (0.03 μm pore, Norit X-Flow) membranes, with space for expansion to up to five modules. The membranes operate at an average flux of 153 LMH normalised to 25°C. The permeate water has mean TSS, BOD and COD concentrations of 4, 7 and 16 mg/L, respectively, with COD residuals generally ranging between 5 and 25 mg/L.

Up to 34 m^3/h of the UF permeate is passed to a two-stage reverse osmosis plant comprising low fouling RO membrane modules arranged as a 3 × 2 array which achieves an overall recovery of 75–80%. The reject stream from the second stage is discharged

Table 5.20 Summary of process treatment scheme, Kanes Foods

Process stage	Characteristics
Preliminary screening	Rotary drum screen, 0.75 mm
Balancing	900 m^3 balance tank c/w mixing/aeration
Secondary screening	Backflushing filter/screen, 0.5 mm
Biological treatment	2 No. 250 m^3 Bioreactors c/w mixing/aeration
UF membrane separation	4 No. Membrane banks, 324 m^2 total membrane area, 34 m^3/h
Reverse osmosis	two stage RO system, 27 m^3/h
UV disinfection	UV disinfection unit, 27 m^3/h
Sludge handling	22 m^3 Sludge tank

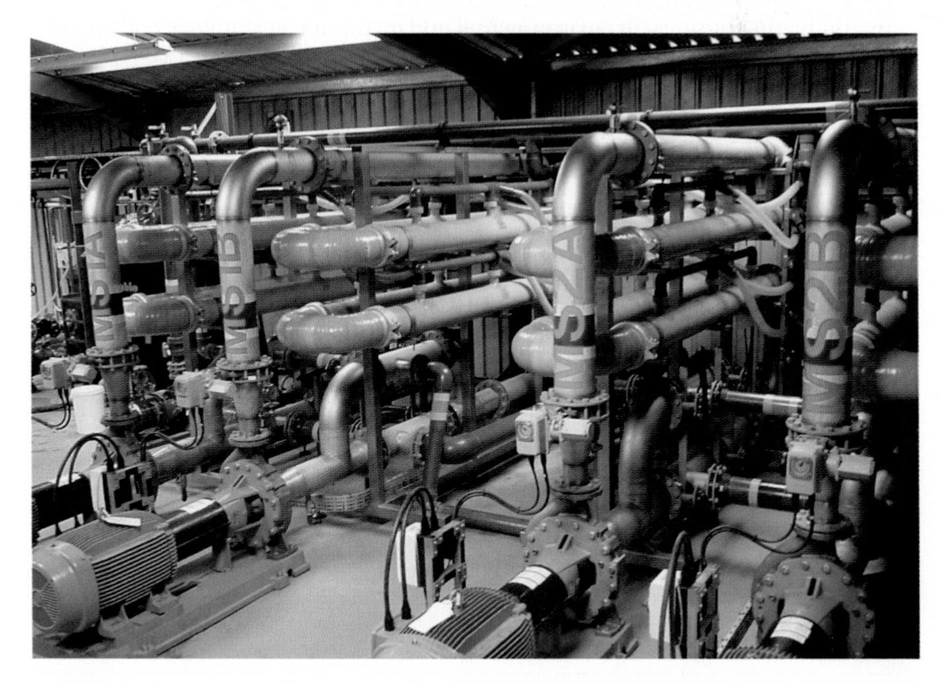

Figure 5.39 UF membrane banks at Kanes Foods

to sewer and the permeate, which typically has a conductivity of 40–100 μS/cm, is passed to the UV disinfection unit which applies a UV dose of >38 W/m^2.

Both the UF membrane and RO systems are provided with CIP systems. For the UF membranes, individual banks can be taken off-line for cleaning with hypochlorite or proprietary cleaners, normally every 120–180 days for around 3 h, while the remaining banks continue to provide the full throughput requirement.

5.4.2.3 Plant performance and cost savings

Regular monitoring of the final reuse water has shown it to be completely clear of total coliforms and total bacteria at all times, as well as suspended solids, and to have a conductivity ranging from 40 to 100 μS/cm^2. Membrane hydraulic performance has been better than the design specification: no long-term deterioration in flux has been noted over 3½ years operation. This has allowed one of the membrane banks to be maintained as a standby and so offering more process flexibility and lower energy use. Costings (Table 5.21) indicate an annual operating saving of £39.3 k (excluding capital costs for upgrading of municipal infrastructure) over conventional supply and discharge, despite a total specific energy demand of around 7 kWh/m^3 for the complete treatment, 3.5 kWh/m^3 relating to the MBR. The municipal upgrade cost (Table 5.21) would have been a tangible cost from the water supplier if the reuse option was not available.

5.4.3 Landfill leachate treatment systems, Wehrle, Germany

Wehrle Environmental is a subsidiary of WEHRLE UMWELT GMBH, a German company conducting business in environmental and membrane technology and an MBR process supplier since 1989. The company specialises in low-flow (<2 MLD) recalcitrant industrial effluents, and landfill leachate in particular, where the economic viability of sidestream systems becomes comparable with that of immersed systems.

Table 5.21 Cost balance, plant operation, Kanes Foods

Item	Basis	Annual expenditure
Saving on capital expenditure for upgrade of local sewerage/ sewage treatment facilities	Nominally £1.4 million differential cost (amortised over 10 years)	−£140.0 K (saving)
Saving on incoming towns water	Average 0.456 MLD (166 ML/annum) @ 48.7 p/m^3	−£81.1 K (saving)
Saving on effluent disposal	Average 0.456 MLD (166 ML/annum) @ 43.7 p/m^3	−£72.7 K (saving)
Maintenance costs including membrane replacement	Normal spares and membrane replacement after 7 years	+£13.6 K (cost)
Electricity costs	5740 kWh/day @ 3.2 p/kWh	+£67.0 K (cost)
Chemical costs	UF and RO cleaning anti-scalant & bisulphite for RO	+£10.5 K (cost)
Sludge disposal costs	156 × 20 tonne tankers per annum	+£23.4 K (cost)
Net saving		**−£179.3 K***

*£39.3k excluding the capital expenditure for upgrading of municipal infrastructure.

The company had installed 85 sMBR plants by 2005, of which 58 were dedicated to treating landfill leachate.

5.4.3.1 Landfill leachate

A landfill site is a large area of ground, normally lined, that is used for tipping and disposal of waste material. As long as the rainfall is greater than the rate of water evaporation, then the liquid level (leachate) within the landfill area will tend to rise. Environmental regulations require that the leachate level be controlled, which means that excess leachate must be removed and disposed of.

Landfill leachate represents one of the most challenging effluents to treat biologically (Alvarez *et al.*, 2004). Whilst nitrification is generally readily achievable, with >95% removal of ammonia reported through the exclusive application of biological techniques, COD removal is considerably more challenging. Removal efficiency values range from over 90% to as low as 20% according to leachate characteristics (origin and, more significantly, age), process type and process operational facets. Treatment process schemes generally comprise some combination of biological and physical and/or chemical treatment, with key operational determinants being organic loading rate and the related HRT. Leachate matrices can be characterised conveniently with reference to BOD/COD ratio, which normally lies within the range 0.05–0.8 and is generally held to be an indicator of leachate age as well as biodegradability (Table 5.22). Considerable variability in actual levels of organic material and ammonia arise, and a low assimilable carbon level combined with a high ammonia concentration may demand dosing with methanol to ensure an adequate carbon supply.

Leachate not only presents a challenge with respect to biodegradability, but the mixed liquor generated from leachate is substantially less filterable than that produced from sewage or other industrial effluents (Fig. 5.40). It is mainly for this reason that leachate biotreatment by MBRs tends to be a high-energy process, with specific energy demands ranging between 1.75 and 3.5 kWh/m^3 for standard sMBRs operating with pumped crossflow.

5.4.3.2 Bilbao

The leachate treatment MBR plant at the landfill in Bilbao (Fig. 5.41) is the largest such one in the world and has been in operation since 2004. The design inflow to the plant is 1.8 MLD, though it has actually been treating peak loads of 2.2 MLD. The process is a classical BIOMEMBRAT® with a pressurised bioreactor tank operating at an HRT of 15 h, an SRT of 53 days and an aeration rate of 4000 Nm^3/h. Following flow balancing, the leachate is treated in two parallel lines, each consisting of a denitrification and two nitrification tanks (Fig. 5.42). The sludge from each line is pumped into two UF plants comprising three streams of 6 modules in series, the permeate being directly discharged. The membrane plant operates at a mean TMP and permeate flux of 3 bar and 120 LMH, respectively.

5.4.3.3 Freiburg

The 0.1 MLD Freiburg plant (Fig. 5.43) represents an early example of an airlift MBR for leachate treatment, having been commissioned in 1999, and is designed for ammonia and COD removal (Table 5.23). The treatment scheme (Fig. 5.44) comprises flow

Table 5.22 Mean leachate quality values from various landfill sites worldwide

Parameters – all mg/L except pH and BOD/COD ratio	Reference							
	Urbini *et al.* (1999)		Tchobanoglous *et al.* (1993)		Robinson (1996)		Irene and Lo (1996)	
	Young leachate (<2 years)	Old leachate (>6.5 years)	New landfills (<2 years)	Mature landfills (>10 years)	Young leachate	Old leachate	Active landfills	Closed landfills
BOD	24 000	150	2500–3000	10–20	11 900	260	1600	160
COD	62 000	3000	3000–60 000	100–500	23 800	1160	6610	1700
TOC	NG	NG	1500–20 000	80–160	8000	465	1565	625
BOD/COD	0.39	0.05	0.05–0.67	0.04–0.1	0.5	0.2	0.24	0.09
$N\text{-}NH_4$	1400	350	10–800	20–40	790	370	1500	2300
pH	5.8	8	4.5–7.5	6.6–7.5	6.2	7.5	5.6–7.3	7.9–8.1

(from Alvarez *et al.*, 2004).

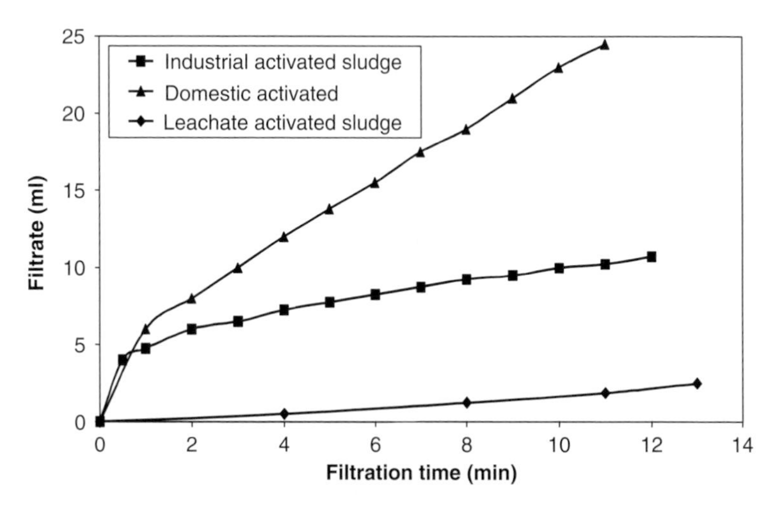

Figure 5.40 Leachate biomass filterability vs. other matrices (Robinson, 2005)

Figure 5.41 The plant at Bilbao showing the bioreactor towers and sidestream membranes

balancing followed by separate denitrification and nitrification, with the sludge being pumped into the airlift UF plant. The UF permeate is directed through an activated carbon vessel before being discharged. The biomass (sludge) recirculation rate is controlled by a variable speed circulation pump where the optimum tubular crossflow

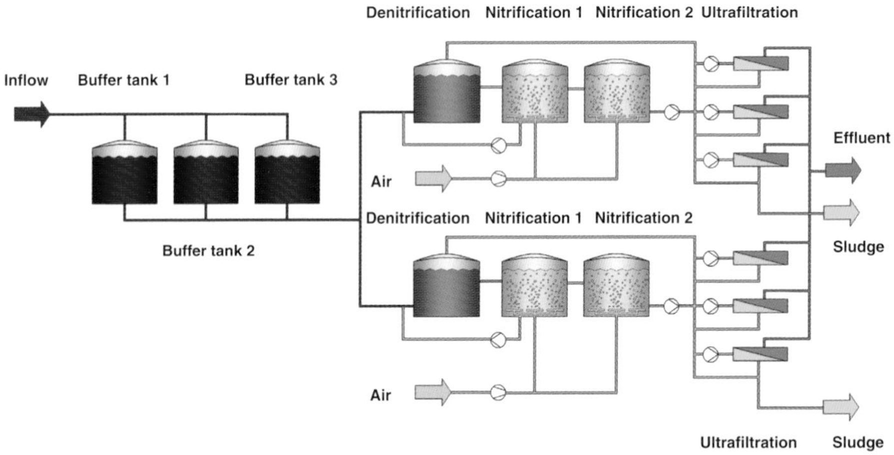

Denitrification Nitrification 1 Nitrification 2 Ultrafiltration
Inflow Buffer tank 1 Buffer tank 3
Effluent
Air
Denitrification Nitrification 1 Nitrification 2
Buffer tank 2
Sludge
Air
Ultrafiltration Sludge

Figure 5.42 The treatment scheme at Bilbao

Figure 5.43 The plant at Freiburg

Table 5.23 Inlet and outlet characteristics, Freiburg

	Inlet	Outlet (consent for discharge to sewer)
COD (mg/L)	1500	400
Ammonia (mg/L)	900	100

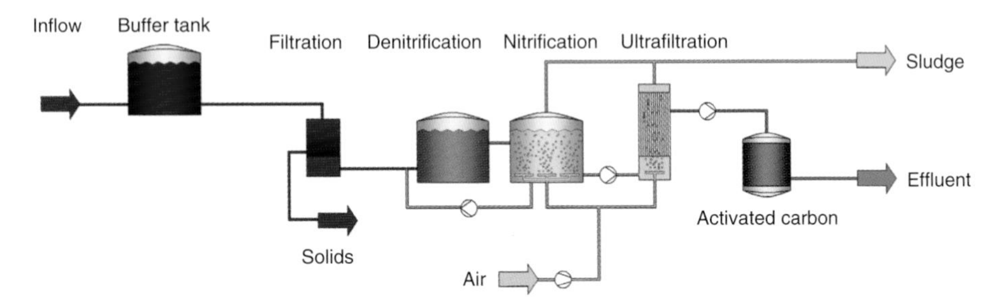

Figure 5.44 *The treatment scheme at Freiburg*

velocity is 2 m/s. This velocity provides fluxes between 50 and 60 LMH at a specific energy demand of 2–3 kW/m^3. Following excessive rainfall, it is sometimes necessary to operate at higher fluxes (up to 90 LMH), which can be achieved by increasing the liquid pumping rate to achieve a CFV of up to 4 m/s. At this flow rate, the specific energy demand increases to 5–6 kW/m^3. During summer months of low rainfall and so reduced volumetric demand on the MBR, the circulation pump can be either switched off or operated at low flow, in which case the energy demand decreases. The MLSS is generally between 12 and 15 g/L.

5.4.3.4 Performance comparison

Given the availability of the data (Robinson, 2005), it is instructive to compare:

- the hydraulic performance characteristics of the pumped and airlift mode sMBRs;
- treatment of leachate vs. dairy effluent feedwaters.

This comparison yields an indication of the relative fouling propensities of the respective mixed liquors generated by the two matrices.

Data for the airlift and pumped configurations (Table 5.24) indicate that the latter operate at almost twice the flux of the airlift type (107 LMH daily average for Bilbao, *cf.* 57 for Freiburg) but at the expense of a 66% higher specific energy demand (5 vs. 3 kWh/m^3 permeate). The projected membrane life for the airlift configuration is also higher, such that the overall specific operating cost for this mode of operation is 32% lower than that of the pumped sMBR. The filterability of the Bilbao leachate sludge is also much higher than that of Freiburg.

Compared with the three food effluent MBRs for which data are provided, the landfill leachate plant operates at a higher cost, associated mainly with a higher energy demand (due to the higher crossflow required to maintain the flux) and a shorter membrane life (brought about by the low filterability/high fouling propensity of the leachate biomass).

5.4.4 Thermophylic MBR effluent treatment, Triqua, the Netherlands

Veiligheidspapierfabriek (VHP) Ugchelen B.V., based near Apeldoorn in the Netherlands, is a subsidiary of the international banknote paper manufacturer Arjo Wiggins.

Table 5.24 Data for various Wehrle leachate and food effluent treatment plant

Customer/site: *Application:*	*Units*	**Freiburg** Leachate *Airlift*	**Bilbao** Leachate *Pumped*	**Kellogg** Food *Pumped*	**Dairy Crest** Dairy *Pumped*	**Dairy Gold** Dairy *Pumped*
Flow rate	MLD	0.12	1.8	1.8	1.2	2.0
Membrane area	m^2	88	700	864	486	648
Flux (year average)	LMH	54	102	71	98	125
Flux (day average)	LMH	57	107	87	103	129
Membrane price total	€	19 360	189 000	198 400	111 600	163 200
Membrane life	years	6	4	4.50	5.00	6.00
Membrane replacement cost	€/a	3227	47 250	44 089	22 320	27 200
Specific energy demand[b]	kWh/m^3	3	5	3.75	3.18	2.14
Specific energy cost	€/kWh	0.06	0.06	0.06	0.06	0.06
Total energy cost	€/a	7490	187 245	121 564	79 347	91 264
Total operating cost[c]	€/a	10 716	234 495	165 653	101 667	118 464
Total specific operating cost	€/m^3	0.26	0.38	0.31	0.24	0.17

[a]Membrane + Biology + Civils
[b]UF stage
[c]Sum of energy and membrane replacement
Robinson (2005).

Banknote paper is made from cotton combers bleached with sodium hydroxide and hydrogen peroxide. The project concerned the use of an MBR, originally developed for landfill leachate treatment, operating under thermophilic conditions for removing colloidal and high molecular weight dissolved material to allow the reuse of wastewater from the bleaching plant. The process, based on 8 mm diameter MT polyvinyllidene difluoride (PVDF) membranes, has been piloted on a variety of high-strength recalcitrant industrial effluents.

A trial was originally conducted between June and October 1999 for the clarification of wastewater from the bleaching plant. Wastewater from the bleaching process has a mean COD of 4500 mg/L, a pH of >12 and a temperature of 75–85°C. The quality of the effluent from these trials, which has a COD of around 400 mg/L, was sufficient to allow it to be reused in the bleaching vessels and led to the decision to install a full-scale plant (Fig. 5.45) with a capacity of 0.24 MLD in November 2000.

The system comprises a buffer tank followed by a dual-chamber DAF process for the removal of cotton flocks and fibres. In the second chamber, waste carbon dioxide from the boiler house is introduced to lower the pH of the solution from >12 to 8–9, an ingenious use of waste gas which reduces CO_2 emissions. The pH-adjusted clarified effluent is then transferred to the MBR aeration tank, 250 m^3 in volume and thus providing an HRT of ~24 h. The MBR operates at between 55°C and 60°C and an MLSS of 20 g/L. Three sidestream MT membrane units are employed (Fig. 5.46), each comprising 6 × 3 m-long modules in series with each module having a membrane area of 4 m^2 and hence providing a total area of 72 m^2. The membrane units operate at a mean flux of >120 LMH, an average TMP of 4 bar, and a mean crossflow velocity of ~4.5 m/s. The permeate COD concentration is 400–500 mg/L, and it is reheated by passing it through a heat exchanger with the influent before being reused

Figure 5.45 The VHP MBR

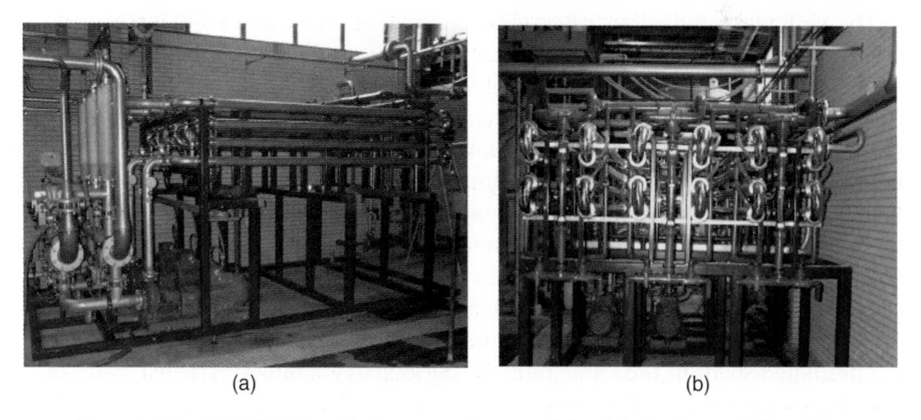

(a) (b)

Figure 5.46 VHP Ugchelen sidestream membrane modules: (a) side view and (b) end view

in bleaching vessels. Pipework is of RVS 304 and 316 stainless steel, since the high solution temperatures and ionic strengths present a fairly corrosive environment.

The total project costs were ~\$1.1 m, and the annual operating costs around \$50 k p.a. for the annual treated volume of 75 000 m^3. This amount includes (a) monitoring of the system (2 h/day) based on routine measurement and maintenance of pumps, DAF, sensors, etc. and (b) membrane replacement based on a lifetime of 2 years. The annual reduction of discharge of wastewater from 10 000 to 5000 p.e.

leads to cost savings of around \$270 k p.a. Furthermore, the total reduction of freshwater intake is around 40 000 m^3/year, and the annual reduction in gas consumption ~700 000 m^3. These reductions provide an additional cost saving of \$100 k p.a. The payback time on the original investment is around 4 years, although a governmental subsidy reduced this period to 3.3 years.

5.4.5 Millenniumpore

5.4.5.1 Eden Project, St Austell, Cornwall, UK

The Eden Project in Cornwall, UK was devised and constructed as part of the UK Millennium programme "to promote the understanding and responsible management of the vital relationship between plants, people and resources leading to a sustainable future for all" (Eden Project, 2006). Major landmarks of the project, which has been constructed in a crater and which views itself as a "living theatre", are the two greenhouses or "biomes" which sustain plants of both the humid tropical and the warm temperate regions of the world. The Eden project is a major tourist attraction, and there are plans to develop the site further in the future, including the construction of a third biome (Eden Project, 2006).

The Millenniumpore MBR is situated at the Waste Recycling Centre and receives liquid waste streams from the toilet and washrooms, the biocomposter effluent, storm water transferred from the site collection lagoon and run-off from the waste recycling compound area. Water is reused for irrigation of the plants, which form the central part of this tourist attraction.

The MBR plant at the Waste Recycling Centre (Figs. 5.47 and 5.48) has been operational since March 2004 and treats up to 40 m^3 of effluent per day. Liquid waste having a COD which varies between 120 and 780 mg/L and containing some solids debris is delivered by gravity to a 6 m^3 capacity underground septic tank where some solids settlement occurs. Supernatant liquid and non-settled solids overflow via underground ducting to a concrete-lined underground buffer tank (a manhole), which also collects storm water when capacity permits. The accumulating waste in the buffer storage tank is pumped for treatment to the downstream MBR process via a 0.5 mm bag filter for removal of gross non-settled solids. The bag filter is periodically removed and cleaned. Waste sludge from the MBR process is fed back to the septic tank where it settles along with the gross solids from the incoming effluent stream. Accumulated solids in the septic tank are periodically removed. Treated water from the MBR process is stored for reuse.

The MBR is configured as a sidestream unit and comprises a 4 m^3-capacity cylindrical bioreactor tank fitted with two vertically oriented PVC-housed MT modules which are each 225 mm in diameter and 1.2 m in length. Each is fitted with epoxy resin-sealed 5.3 mm internal diameter (i.d.) lumens of pore size 0.05 μm, providing an area of 8.6 m^2 per module. A 3 kW centrifugal pump produces a flow of ~70 m^3/h, yielding a mean crossflow velocity of 2 m/s, at a TMP of ~0.8 bar. This produces a mean operating flux of 16.5 LMH (peaking at 30), equating to a specific energy demand of 1.8 kWh/m^3 for membrane permeation alone. The MLSS is maintained at around 12–15 g/L by wasting 2 m^3 of sludge every 2 months to produce an SRT of ~120 days.

Figure 5.47 The Eden Project MBR

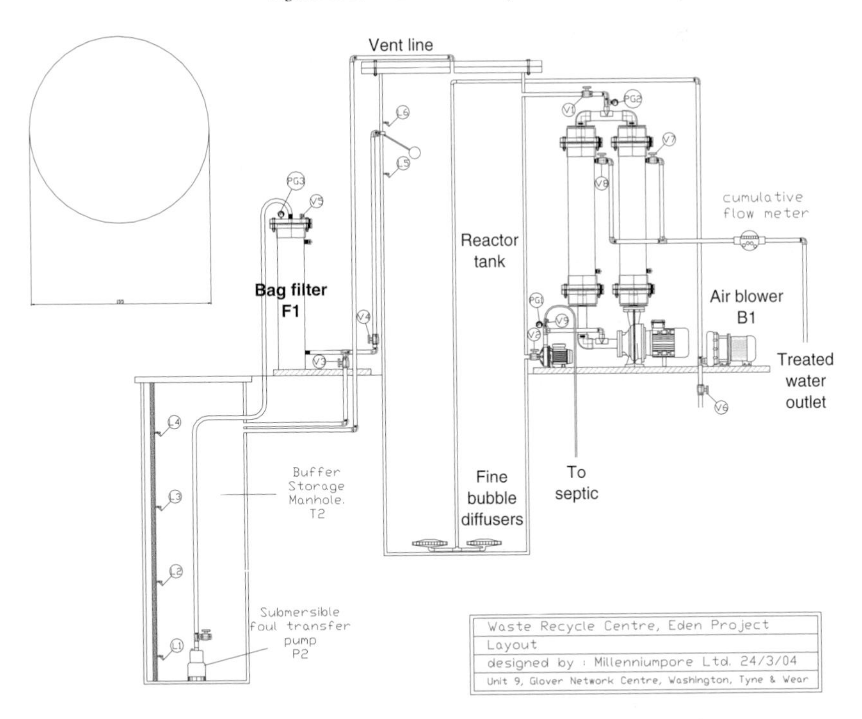

Figure 5.48 The Eden Project wastewater recycling treatment scheme

5.4.5.2 Other Millenniumpore plant

A Millenniumpore MBR plant of very similar design to that used at the Eden Project site was installed for treating highly concentrated organic waste from a beverage production factory, owed by VitaSoy International Holdings Limited, in Hong Kong in 2003. The effluent, which has a COD varying between 500 and 20 000 mg/L, is partially clarified by coagulation/DAF to produce a feed to the MBR of 320–17 000 mg/L COD. The MBR is operated at an MLSS of 7–9 g/L, and six MT modules provide sufficient capacity for the flow of 88 m^3/day. The effluent COD level is between 11 and 93 mg/L.

5.4.6 Novasep Orelis

5.4.6.1 The Queen Mary II WWTP

Background Legislation concerning water discharged to sea has recently been tightened. In the past, the practice had been to employ conventional wastewater treatment for foul waters and discharge grey water to the sea. The latter is no longer considered acceptable; a number of pieces of legislation applied to different geographical regions have been promulgated which effectively disallow grey water discharge, the most restrictive being the Alaskan regulations which limit both TSS and BOD to 30 mg/L, faecal coliforms to 20 per 100 ml and undetectable levels of heavy metals. Some commercial cruise ship companies have adopted targets which are even stricter than the legislation, since (a) the marine conservation lobby is likely to bring about ever more stringent standards and (b) only slightly more rigorous treatment would permit water reuse for general purposes such as cooling and sluicing down.

Ship-board wastewater treatment presents a number of challenges. The deck height, the floor-ceiling distance in which the MBR has to be located, is normally around 2.5 m. This limits the aeration tank depth and prohibits inserting the membrane module into or lifting it from the tank *in situ*. The plant has to run with very little maintenance, since marine crew generally have little or no experience of wastewater treatment. Full plant automation is thus required, and in the case of the Queen Mary II (QM2) staff training has also been provided. Lastly, and perhaps most importantly, accurate effluent water data is at a premium in the ship industry.

The QM2 is the world's largest passenger ship and is owned by the Carnival Group. A total of around 1.1 MLD of wastewater is generated on board the ship from four main sources, these being:

- Grey water from accommodation (0.65 MLD) and laundering operations (0.150 MLD).
- Galley water (0.2 MLD) from the kitchens and food waste drainage.
- Black (or foul) water (0.100 MLD).
- Recreational pool and sauna wastewater (<0.01 MLD).

These flows can be segregated into two streams of black and grey water (Table 5.25). Carnival has set targets of 10 mg/L TSS and undetectable faecal coliforms for the treated water, thereby providing the option of reuse of the grey water stream.

Table 5.25 Basis for process design: wastewater quality on board the QM2

Parameter	Blackwater	Grey water
Flow (m^3/day)	300	800
COD (average)	3500	600
TSS (mg/L)	1000	1200
BOD_5 (mg/L)	1200	200
N-NTK (mg/L)	Not defined	Not defined
P-tot (mg/L)	Not defined	Not defined
pH	4.5–8	4.5–7.5
Temperature (°C)	Not defined	Not defined
FOG (mg/L)	200	20
TDS (mg/L)	500	500
Phenol (mg/L)	0.002	0.018 g/L
Arsenic (mg/L)	Not defined	0.06 mg/L
Copper (mg/L)	Not defined	2.00 mg/L
Lead (mg/L)	Not defined	0.07 mg/L
Zinc (mg/L)	Not defined	2.90 mg/L

Treatment scheme The full treatment scheme (Fig. 5.49) comprises a prefiltration unit using a 1 mm drum filter with hydrocyclones for sand removal. This is followed by biotreatment in a 150 m^3 tank fitted with two blowers each providing 400 Nm^3/h of air which maintains the DO between ~0.5 and ~2 mg/L. The tank height is 4.5 m installed within two decks: there is a 0.5 m clearance between the top of the tank and the floor of the deck above. The MLSS in the bioreactor is kept between ~8 and ~12 g/L and the reactor temperature between 20°C and 35°C. The sludge production rate is 0.15 kg MS/kg COD.

Filtration is achieved by two lines of five skid-mounted Pleiade® UF modules (Fig. 5.50) containing 200 membrane elements providing 70 m^2 total membrane area per module, and thus 700 m^2 of total MF area. The skid has a footprint of 9 m × 4 m and is 2.5 m high, and each line is fitted with a cleaning tank providing an automated CIP every 3–5 weeks to supplement the water flush every 2–4 h. The skid is fed with two feed pumps (duty and standby) and five circulation pumps, providing a mean TMP of 2 bar and a flux of 75 LMH. The grey water line is fitted with activated carbon and UV irradiation downstream of the membrane stack to enable reuse. The wasted sludge from both the grey and blackwater lines is delivered to a centrifuge.

A number of challenges were presented in the early stages of operation. The blackwater was out of specification, causing "balling" of solids (formation of gross solid ~10 mm particles) in the bioreactor from paper fibrous material which caused blockage of the membrane flow channels. This necessitated disassembling of the modules for manual cleaning of the membranes. A number of remedial modifications were investigated to overcome this problem, including changing the toilet paper to one with longer cellulose fibres and dosing with enzymes to encourage the breakdown of cellulosic matter. Successful demonstration led to the installation of a supplementary prefilter. The discharge from the kitchen waste was also out of specification (Table 5.26) downstream of the grease traps, causing foaming which demanded dosing with antifoaming agent. The greatest challenge experienced,

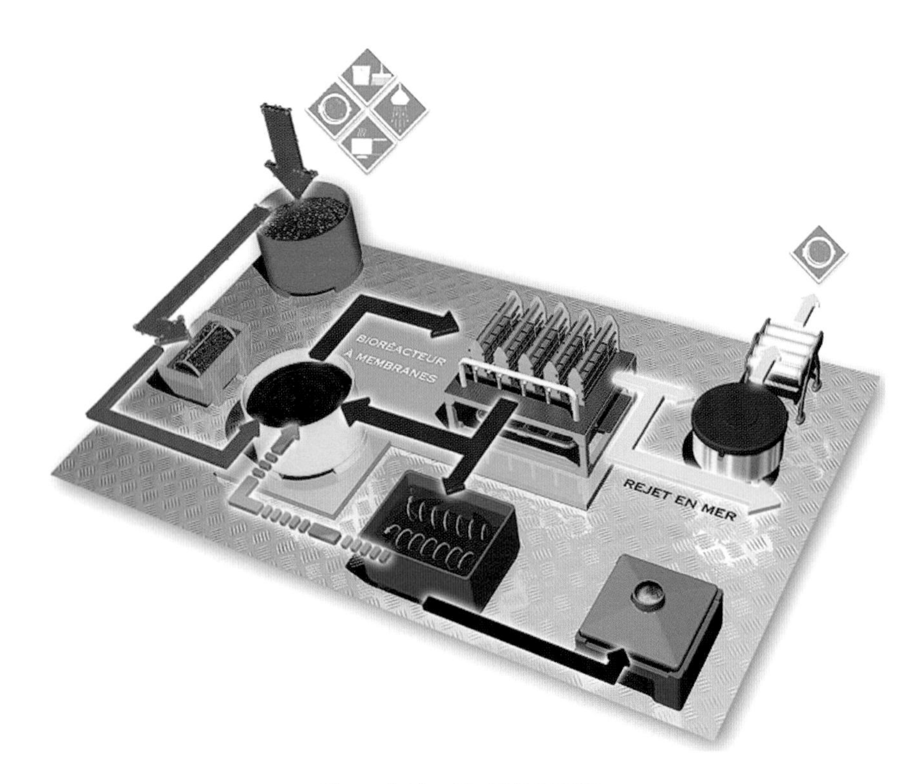

Figure 5.49 The QM2 WWTP

Figure 5.50 Pleiade® skid on the Queen Mary 2

though, was with engaging the crew, who are changed every 3 months making it difficult for them to be familiarised with the technology.

The plant has been in operation since January 2003 and is now operating well, thanks to better information regarding the quality of the various effluent streams and a constructive relationship with the crew.

5.4.7 Other Orelis plant

There are a large number of relatively small Orelis MBR plant for industrial effluent treatment. A classic example is a plant which treats effluent from a sauce preparation facility in France, where a flow of 0.2 MLD containing high organic levels (8000 mg/L COD) has to be treated down to below 125 and 35 mg/L COD and BOD5, respectively, prior to discharge, as well as providing the prospective option of recycling by retrofitting of a reverse osmosis system at some point in the future. The system fitted (Fig. 5.51) comprised two RP2085 Pleiade® membrane modules of 56 m^2 each, thereby operating at a mean flux of ~150 LMH.

Table 5.26 Actual wastewater quality on board QM2: averaged data from three samples of Food waste drainage (FWD) and two samples of Galley grey water (GGW)

Parameter (mg/L)	FWD			GGW	
COD	52 000	880	2000	25 000	2000
Total organic carbon	3 480	270	580	3000	530
BOD_5		350	33 000	23 000	1300
TSS	21 100	110	39 000	3400	300
Ammonia nitrogen	63	1	12	5.8	ND
Total phosphorus		10	26	51	26
FOG		81	200		210
PH	3.44	5.59	3.42	3.79	5
Temperature				31	

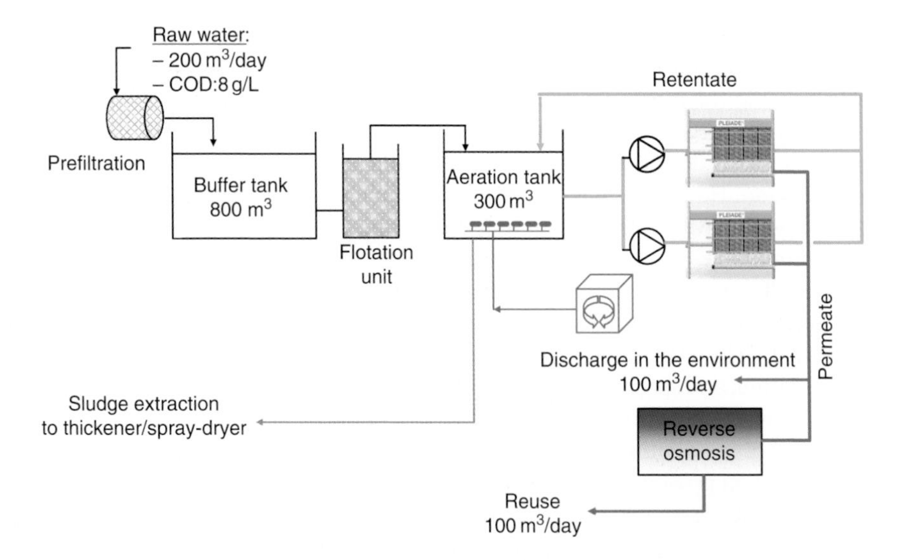

Figure 5.51 Orelis Novasep MBR treatment scheme for sauce preparation effluent treatment

5.5 MBRs: prognosis

A number of case studies have been provided from 14 commercial technology providers. Design features of each technology have been reviewed in Chapter 4, and O&M data from the full-scale case studies, along with comparative pilot study data, are analysed in Section 3.3.

MBRs are generally regarded by the water industry as a whole as being a novel technology (although even the new submerged technology is now over 15 years old), with a degree of risk associated with their safe operation and satisfactory performance. They are also comparatively expensive, both due to the additional cost of the membrane over that of a conventional activated sludge or SBR and the supplementary membrane aeration requirement.

There can be little doubt that the capital cost of the MBR technology itself is likely to remain higher than that of a conventional ASP. However, the added value provided by the decreased footprint and higher-purity product of the MBR cannot be discounted. Indeed, for some older STWs based near residential areas within certain urban conurbations, it is conceivable that the land made available by installing an MBR may be of sufficient financial value to cover the capital cost of the MBR. It is a reflection of both the increasingly stringent demands placed on discharged effluent quality and the decreasing costs of the technology (Figs. 1.4 and 1.5) that MBRs are increasing in capacity, the largest having a capacity of around 50 MLD as of end 2005. Moreover, there are an increasing number of players in the market (Table 4.4), particularly from South-East Asia, with more reference sites providing an increasing knowledge base – especially with respect to the crucially important parameter of membrane life (Table 5.2).

The perception of MBRs as being high risk is perhaps based on the limited tolerance of the membrane itself. All membranes foul, and MBR membranes are no exception. By conventional membrane technology standards, the medium undergoing filtration by the membrane (8–18 g/L solids) is a very challenging one, and it is perhaps unsurprising that fouling problems can arise. There are also issues of foaming, as indeed can occur with all aerobic treatment processes. MBRs are possibly more susceptible to this problem due to the more vigorous aeration demanded, and foaming can be severe if the biomass undergoes rapid dilution. Membrane channel clogging can also arise at higher solids concentrations, coupled with insufficient or unevenly distributed membrane aeration. However, steps can be taken to ameliorate such problems. It is generally the case that MBRs can and have been operated successfully when the plant has been appropriately, and possibly, slightly conservatively, designed and operated. MBRs are possibly less tolerant hydraulically than conventional processes due to the vagaries of membrane fouling, but are much more tolerant with respect to maintaining treated water quality. This is manifested as an increasing number of effluent recycling plants based on MBRs, and specifically MBRs with downstream RO (Section 5.3.1.5).

Whilst it would be wrong to regard MBRs as a panacea, it is definitely the case that the number of installations has increased at an accelerated rate as their acceptance widens based on proven performance and process economics. There is substantially less process risk associated with the process than existed ten years ago, and a wider range of technologies than ever is now available, albeit predominantly based on just two membrane module configurations. A number of process improvements have been introduced within the past 5–10 years to increase reliability and also to reduce membrane module fabrication and process O&M costs. As a result, the MBR process is becoming increasingly competitive, and few would predict that their increasing market penetration is likely to wane in the near future.

References

Alvarez-Vazquez, H., Jefferson, B. and Judd, S.J. (2004) Membrane bioreactors vs. conventional biological treatment of landfill leachate: a brief review. *J. Chem. Technol. Biotechnol.*, **79**(10), 1043–1049.

Catley, G. and Goodwin, S. (2004) Industrial water re-use, full scale studies in the food industry: technical performance and cost savings, *CIWEM 2nd National Conference*, September, Wakefield, UK.

Chen, T.K., Ni, C.H., Chen, J.N. and Lin, J. (2003) High-strength nitrogen removal of opto-electro industrial wastewater in membrane bioreactor – a pilot study. *Wat. Sci. Techol.*, **48**(1), *pp*. 191–198.

Churchouse, S. and Brindle, K. (2003) Long-term operating experience of membrane bioreactors, *MBR4*, April, Cranfield University, UK.

Côté, P., Belli, L., Rondi, S. and Meraviglia, I. (2004) Brescia large-scale membrane bioreactor: case study, *Paper Presented at UK MAIN*, May, Cranfield University, UK.

Eden Project www.edenproject.com accessed February 2006.

Irene, M. and Lo, C. (1996) Characteristics and treatment of leachates from domestic landfills. *Environ. Intern.*, **4**, 433–442.

Jeavons, J., Stokes, L., Upton, J. and Bingley, M. (1998) Successful sidestream nitrification of digested sludge liquors. *Wat. Sci. Techol.*, **38**(3), 111–118.

Judd, S. (1997) MBRs – why bother? *Proceedings MBR1*, Cranfield University, March.

Judd, S. and Jefferson, B. (2003) *Membranes for Industrial Wastewater Recovery and Re-use*. Elsevier, Oxford.

Kennedy, S. and Churchouse, S.J. (2005) Progress in membrane bioreactors: new advances, *Proceedings of the Water and Wastewater Europe Conference*, Milan, June.

Lozier, J. and Fernandez, A. (2001) Using a membrane bioreactor/reverse osmosis system for indirect potable reuse, *Water Suppl* 5–6(1) 303.

Mitsubishi Rayon (2000). Sterapore Water Treatment, Case Reports 1 & 2: Dairy and ume plum processing wastewater treatment system, and Livestock wastewater treatment and reuse system

Mulder, J.W. and Kramer, F. (2005) *Hybrid MBR – the Perfect Upgrade for Heenvliet*, *H_2O*, Rinus Visser, the Netherlands, *pp*. 66–68.

Paranjape, S. Reardon, R. and Foussereau, X. (2003) Pretreatment technology for reverse osmosis membrane used in wastewater reclamation application – past, present and future literature review, *Proceedings of the 76th Water Environment Federation Technical Exposition and Conference* (WEFTEC), Oct., Los Angeles.

Qin, J.J., Kekre, K.A., Tao, G., Oo, M.H., Wai, M.N., Lee, T.C., Viswanath, B. and Seah, H. (2006) New option of MBR-RO process for production of NEWater from domestic sewage *J. Membrane Sci.*, in press.

Robinson, H.D. (1995) A review of the composition of leachates from domestic wastes in landfill sites. Report Prepared for the *UK Department of the Environment, Contract PECD 7/10/238*, ref: DE0918A/FR1.

Robinson, A. (2005) Landfill leachate treatment, *Proceedings MBR5*, January 2005, Cranfield University.

Symposium. S. Margherita di Pula, Cagliari, Italy, CISA. Environmental Sanitary Engineering Centre Cagliari, Italy, pp. 73–80.

Tchobanoglous, G., Theisen, H. and Vigil, S.A. (1993) Integrated Solid Waste Management: Engineering Principles and Management Issues. McGraw-Hill International Editions, p. 978.

Trivedi, H.K. and Heinen, N. (2000) Simultaneous nitrification/denitrification by monitoring NADH fluorescence in activated sludge, *Proceedings of the 73rd Water Environment Federation Technical Exposition and Conference*, Anaheim, CA.

Urbini, G., Ariati, L., Teruggi, S. and Pace, C. (1999) Leachate Quality and Production from Real Scale MSW Landfills, *Proceedings of the Seventh Management and Landfill Conference*, Jun., Sardina.

Appendix A

Blower Power Consumption

An air blower is a mechanical process that increases the pressure of air by applying shaft work. According to the first law of thermodynamics:

$$\Delta Q = \Delta E + \Delta W \tag{A.1}$$

where Q = heat transferred, E = system energy and W = work done by the system. The system energy comprises kinetic energy ($\frac{1}{2}mu^2$, m and u being mass and velocity, respectively), potential energy (mgh, h being height and g being acceleration due to gravity) and internal energy (e). The work done by the system is made up of external work (W) and work performed against forces due to pressure, p/ρ. From this information it is possible to derive Equation (A.2) for a fluid moving at steady flow between points 1 and 2 as illustrated in Fig. A.1.

$$\frac{\delta Q}{\delta m} = \left(e_2 - e_1\right) + \frac{\left(u_2^2 - u_1^2\right)}{2} + g\left(z_2 - z_1\right) + \frac{\delta W}{\delta m} + \frac{\left(p_2 - p_1\right)}{\rho} \tag{A.2}$$

This equation is known as the *steady-flow* energy equation. In this case the change in internal, kinetic and potential energy of the fluid are negligible, such that associated terms in Equation (A.2) may be ignored. Ignoring, in the first instance, frictional losses, manifested as heat, $\delta Q = 0$. Because a blower exerts work on the fluid the work term will be negative in this case and to simplify the equation this will be ignored. The fluid density in this system is not constant and the conditions are said to be adiabatic (change in density with no heat transferred to the gas):

$$dw = \frac{dp}{\rho} \tag{A.3}$$

where w is the work done per unit mass of fluid. For an ideal gas under adiabatic conditions

$$\frac{p}{\rho^{\lambda}} = \text{const}$$

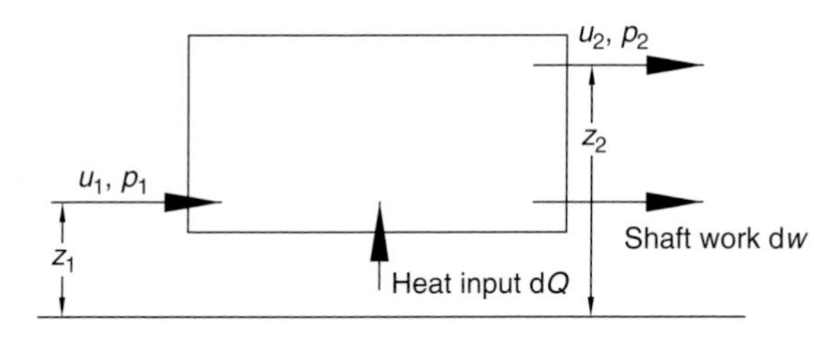

Figure A.1 Steady flow system

or

$$\frac{p_1}{\rho_1^{\lambda}} = \frac{p}{\rho^{\lambda}}$$

or

$$\rho = \frac{\rho_1}{p_1^{1/\lambda}} p^{1/\lambda} \tag{A.4}$$

where λ is the ratio of specific heat capacity at constant pressure to specific heat capacity at constant volume (c_p/c_v) and is assumed to be constant. Substituting Equation (A.4) into Equation (A.3) and integrating between points 1 and 2:

$$w = \frac{p_1^{1/\lambda}}{\rho_1} \int_{p_1}^{p_2} \frac{dp}{p^{1/\lambda}} = \frac{p_1^{1/\lambda}}{\left(1 - 1/\lambda\right)\rho_1} \left(p_2^{1-1/\lambda} - p_1^{1-1/\lambda}\right)$$

This equation can be manipulated to give:

$$w = \frac{p_1 \lambda}{\left(\lambda - 1\right)\rho_1} \left[\left(\frac{p_2}{p_1}\right)^{1-1/\lambda} - 1\right] \tag{A.5}$$

This equation gives the work done per unit mass by a blower that is 100% efficient. No mechanical process is 100% efficient so a term for the mechanical efficiency of the system (ξ) is added; this accounts for any losses in the system, including friction. w is work per unit mass and power is total work per second. Equation (A.5) can be converted into power consumption in (W) by dividing by time. The density of air varies with temperature according to:

$$\rho_t = \rho_0 \frac{T_0}{T_t} \tag{A.6}$$

Where T_0 is the standard temperature (273°K). The power consumed, in kW, thus becomes:

$$\frac{P}{m} = \frac{p_1 \lambda}{1000\xi\left(\lambda - 1\right)\rho_0 t} \frac{T_1}{273} \left[\left(\frac{p_2}{p_1}\right)^{1-1/\lambda} - 1\right] \tag{A.7}$$

but

$$V = qt \tag{A.8}$$

where V is volume, q is volumetric flow and t is time. Also:

$$\rho = \frac{m}{V} = \frac{m}{qt} \tag{A.9}$$

Substituting Equation (A.9) into (A.7) provides theoretical blower power of:

$$\text{Power} = \frac{p_1 T_1 \lambda q}{2.73 \times 10^5 \xi (\lambda - 1)} \left[\left(\frac{p_2}{p_1} \right)^{1-1/\lambda} - 1 \right] \tag{A.10}$$

Note: all values are in SI units unless otherwise stated.

Appendix B

MBR Biotreatment Base Parameter Values

Parameter	Value	Units	Reference	Process	Wastewater tested and additional comments
$k_{d,n}$	0.21	per day	Dinçer and Kargi (2000)	ASP with denitrification	Synthetic wastewater (100:100 COD:NH_4-N)
k_{dn}	0.05–0.15	per day	Metcalf and Eddy (2003)		0.08 typical
k_e	0.06–0.2	per day	Metcalf and Eddy (2003)		0.12 typical
k_e	0.067	per day	Yenkie (1992)	High-rate CAS	22°C, synthetic wastewater (1090:872 COD:BOD), SRT 0.3 day
k_e	0.050	per day	Fan *et al.* (1996)	MBR	Municipal wastewater, 30°C, 411–72 COD: 26–53 NH_4-N, 20 days SRT
k_e	$0.85\theta_x^{-0.62}$	per day	Huang *et al.* (2001)	MBR	Domestic wastewater (~250:20:170 COD:NH_3-N:SS), SRT 5–40 days
k_e	0.023	per day	Liu *et al.* (2005)	MBR	Synthetic wastewater (220–512 mg/L COD, 36–72 mg/L NH_4-N), infinite SRT
k_e	0.08	per day	Wen *et al.* (1999)	MBR	30°C, urban wastewater (~500 COD), SRT 5–30 days
k_e	0.025–0.075	per day	Xing *et al.* (2003)	MBR	Variable (30–2234 mg/L COD) municipal wastewater, SRT 5–30 days
k_e	0.048	per day	Yilditz *et al.* (2005)	MBR	Synthetic wastewater, 26°C, 1090 COD, SRT 0.3 day
$k_{d,n}$	0.12	per day	Harremoës and Sinkjaer (1995)	ASP with denitrification	Municipal wastewater, 20°C, 397–256 COD: 40–35 TKN, 18–21 days SRT
K_n	0.1–0.4	g/m^3	Harremoës and Sinkjaer (1995)	ASP with denitrification	Municipal wastewater, 20°C, 397–256 COD: 40–35 TKN, 18–21 days SRT
K_n	0.5–1	g/m^3	Metcalf and Eddy (2003)		0.74 typical
K_n	0.1–0.15	g/m^3	Manser *et al.* (2005)	MBR & CAS in parallel	Domestic wastewater (quality not given), SRT 20 days
K_n	0.85	g/m^3	Wyffels *et al.* (2003)	MBR with low DO	30°C, ~sludge digester supernatant 605:931 COD:TAN (total ammonium nitrogen), SRT >650 days
K_n	0.01–0.34	g/m^3	Gröeneweg *et al.* (1994)	Nitrifying chemostat	Synthetic wastewater, 30°C, 392 mg NH_x-N/L. Values were dependant on pH, temp and bacterial species
K_s	5–40	g/m^3	Metcalf and Eddy (2003)		20 typical
K_s	80	g/m^3	Yenkie (1992)	High-rate CAS	22°C, synthetic wastewater (1090:872 COD:BOD), SRT 0.3 day
K_s	192	g/m^3	Yilditz *et al.* (2005)	MBR	Synthetic wastewater, 26°C, 1090 COD, SRT 0.3 day
Y	0.3–0.5	g VSS/g bCOD	Metcalf and Eddy (2003)		0.4 typical
Y	0.44	per day	Yenkie (1992)	High-rate CAS	22°C, synthetic wastewater (1090:872 COD:BOD), SRT 0.3 day

Y	0.61	g VSS/g COD	Fan *et al.* (1996)	MBR	Municipal wastewater, 30°C, 411–72 COD: 26–53 NH_4-N, 20 days SRT
Y	0.28–0.37	g VSS/g COD	Huang *et al.* (2001)	MBR	Domestic wastewater (~250:20:170 COD:NH_3-N:SS), SRT 5–40 days
Y	0.288	g VSS/g COD	Liu *et al.* (2005)	MBR	Synthetic wastewater (220–512 mg/L COD, 36–72 mg/L NH_4-N), infinite SRT
Y	0.40–0.45	g VSS/g BOD	Lübbecke *et al.* (1995)	MBR	Synthetic wastewater (8500–17 600 mg/L COD, 36–72 mg/L NH_4-N), SRT 1.5–8 days
Y	0.56	g VSS/g COD	Wen *et al.* (1999)	MBR	30°C, urban wastewater (~500 COD), SRT 5–30 days
Y	0.25–0.40	g VSS/g COD	Xing *et al.* (2003)	MBR	Variable (30–2234 mg/L COD) municipal wastewater, SRT 5–30 days
Y	0.58	g VSS/g COD	Yilditz *et al.* (2005)	MBR	Synthetic wastewater, 26°C, 1090 COD, SRT 0.3 day
Y_n	0.34	g TSS/g N	Dinçer and Kargi (2000)	ASP with denitrification	Synthetic wastewater (100:100 COD:NH_4-N)
Y_n	0.16	g VSS/g N	Harremoës and Sinkjaer (1995)	ASP with denitrification	Municipal wastewater, 7.5°C, 397–256 COD: 40–35 TKN, 18–21 days SRT
Y_{obs}	0.31–0.36	g VSS/g COD	Tao *et al.* (2005)	3 MBR plants in parallel	SRT 14–28 days, settled sewage, 265 COD
Y_{obs}	0.11	g VSS/g COD	Liu *et al.* (2005)	MBR	Synthetic wastewater (220–512 mg/L COD, 36–72 mg/L NH_4-N), infinite SRT
Y_{obs}	0.16–0.38	g VSS/g COD	Wen *et al.* (1999)	MBR	30°C, urban wastewater (~500 COD), SRT 5–30 days
μ_{max}	3–13.2	per day	Metcalf and Eddy (2003)		6 typical
μ_{max}	0.125	per day	Yenkie (1992)	High-rate CAS	22°C, synthetic wastewater (1090:872 COD:BOD), SRT 0.3 day
μ_{max}	3.24	per day	Yilditz *et al.* (2005)	MBR	Synthetic wastewater, 26°C, 1090 COD, SRT 0.3 day
μ_{max}/K_s	0.001–0.01	per day	Wen *et al.* (1999)	MBR	Lower than ASP. 30°C, urban wastewater (~500 COD), SRT 5–30 days
$\mu_{n,max}$	0.1–0.2	per day	Fan *et al.* (1996)	MBR	Municipal wastewater, 30°C, 411–72 COD: 26–53 NH_4-N, 20 days SRT
$\mu_{n,max}$	0.2–0.9	per day	Metcalf and Eddy (2003)		0.75 typical
$\mu_{n,max}$	2.02	per day	Wyffels *et al.* (2003)	MBR with low DO	30°C, ~sludge digester supernatant 605:931 COD:TAN (total ammonium nitrogen), SRT >650 days

References

Dinçer, A. and Kargi, F. (2000) Kinetics of sequential nitrification and denitrification processes. *Enzyme Microb. Tech.*, **27**, 37–42.

Fan, X-J., Urbain, V., Qian, Y. and Manem, J. (1996) Nitrification and mass balance with a membrane bioreactor for municipal wastewater treatment. *Water Sci. Technol.*, **34**(1–2), 129–136.

Gröeneweg, J., Sellner, B. and Tappe, W. (1994) Ammonia oxidation in *Nitrosomonas* at NH_3 concentrations near Km: effects of pH and temperature. *Water Res.*, **28**(12), 2561–2566.

Harremoës, P. and Sinkjaer, O. (1995) Kinetic interpretation of nitrogen removal in pilot scale experiments. *Water Res.*, **29**(3), 899–905.

Huang, X., Gui, P. and Qian, Y. (2001) Effect of sludge retention time on microbial behaviour in a submerged membrane bioreactor. *Process Biochem.*, **36**, 1001–1006.

Lübbecke, S., Vogelpohl, A. and Dewjanin W. (1995) Wastewater treatment in a biological high performance system with high biomass concentration. *Water Res.*, **29**(3), 793–802.

Liu, R., Huang, X., Jinying, X. and Quan, Y. (2005) Microbial behaviour in a membrane bioreactor with complete sludge retention. *Process Biochem.*, **40**, 3165–3170.

Manser, R., Gujer, W. and Hansruedi, S. (2005) Consequences of mass transfer effects on the kinetics of nitrifiers. *Water Res.*, **39**, 4633–4642.

Metcalf and Eddy (2003) *Wastewater Engineering – Treatment and Reuse* (3rd edn.). McGraw-Hill, New York.

Tao, G., Kekre, K., Wei, Z., Lee, T.-C., Viswanath, B. and Seah, H. (2005) Membrane bioreactors for water reclamation. *Water Sci. Technol.*, **51**(6–7), 431–440.

Wen, X.-H., Xing, C.-H. and Qian, Y. (1999) A kinetic model for the prediction of sludge formation in a membrane bioreactor. *Process Biochem.*, **35**, 249–254.

Wyffels, S., Van Hulle, S., Boeckx, P., Volcke, E., Van Cleemput, O., Vanrolleghem, P. and Verstraete, W. (2003) Modelling and simulation of oxygen-limited partial nitritation in a membrane assisted bioreactor (MBR). *Biotechnol. Bioeng.*, **86**(5), 531–542.

Xing, C.-H., Wu, W.-Z., Qian, Y. and Tardieu, E. (2003) Excess sludge production in membrane bioreactors: a theoretical investigation. *J. Environ. Eng.*, **129**(4), 291–297.

Yenkie, M., Gerssen, S. and Vogelphol, A. (1992) Biokinetics of wastewater treatment in the high performance compact reactor (HCR). *Chem. Eng. J.*, **49**, B1–B12.

Yilditz, E., Keskinler, B., Pekdemir, T., Akay, G. and Nuhoglu, A. (2005) High strength wastewater in a jet loop membrane bioreactor: kinetics and performance evaluation. *Chem. Eng. Sci.*, **60**, 1103–1116.

Appendix C

Hollow Fibre Module Parameters

C.1 Separation distance δ

Consider three equally spaced hollow fibres (HFs) in an element of internal volume V (Fig. C.1). The fibres of diameter d are separated by a distance of D, where D represents the diameter of a virtual cylinder surrounding each HF. The total volume of these virtual cylinders is equal to V minus the interstitial volume created by three touching cylinders. Each HF is then associated with one interstitial volume.

Given the axial symmetry and from basic trigonometry, the % volume occupied by the interstitial volume is given by the ratio:

$$R = \frac{\text{area of equilateral triangle of sides } D - \text{area of semicircle of side } D}{\text{area of equilateral triangle of sides } D}$$

Thus:

$$R = \frac{(D/2)D\sin 60 - (\pi/8)D^2}{(D/2)D\sin 60} = \frac{0.886/2 - 0.3923}{0.886/2} = 0.0460 \tag{C.1}$$

The volume occupied and the surface area provided by the fibres are given by:

$$V_{\text{f}} = \frac{\pi}{4}d^2 NL \tag{C.2}$$

$$A_{\text{f}} = \pi dNL \tag{C.3}$$

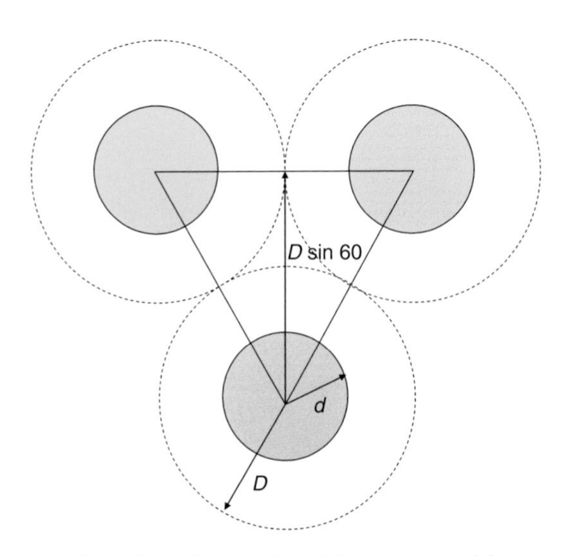

Figure C.1 Cross-section of three HFs in module

L being the internal module length and N being the number of fibres per element. The total cross-sectional area associated with each fibre is thus given by:

$$X_{\mathrm{f}} = \frac{V_{\mathrm{f}}}{LN} \tag{C.4}$$

Substituting Equation (C.3) into Equation (C.4) yields:

$$X_{\mathrm{f}} = \frac{\pi d V_{\mathrm{f}}}{A_{\mathrm{f}}} = \frac{\pi d}{\phi} \tag{C.5}$$

where ϕ is the fibre surface area per unit module volume ($A_{\mathrm{f}}/V_{\mathrm{f}}$). So, combining Equation (C.1) and Equation (C.5) yields:

$$D = \sqrt{\frac{4(1-0.046)d}{\phi}} \tag{C.6}$$

The separation, δ is thus given by:

$$\delta = 1.95\sqrt{\frac{d}{\phi}} - d \tag{C.7}$$

In fact, this separation differs by only 2.5% from the value calculated on the basis of ignoring the excluded area, for which the coefficient in the above equation becomes 2.

C.2 Free area A_{x}

The total free area can be obtained by subtracting the cross-sectional area of the fibres from that of the module:

$$A_{\mathrm{x}} = \frac{V}{L} - \frac{\pi d^2 N}{4} \tag{C.8}$$

Substituting for V and N with ϕ and A_{f}, from Equation (C.3), produces:

$$A_{\mathrm{x}} = \frac{A_{\mathrm{f}}}{\phi L} - \frac{A_{\mathrm{f}} d}{4L} \tag{C.9}$$

Thus:

$$A_x = \frac{A_f}{L}\left(\frac{1}{\phi} - \frac{d}{4}\right) \tag{C.10}$$

The free area per filament is simply given by:

$$A_{x,filament} = \frac{\pi}{4}\left(\frac{D^2}{1-R} - d^2\right) \tag{C.11}$$

and hence:

$$A_{x,filament} = \frac{\pi}{4}\left(\frac{4d}{\phi} - d^2\right) \tag{C.12}$$

Appendix D

Membrane Products

On the following pages are listed the membrane products which have either been used for membrane bioreactor (MBR) duties or are being recommended for this purpose by the suppliers (Table D.1). Those products in which no specifications are presented in the subsequent sections are italicised. These products are either not commercially available or, as yet, have not been employed for MBR duties. Some multitube (MT) products may be used by process suppliers interchangeably in their processes provided they are a standard module size (typically 8″ diameter modules). Note that a number of generic terms and definitions exist for the three MBR membrane configurations of flat sheet (FS), hollow fibre (HF) and MT (Table D.2). The membrane module details of FS, HF and MT are given in Tables D.3–D.5.

The specific membrane area (membrane area per unit module volume) refers to the area based on the FS panel or HF/MT module. The area per module (FS) or rack (HF) is given in parenthesis, where provided by the supplier. Abbreviations for membrane polymeric materials are:

PAN polyacrylonitrile
PE polyethylene
PES polyethylsulphone
PS polysulphone
PVDF polyvinylidine difluoride

Table D.1 Membrane products

FS	HF	MT
Kubota	Asahi Kasei	Berghof
Huber	**Siemens Water Technologies** – Memcor	Norit X-Flow
Orelis Pleiade®	*Kolon*	*Orelis Kerasep*
Colloide	Mitsubishi Rayon	Millenniumpore
Brightwater	Motimo	
ITRI non-woven	Polymem	
Microdyn-Nadir	Puron	
Han-S	Zenon	
	Ultraflo	

Table D.2 Membrane configuration definitions

Generic term	FS	HF or MT
Membrane	Single sheet, or part thereof	Single filament/fibre or tube
Element	Panel	Module
Multiple element	Module, cassette or stack	Cassette, stack or rack
Multiple multi-element	Train	Train

Table D.3 Membrane module details of FS

Supplier	**Brightwater Engineering**
Membrane or module proprietary name (model)	Membright
Membrane material	PES
Pore size (μm) or MWCO (kDa)	150 kDa
Panel dimensions, length × width × depth (mm)	950 × 950 × 7
Panel area (m^2)	1.84
Membrane separation (mm)	~9
Module dimensions, length × width × depth (mm)	1120 × 1215 × 1450 (50-panel unit) 1120 × 715 × 1450 (25-panel unit)
Number of panels per module	50 or 25
Total membrane area per module (m^2/m^3)	110 (47)
Clean water permeability (LMH/bar)	–
Supplier	**Colloide Engineering Systems**
Membrane or module proprietary name (model)	Sub snake
Membrane material	PES
Pore size (μm) or MWCO (kDa)	0.04 μm
Panel dimensions, length × width × depth (mm)	1000 width × 5
Panel area (m^2)	10
Membrane separation, mm	10
Module dimensions, length × width × depth (mm)	–
Number of panels per module	5
Total membrane area per module (m^2/m^3)	160
Clean water permeability (LMH/bar)	–
Supplier	**Huber Technology**
Membrane or module proprietary name (model)	VRM
Membrane material	PES
Pore size (μm) or MWCO (kDa)	0.038
Panel dimensions, length × width × depth (mm)	–
Panel area (m^2)	0.75 for a single four-panel trapezoid element
Membrane separation (mm)	6 (6 mm membrane thickness)
Module dimensions, length × width × depth (mm)	Diameter = 2 or 3 m
Number of panels per module	Four plates/element; 6–8 elements per complete panel
Total membrane area per module (m^2/m^3)	160 (80–90)
Clean water permeability (LMH/bar)	–
Supplier	**Kubota**
Membrane or module proprietary name (model)	Kubota (ES, single-deck; EK, double 510)
Membrane material	Chlorinated PE
Pore size (μm) or MWCO (kDa)	0.4 μm
Panel dimensions, length × width × depth (mm)	1000 × 500 × 6
Panel area (m^2)	0.8
Membrane separation (mm)	7
Module dimensions, length × width × depth (mm)	1000 × 500 × 1300
Number of panels per module	100

(Table continued on next page)

Table D.3 *(continued)*

Supplier	**Kubota**
Total membrane area per module (m^2/m^3)	115
Clean water permeability (LMH/bar)	1110

Supplier	**Microdyne-Nadir**
Membrane or module proprietary name (model)	BioCel
Membrane material	PES
Pore size (μm) or MWCO (kDa)	150 kDa
Panel dimensions, length × width × depth (mm)	–
Panel area (m^2)	–
Membrane separation (mm)	8 (2 mm membrane thickness)
Module dimensions, length × width × depth (mm)	1200, 1340 × 650, 1140 × 1800, 2880
Number of panels per module	–
Total membrane area per module (m^2/m^3)	70 (90)
Clean water permeability (LMH/bar)	450–550

Supplier	**Novasep Orelis**
Membrane or module proprietary name (model)	Pleiade®
Membrane material	PAN
Pore size (μm) or MWCO (kDa)	40 kDa
Panel dimensions, length × width × depth (mm)	2610 × 438
Panel area (m^2)	~2
Membrane separation (mm)	3
Module dimensions, length × width × depth (mm)	2610 × 438 × 1710
Number of panels per module	–
Total membrane area per module (m^2/m^3)	(36)
Clean water permeability (LMH/bar)	1250 ± 500

Supplier	**Toray Industries**
Membrane or module proprietary name (model)	Toray (TRM140-100S)
Membrane material	PVDF
Pore size (μm) or MWCO (kDa)	0.08 μm
Panel dimensions, length × width × depth (mm)	1608 × 515 × 13.5 (inc. 6 mm separation)
Panel area (m^2)	1.4
Membrane separation (mm)	6
Module dimensions, length × width × depth (mm)	2100 × 810 × 1620
Number of panels per module	50 , 100 or 200
Total membrane area per module (m^2/m^3)	135
Clean water permeability (LMH/bar)	–

Table D.4 Membrane module details of HF

Supplier	**Asahi-Kasei**
Membrane or module proprietary name (model)	Microza
Membrane material	PVDF
Pore size (μm) or (MWCO) (kDa)	0.1 μm
Filament diameter (mm)	1.3
Module dimensions, length × diameter (mm)	2000 × 150

Table D.4 *(continued)*

Supplier	**Asahi-Kasei**
Module area (m^2)	25
Number of filaments per element	765
Rack dimensions, length × width × depth (mm)	2700 × 800 × 700
Number of modules per rack	4
Total membrane area per module (rack) (m^2/m^3)	710 (89)
Clean water permeability (LMH/bar)	–

Supplier	**Koch Membrane Systems**
Membrane or module proprietary name (model)	Puron
Membrane material	PES
Pore size (μm) or MWCO (kDa)	0.04 μm
Filament diameter	2.6 mm
Module dimensions, length × diameter (mm)	–
Module area (m^2)	30, 235 and 500
Number of filaments per element	–
Rack dimensions, length × width × depth (mm)	2000 × 1000 × 2600
Number of modules per rack	–
Total membrane area per module (rack) (m^2/m^3)	260 (96)
Clean water permeability (LMH/bar)	–

Supplier	**Mitsubishi Rayon Engineering**
Membrane or module proprietary name (model)	Sterapore SUR
Membrane material	PE
Pore size (μm) or MWCO (kDa)	0.4 μm
Filament diameter	0.54 mm
Module dimensions, length × width × depth (mm)	1035 × 13 × 524 446
Module area (m^2)	Up to 210
Number of filaments per element	–
Cassette dimensions, length × width × depth (mm)	1442 × 1538 × 725
Number of modules per cassette	70
Total membrane area per module (m^2/m^3)	485
Clean water permeability (LMH/bar)	–

Supplier	**Motimo Membrane Technologies Ltd.**
Membrane or module proprietary name (model)	Flat plat (FP AIV)
Membrane material	PVDF
Pore size (μm) or MWCO (kDa)	0.1–0.2 μm
Filament diameter	1.0 (0.6 ID)
Element area (m^2)	25
Module dimensions, length × width × depth (mm)	1510 × 530 × 450
Module area (m^2)	10 000
Number of filaments per element	–
Total membrane area per module (m^2/m^3)	1100
Number of modules per rack	360

Supplier	**Polymem**
Membrane or module proprietary name (model)	Immem (WW120)
Membrane material	PS

(Table continued on next page)

Table D.4 *(continued)*

Supplier	**Polymem**
Pore size (μm) or MWCO (kDa)	300 kDa; 0.08 μm
Filament diameter	0.7 to 1.4 mm
Module dimensions, length × diameter (mm)	1000 to 1500 × 315
Module area (m^2)	60–100 m^2
Number of filaments per element	a few tens of thousands
Number of modules per rack	na
Total membrane area per module (m^2/m^3)	800
Clean water permeability (LMH/bar)	500

Supplier	**Siemens Water Technologies**
Membrane or module proprietary name (model)	Memcor *MemJet*® (B10R, B30R)
Membrane material	PVDF
Pore size (μm) or MWCO (kDa)	0.04 μm
Filament diameter	1.3
Module dimensions, length × diameter (mm)	1610 × 154 (B10R); 203 × 203 (B40R)
Module area (m^2)	10 (B10R) or 38 (B30R)
Number of filaments per element	–
Number of modules per rack	40 (B10R) or 16 (B30R)
Total membrane area per module (m^2/m^3)	334
Clean water permeability (LMH/bar)	300

Supplier	**Zenon Environmental**
Membrane or module proprietary name (model)	ZeeWeed 500 c–d
Membrane material	PVDF
Pore size (μm) or MWCO (kDa)	0.04 μm
Filament diameter (mm)	1.9 (0.8 ID)
Module dimensions, length × width × depth (mm)	1940 × 678 × 60 (c); 2198 × 844 × 56 (d)
Module area (m^2)	23.2 (c), 31.6 (d)
Number of filaments per element	~2700
Stack dimensions, length × width × depth (mm)	–
Number of modules per stack	–
Total membrane area per module (m^2/m^3)	294 (c), 304 (d)

Supplier	**Ultrafo**
Membrane or module proprietary name (model)	Ultraflo (SS60)
Membrane material	PAN
Pore size (μm) or MWCO (kDa)	0.01–0.1
Filament diameter	2.1 mm OD (1.2 mm ID)
Element dimensions, length × diameter (mm)	1515 × 152
Element area (m^2)	28
Number filaments per element	–
Number of elements per module	na
Total membrane area per module (m^2/m^3)	1020
Clean water permeability (LMH/bar)	500

Table D.5 Membrane module details of MT

Supplier	**Norit X-Flow**
Web site/contact	
Membrane or module proprietary name (model)	F4385, F5385
Membrane material	PES
Pore size (μm) or MWCO (kDa)	0.03 μm
Tube diameter	5.2/5.3
Element dimensions, length × diameter (mm)	3000 × 200
Element area (m^2)	29
Number of tubes per element	620
Total membrane area per module (m^2/m^3)	307.7
Clean water permeability (LMH/bar)	>750
Supplier	**Berghof Membrane Technology**
Membrane or module proprietary name (model)	HyPerm-AE, HyperFlux
Membrane material	PVDF or PES
Pore size (μm) or MWCO (kDa)	250 kDa or 20–150 kDa
Tube diameter	Maximum 11.5 mm
Element dimensions, length × diameter, mm	3000 × 225 or 200
Element area (m^2)	11.8, 27.2 or 21.2
Number of tubes per element	112, 224 or 365
Total membrane area per module (m^2/m^3)	110@11.5 mm OD; maximum 280
Clean water permeability (LMH/bar)	PVDF >750; PES >300
Supplier	**Millenniumpore**
Membrane material	PES
Pore size (μm) or MWCO (kDa)	20–500 kDa UF; 0.05–0.5 μm MF
Tube diameter	2.5–15 mm
Element dimensions, length × diameter (mm)	To specification
Element area (m^2)	As dictated by specification
Numberof tubes per element	To specification
Total membrane area per module (m^2/m^3)	As dictated by specification
Clean water permeability (LMH/bar)	3000@0.05 μm; 600@20 kDa

Appendix E

Major Recent MBR and Wastewater Conferences

Listed below is a selection of membrane bioreactor (MBR) and Wastewater International Conferences since 2000. While every effort has been made to include information which is as comprehensive and accurate as possible, this is for guidance only and no responsibility is taken for any omissions or errors. Details were correct at the time of publication and websites were accessed in February 2006. Future conferences are listed where details are available. Conferences relating to desalination have not been included.

Conference title meet	Dates	Location	Web address
Department of Chemical Engineering, Aachen University			
Aachen Membrane Colloquium (AMC)			
8th AMC 2001	27–29 March 2001	Aachen, Germany	www.ivt.rwth-aachen.de/Amk/Eng/index_eng.html
9th AMC 2003	18–20 March 2003	Aachen, Germany	www.ivt.rwth-aachen.de/Amk/Eng/index_eng.html
10th AMC 2005	16–17 March 2005	Aachen, Germany	www.ivt.rwth-aachen.de/Amk/Eng/index_eng.html
11th AMC 2007	27–29 March 2007	Aachen, Germany	www.ivt.rwth-aachen.de/Amk/Eng/index_eng.html
Cranfield University, UK			
Membrane Bioreactors for Wastewater Treatment			
MBR3	16 May 2001	Cranfield, UK	www.cranfield.ac.uk/sims/water
MBR4	09 April 2003	Cranfield, UK	www.cranfield.ac.uk/sims/water
MBR5	28 January 2005	Cranfield, UK	www.cranfield.ac.uk/sims/water
The European Membrane Society; Institute for Membrane Technology			
Euromembrane 2000	24–27 September 2000	Jerusalem, Israel	
Euromembrane 2004	28 September–1 October 2004	Hamburg, Germany	
Euromembrane 2006	24–28 September 2006	Taormina, Italy	euromembrane2006.itm.cnr.it/primaryindex.asp
International Conference on Application of Membrane Technology			
2nd Conference	27–29 September 2002	Beijing, China	
International Conference and Exhibition for Filtration and Separation Technology (FILTEC)			
Filtech Europa 2001	16–18 October 2001	Dusseldorf, Germany	www.filtecheuropa.com
Filtech Europa 2003	21–23 October 2003	Dusseldorf, Germany	www.filtecheuropa.com
Filtech Europa 2005	11–13 October 2005	Wiesbaden, Germany	www.filtecheuropa.com
Filtech Europa 2007	27 February–1 March 2007	Wiesbaden, Germany	www.filtecheuropa.com
International Congress on Membranes and Membrane Processes (ICOM)			
ICOM 2002	7–12 July 2002	Toulouse, France	
ICOM 2005	21–26 August 2005	Seoul, Korea	www.icom2005.org
International Membrane Science and Technology Conference (IMSTEC)			
IMSTEC'03	10–14 November 2003	Sydney, Australia	
International Water Association			
Water Environment Membrane Technology (WEMT)			
WEMT 2004	7–10 June 2004	Seoul, South Korea	www.iwahq.org.uk

Conference title meet	Dates	Location	Web address
International Water Association Leading Edge Technology (LET) Conferences			
1st LET Conference	26–28 May 2003	Amsterdam, The Netherlands	
2nd LET Conference	1–4 June 2004	Prague	
3rd LET Conference and Exhibition on Water and Wastewater Treatment Technologies	6–8 June 2005	Sapporo, Japan	www.iwahq.org.uk
International Water Association Membranes for Water and Wastewater Treatment			
4th International Membranes Conference	13–16 May 2007	Harrogate, UK	www.iwamembranes.info
International Water Association World Water Congress (WWC)			
2nd WWC	15–19 October 2001	Berlin, Germany	www.iwahq.org.uk
3rd WWC	7–12 April 2002	Melbourne, Australia	
4th WWC	19–24 September 2004	Marrakech, Morocco	
5th WWC	10–14 September 2006	Beijing, China	
Membrane Industry Association of China (and others) Water and Membrane China (WMC)			
WMC 2004	27–29 September 2004	Beijing, China	www.grandexh.com
WMC 2005	14–16 September 2005	Beijing, China	
WMC 2005	29–31 March 2005	Shanghai, China	
WMC 2006	29–31 March 2006	Shanghai, China	
Network Young Membrains (NYM)			
NYM 2	7–8 September 2000	Aachen, The Netherlands	www.itm.cnr.it/NYM8/site/history.htm
NYM 3	6–7 September 2001	Delft, The Netherlands	
NYM 4	5–7 July 2002	Toulouse, France	
NYM 5	1–3 October 2003	Barcelona, Spain	
NYM 6	22–24 September 2004	Hamburg, Germany	
NYM 7	22–24 June 2005	Enschede, The Netherlands	
NYM 8	21–23 September 2006	Rende, Italy	www.itm.cnr.it/NYM8
North American Membrane Society (NAMS)			
11th Annual Meeting	23–27 May 2000	Colorado, USA	www.membranes.org
12th Annual Meeting	15–20 May 2001	Kentucky, USA	
13th Annual Meeting	11–15 May 2002	California, USA	
14th Annual Meeting	17–21 May 2003	Wyoming, USA	
15th Annual Meeting	26–30 June 2004	Hawaii, USA	
16th Annual Meeting	11–15 June 2005	Rhode Island, USA	
Slovak Society of Chemical Engineering Membrane Science and Technology Conference of Visegrad Countries (PERMEA)			
PERMEA 2003	7–11 September 2003	Slovakia	sschi.chtf.stuba.sk/permea
PERMEA 2005	18–22 September 2005	Poland	

Conference title meet	Dates	Location	Web address
Water Environment Federation			
Water Environment Federation's Technical Exhibition and Conference (WEFTEC)			
WEFTEC.00	October 2000	Anaheim, USA	www.weftec.org
WEFTEC.01	13–17 October 2001	Atlanta, USA	
WEFTEC.02	28 September–2 October 2002	Chicago, USA	
WEFTEC.03	11–15 October 2003	Los Angeles, USA	
WEFTEC.04	2–6 October 2004	New Orleans, USA	
WEFTEC.05	29 October–2 November 2005	Washington DC, USA	
WEFTEC.06	21–25 October 2006	Dallas, USA	
WEFTEC.07	13–17 October 2007	San Diego, USA	
WEFTEC.08	18–22 October 2008	Chicago, USA	
WEFTEC.09	17–21 October 2009	Orlando, USA	
WEFTEC.10	2–6 October 2010	New Orleans, USA	
Water Quality Association; Amsterdam RAI			
Aquatech			
Aquatech Amsterdam 2004	28 September–1 October 2004	Amsterdam, The Netherlands	www.wqa.org
Aquatech USA 2005	29 March–2 April 2005	Las Vegas, USA	
Aquatech USA 2006	29–31 March 2006	Chicago, USA	
Aquatech Amsterdam 2006	26–29 September 2006	Amsterdam, The Netherlands	
Aquatech USA 2007	28–30 March 2007	Orlando, USA	www.eshow2000.com/wqa
World Filtration Congress (WFC)			
WFC8	3–7 April 2000	Brighton, UK	
WFC9	18–22 April 2004	New Orleans, USA	
WFC10	14–18 April 2008	Leipzig, Germany	www.wfc10.com

In addition to the above, the following related conferences and workshops have also been held – again, this list is not comprehensive but represents a flavour of the events which are available:

Event	Last held	Location	Website
Advanced Membrane Technology	23–28 May 2004	Irsee, Germany	www.engconfintl.org/calendar.html#environtech
Annual Membrane/Separations Technology Planning Conference	December 2003	USA	www.bccresearch.com/conferences
CITEM (Ibero American Congress on Membrane Science and Technology)	6–8 July 2005	Valencia, Spain	

Event	Last held	Location	Website
Fouling and Critical Flux: Theory and Applications	16–18 June 2004	Finland	www.lut.fi/cst/fouling2004
IWEX (International Water and Effluent Treatment Exhibition)	October 2005	Birmingham UK	www.iwex.co.uk
Regional Symposium on Membrane Science and Technology	26–27 April 2005	Bandung, Indonesia	www.mst2006.com
WateReuse 20th Annual Symposium	18–21 September 2005	Denver, USA	www.watereuse.org/2005Symposium
Water Sciences and Technology Association – 7th Gulf Water Conference	19–23 November 2005	Kuwait	
Water and Wastewater Europe	28–30 June 2005	Milan, Italy	wwe05.events.pennnet.com
Water and Wastewater Treatment with Hybrid Membrane Processes and Membrane Bioreactors – Recent Advances	2 November 2004	Singapore	www.iese.ntu.edu.sg/eventsconf.htm#bioreactor2
Watermex Asia	29 November–2 December 2005	Singapore	www.environmexasia.com

Appendix F

Selected Professional and Trade Bodies

Listed below is a selection of major membrane-related professional and trade bodies, societies and associations. This list is for guidance only, and no responsibility is taken for any omissions or errors. Details were correct at the time of publication and websites were accessed in February 2006. Associations relating specifically to desalination have not been included.

Name of body, society or association	Based	Web address
American Filtration and Separations Society	Richfield, MN, USA	www.afssociety.org
American Membrane Technology Association (AMTA)	Florida, USA	www.membranes-amta.org
American Water Works Association (AWWA)	Denver, USA	www.awwa.org
Association of State Drinking Water Administrators (ASDWA)	Washington DC, USA	www.asdwa.org
Australian Water Association (AWA)	Artarmon, Australia	www.awa.asn.au
British Water	London, UK	www.britishwater.co.uk
Canadian Water and Wastewater Association (CWWA)	Ottawa, Canada	www.cwwa.ca
Co-operative Research Centre for Water Quality and Treatment	Adelaide, Australia	www.waterquality.crc.org.au
DECHEMA, Subject Division Membrane Technology	Frankfurt, Germany	www.dechema.de
European Membrane Society (EMS)	Toulouse, France	www.ems.cict.fr
The Filtration Society	UK	www.lboro.ac.uk/departments/cg/research/filtration
Foundation for Water Research (FWR)	Marlow, UK	www.fwr.org
Institute for Membrane Technology (IMT)	Rende, Italy	www.itm.cnr.it/english
International Water Association (IWA)	London, UK	www.iwahq.org.uk

Japan Water Works Association (JWWA)	Tokyo, Japan	www.jwwa.or.jp (Japanese)
Membrane Academia Industry Network (MAIN)	Cranfield, UK	www.cranfield.ac.uk/sims/water/memnet/mise.html
Membrane Industry Association of China (MIAC)	Beijing, China	www.membranes.com.cn (Chinese)
Membrane Society of Japan	Tokyo, Japan	www.soc.nii.ac.jp/membrane/index_e.html
Membrane Society of Korea	Seoul, Korea	www.membrane.or.kr/index_eng.asp
Middle East Water Information Network		www.water1.geol.upenn.edu/index.html
North American Membrane Society (NAMS)	Toledo, Ohio, USA	www.membranes.org
Society of British Water and Wastewater Industries (SBWWI)	Warwickshire, UK	www.sbwi.co.uk
South African Water Research Commission	Pretoria, South Africa	www.wrc.org.za
UK Water Industry Research Ltd (UKWIR)	London, UK	www.ukwir.co.uk
Water Environment Federation (WEF)	Alexandria, USA	www.wef.org
Water Quality Association (WQA)	Lisle, IL, USA	www.wqa.org
WateReuse Association	Alexandria, VA	www.watereuse.org
Water Sciences and Technology Association (WSTA)	Manama, Bahrain	www.wsta.org.bh
Water Technology	London, UK	www.water-technology.net
Water UK	London, UK	www.water.org.uk

Nomenclature

δ	Separation (m)
Φ	Temperature correction constant
η	Viscosity (Pa/s)
ψ	Resistance per cake depth (1/m)
κ	Membrane geometry constant
ϕ	Membrane packing density (m^2/m^3)
μ_m	Maximum specific growth rate (g VSS/(g VSS/day))
ΔP	Pressure drop (Pa)
ΔP_h	Hydrostatic pressure (kPa)
ΔP_m	Transmembrane pressure or TMP (bar)
a	Gas-liquid or water-air interface surface area per unit volume (1/m)
A_f	Membrane filament area (m^2)
A_m	Membrane area (m^2)
A_t	Cross-sectional area of tube (m)
A_x	Membrane open cross sectional area (m^2)
$A_{x,filament}$	Membrane filament open cross sectional area, HF (m^2)
$A_{x,panel}$	Membrane panel open cross sectional area, FS (m^2)
C	Dissolved oxygen concentration (kg/m^3)
C^*	Saturated oxygen concentration (kg/m^3)
c_c	Cleaning reagent concentration (kg/m^3)
d	Diameter (m)
E	Process efficiency
F	Membrane replacement (years)
f_d	Fraction of the biomass that remains as cell debris (g VSS/g BOD)
g	acceleration due to gravity (9.81 m/s^2)
H	Pump head (m of water)
J	Flux (LMH)
J'	Temperature pressure corrected flux (LMH)
J_b	Backflush flux (LMH)
J_c	Critical flux (LMH)
$J_{CIP\ (CEB)}$	Cleaning flux (chemically enhanced backflush (LMH))
J_{net}	Net flux (LMH)
K	Permeability (LMH/bar)
k	Maximum specific substrate utilisation rate (1/d)

K'	Temperature pressure corrected permeability (LMH/bar)
k_e	Death coefficient or death rate constant (g VSS/(g VSS/day))
$k_{e,n}$	Death rate constant for nitrifying bacteria (g VSS/(g VSS/day))
k_L	Mass transfer coefficient (m/s)
k_La	Volumetric mass transfer coefficient (1/s)
K_n	Saturation coefficient for nitrogen (g/m^3)
K_s	Saturation coefficient (g/m^3)
L	Length of membrane fibre (m)
m_A	Mass flow of air (kg/s)
$m_{A,b}$	Mass flow of air for biological aeration (kg/day)
$m_{A,m}$	Mass flow of air for membrane scour (kg/day)
M_c	Mass of chemical per unit permeate (kg/m^3)
m_o	Total oxygen required (g/day)
n	Number of physical cleans per chemical clean
N	Total nitrogen in the influent or TKN (g/m^3)
N_e	Effluent nitrogen concentration (g/m^3)
NO_x	Concentration of NH_4-N that is oxidised (g/m^3) to form nitrate
OTE	Oxygen transfer efficiency
OTR	Oxygen transfer rate (kg O_2/s)
P	Pressure (Pa, m or bar)
$P_{A,1}$	Inlet blower pressure as an absolute pressure (Pa)
$P_{A,2}$	Inlet blower pressure as an absolute pressure (Pa)
P_{max}	A threshold pressure beyond which operation cannot be sustained (bar)
$Power_b$	Air blower power demand (kW)
P_x	Biomass sludge production (g/day), where $P_{x,het}$ is heterotrophic sludge production and $P_{x,aut}$ is the sludge production by autotrophic organisms
Q	Average feed flow rate (m^3/day)
$Q'_{A,m}$	Temperature pressure corrected airflow to the membrane (m^3/day)
$Q_{A,b}$	Airflow to the biomass (m^3/day)
$Q_{A,m}$	Airflow to the membrane (m^3/day)
Q_{dn}	Denitrification recycle flow rate (m^3/day)
Q_e	Effluent flow rate (m^3/day)
Q_P	Permeate flow rate (m^3/day) note $Q_P = Q_e$
Q_{peak}	Peak feed flow rate (m^3/h)
Q_{pump}	Pumped flow rate (m^3/day)
Q_R	Retentate flow rate (m^3/day)
Q_w	Sludge wastage rate (m^3/day)
r	Denitrification recycle ratio as a fraction of Q
R	Resistance (1/m)
R_{as}	Resistance of the activated sludge (1/m)
R_b	Biomass aeration per m^3 permeate produced (m^3/m^3)
R_c	Cake resistance (1/m)
R_{col}	Hydraulic resistances for colloidal matter in the biomass (1/m)
R_M	Membrane aeration per m^2 membrane area (m^3/m^2/day)
R_{sol}	Hydraulic resistances for soluble matter in the biomass (1/m)
R_{ss}	Hydraulic resistances for suspended solids (1/m)

R_{sup}	Hydraulic resistances for colloidal and soluble matter in the biomass (1/m)
R_{tot}	Total hydraulic resistances (1/m)
R_V	Membrane aeration per m^3 permeate produced (m^3/m^3)
S	Limiting substrate concentration (g/m^3)
S_e	Effluent substrate concentration (g/m^3)
T	Temperature (°C)
t	Time (s)
t_c	Period between chemical cleans (h)
t_{crit}	Time over which low-fouling operation is maintained (h)
T_K	Temperature (K)
t_p	Period between physical cleans (h)
TSS	Total suspended solids (g/m^3)
T_w	Wastewater temperature (°C)
U_G	Gas velocity (m/s)
U_L	Liquid crossflow velocity (m/s)
U_R	Retentate velocity (m/s)
V	Aeration tank volume (m^3)
V_{an}	Anoxic tank volume (m^3)
v_c	Cleaning reagent volume (m^3)
VSS	Volatile suspended solids (g/m^3)
w	Panel width (m)
W_b	Biological aeration blower power demand per unit permeate (kWh/m^3)
$W_{b,m}$	Blower power demand per unit membrane area (kW/m^2)
$W_{b,V}$	Membrane aeration blower power demand per unit permeate (kWh/m^3)
W_h	Permeate pumping energy demand per unit permeate (kWh/m^3)
W_p	Pumping power demand per unit permeate (kWh/m^3)
X	Biomass concentration (g/m^3)
X_0	Non-biodegradable sludge production (g/day)
X_e	Effluent solids concentration (g/m^3)
X_w	Waste stream suspended solids concentration (g/m^3)
Y	Biomass yield for heterotrophs (g VSS/g BOD)
Y_n	Biomass yield for nitrification (g VSS/g NH_4-N)
Y_{obs}	Observed yield (g VSS/g BOD)
α	Difference in mass transfer (k_La) between clean and process water
β	Correction factor for the influence of wastewater constituents on C^*
γ	Shear rate (1/s)
ΔK	Change in permeability (LMH/bar)
ΔP_R	Pressure loss in retentate channel (Pa)
Δt	Change in time (s)
Θ	Conversion
θ_x	Solids retention time or sludge age (day)
λ	Ratio of specific heat capacity at constant pressure to constant volume (1.4 for air)
μ	Growth rate (day)
μ_n	Specific growth rate for nitrifying bacteria (g VSS/(g VSS/day))
$\mu_{n,m}$	Maximum specific growth rate for nitrifying bacteria (g VSS/(g VSS/day))

ξ	Mechanical efficiency
ρ	Density (kg/m^3) where subscript p refers to permeate, b relates to biomass and a to air
τ_c	Duration of chemical clean (h)
τ_p	Duration of physical clean (h)
φ	Temperature correction factor for aeration

Abbreviations

The following lists key abbreviations used in the book. Further definitions are given in the Glossary of Terms. Note that proprietary abbreviations are not listed.

ABR	Anaerobic baffled reactor
AD	Anaerobic digestion
ADUF	Anaerobic digester ultrafiltration
Alum	Aluminum [aluminium?] sulphate
AN	Anaerobic
anMBR	Anaerobic membrane bioreactor
AOC	Assimilable organic carbon
ASP	Activated sludge process
AX	Anoxic
BAC	Biologically activated carbon
BAF	Biological aerated filter
BPA	Biological potential activity
BER	Biofilm-electrode reactor
BNR	Biological nutrient removal
BOD	Biochemical oxygen demand
CEB	Chemically-enhanced backwash
CF	Crossflow
CFV	Crossflow velocity
CIL	Cleaning in line
CIP	Cleaning in place
COD	Chemical oxygen demand
CP	Concentration polarisation
CPR	Chemical phosphorous removal
CST	Capillary suction time
CT	Capillary tube
Da	Dalton
DE	Dead-end (or full flow)
dMBR	Diffusive membrane bioreactor
DO	Dissolved oxygen
DOC	Dissolved organic carbon
DS	Dry solids
EBPR	Enhanced biological phosphate removal
ED	Electrodialysis
(e)EPS	(Extracted) extracellular polymeric substances
EGSB	Expanded granular sludge bed

eMBR	Extractive membrane bioreactor
EPS	Extracellular polymeric substances
EPSc	Extracellular polymeric substances (carbohydrate)
EPSp	Extracellular polymeric substances (protein)
FBDA	Fine bubble diffusion aeration
F:M Ratio	Food-to-micro-organism ratio
FC	Filter cartridge
Flocs	Flocculated particles
FS	Flat sheet (or plate-and-frame)
GAC	Granular activated carbon
GLD	Gigalitres per day
GT	Gas transfer
HF	Hollow fibre
HPSEC	High performance size exclusion chromatography
HRT	Hydraulic retention time
ID	Internal diameter
iMBR	Immersed membrane bioreactor
kDa	kiloDalton
LMH	Litres per m^2 per hour
LMH/bar	Litres per m^2 per hour per bar
MABR	Membrane aeration bioreactor
MCE	Mixed cellulose esters
ME	Membrane extraction
MF	Microfiltration
MHBR	Membrane hydrogenation bioreactor
MLD	Megalitres per day
MLSS	Mixed liquor suspended solids
MLVSS	Mixed liquor volatile suspended solids
MPE	Membrane performance enhancer
MT	Multitube
MW	Molecular weight
MWCO	Molecular weight cut-off
NADH	Nictotinamide adenine dinucleotide hydrogenase
NF	Nanofiltration
NOM	Natural organic matters
O&M	Operation & Maintenance
OC	Organic carbon
OD	Outer diameter
OLR	Organic loading rate
ON	Organic nitrogen
OTR	Oxygen transfer rate
OUE	Oxygen utilisation efficiency
PAC	Powdered activated carbon
PAN	Polyacrylonitrile
p.e.	Population equivalent
PE	Polyethylene

PES	Polyethylsulphone
PP	Polypropylene
PV	Pervaporation
PVDF	Polyvinylidene difluoride
RBC	Rotating biological contactor
Rc	Cake resistance
Redox	Reduction-oxidation
rMBR	(Biomass) rejection membrane bioreactor
RO	Reverse osmosis
SAD	Specific aeration demand
SAD_m	Specific aeration demand – membrane area (in Nm^3 air/(hr·m^2))
SAD_p	Specific aeration demand – permeate volume (in Nm^3 air/m^3 permeate)
SAE	Standard aeration efficiency (kgO_2/kWh)
SBR	Sequencing batch reactor
SCADA	Supervisory control and data acquisitions
SDI	Silt density index
sMBR	Sidestream membrane bioreactor
SME	small-to-medium sized enterprise
SMP	Soluble microbial product
SMP_c	Soluble microbial product (carbohydrate)
SMP_p	Soluble microbial product (protein)
SNdN	Simultaneous nitrification/denitirification
SRF	Specific resistance to filtration
SRT	Solids retention time
SUVA	Specific UV absorbance per unit organic carbon concentration
SVI	Sludge volume index
SW	Spiral-wound
TDS	Total dissolved solids
TF	Trickling filter
THMFP	Tri-halo methane formation potential
TKN	Total Kjedldahl nitrogen
TMDL	Total maximum daily load
TMP	Transmembrane pressure
TOC	Total organic carbon
TSS	Total suspended solids
UASB	Upflow anaerobic sludge blanket
UF	Ultrafiltration
VRM	Vacuum rotating membrane
VSS	Volatile suspended solids
WRP	Water recycling (or reclamation) plant
WWTP	Wastewater treatment plant

Glossary of Terms

A number of key terms used in the book are defined below. Proprietary names and processes are not included.

Aerobic	Conditions where oxygen acts as electron donor for biochemical reactions
Allochthonous	Of terrestrial origin
Anaerobic	Conditions where biochemical reactions take place in absence of oxygen
Anisotropic	Having symmetry only in one plane
anMBR	Anaerobic membrane bioreactor
Annular flow	Flow through an annulus (or gap created concentric cylinders)
Anoxic	Conditions where some species other than oxygen acts as the electron donor for biochemical reactions
Anthropogenic	Of human origin or derived from human activity
Autochthonous	Of microbial origin
Autotrophic	Using carbon dioxide as sole carbon source for growth and development
Backflushing	Reversing flow through a membrane to remove foulants
Biofilm	Film or layer of biological material
Biological treatment (or biotreatment)	Process whereby dissolved organic chemical constituents are removed through biodegradation
Biomass	Viable (living) micro-organisms used to achieve removal of organics through biotreatment
Bubble flow	Air/liquid two-phase flow where the liquid is the continuum
Cake	Solid material formed on the membrane during operation
Cassette	See Appendix D
Churn flow	Air/liquid two-phase flow at high air/liquid ratio

Clogging/ sludging	The accumulation of solids within the membrane channels
Concentration polarisation (CP)	Tendency of solute to accumulate at membrane: Solution interface within concentration boundary layer, or liquid film, during crossflow operation
Conditioning fouling	First stage of membrane fouling through adsorption of material
Critical flux	Flux below which permeability decline is considered negligible
Critical suction pressure	Threshold pressure arising during sub-critical flux fouling
Crossflow	Retentate flow parallel to the membrane surface
Cyclic aeration	Aeration on *n* s on/*n* s off basis, where *n* is normally between 5 and 30 s
Dalton (Da)	Molecular mass relative to that of a hydrogen atom
Dead-end or full-flow	Flow where all of the feed is converted to permeate
Death coefficient	A biokinetic parameter
Denitrification	Biochemical reduction of nitrate to nitrogen gas
Dense membrane	Membrane of high selectivity attained by specific physicochemical interactions between solute and membrane
Electrodialysis	Membrane separation process by which ions are removed via ion-exchange membranes under the influence of an electromotive force (voltage)
Electron donor	Species capable of donating an electron to a suitable acceptor, and is oxidised as a result
Element (membrane)	See Appendix D
EMBR	Extractive MBR: MBR configured so that priority pollutants are selectively extracted into the bioreactor via the membrane
Endogenous metabolism	Developing or originating within, or part of, a micro-organism or cell
Exogenous	Originating outside the micro-organism or cell
F:M (ratio)	Food-to-micro-organism (ratio): Rate at which substrate is fed to the biomass compared to the mass of biomass solids
Filament	Single hollow fibre or capillary tube
Filamentous index	Parameter indicating relative presence of filamentous bacteria in sludge
Fixed film process	Process configured with the biofilm attached to a solid medium (which may be a membrane)

Floc	Aggregated solid (biomass) particle
Flux (or permeate velocity)	The quantity of material passing through a unit area of membrane per unit time
Flux-step	Critical flux identification method whereby flux is incrementally increased and the TMP or permeability response recorded
Fouling	Processes leading to deterioration of flux due to surface or internal blockage of the membrane
Gas/air sparging	Introduction of gas/air bubbles
Gas/air-lift	Lifting of liquid using gas/air
Gel layer formation	Precipitation of sparingly soluble macromolecular species at membrane surface
Heterotrophic	Requiring an organic substrate to get carbon for growth and development
Humic matter	Organic matter of terrestrial origin
Hydraulic loading rate	Rate at which water enters the reactor
Hydrogenotrophic	Feeding of hydrogen
Hydrophilic	Water-absorbent
Hydrophobic	Water-repellent
Inoculum	Medium containing micro-organisms initially introduced into a reactor to establish new populations and start the biotreatment process
Intensive/ recovery clean	Cleaning with aggressive chemicals to recover membrane permeability
Irrecoverable fouling	Fouling which is not removed by physical or chemical cleaning
Irreversible/ permanent fouling	Fouling which is removed by chemical cleaning
Lumen-side	Inside the fibre/filament/lumen
Macropore	Pore with diameter above 50 mm
Maintenance cleaning	Cleaning with less aggressive chemicals to maintain membrane permeability
Membrane packing density	Membrane area per unit volume
Mesophilic	Thriving at intermediate temperatures (20–45°C, 15°C optimum)

Mesopore	Pore with diameter between 2 and 50 mm
Methanogens	Microorganisms producing methane as a metabolic byproduct
Micropore	Pore with diameter below 2 mm
Mist flow	Air/liquid two-phase flow where the air is the continuum
Modularisation	Based on modules: Using more modules at higher flows, rather than increasing the unit process size
Module (membrane)	See Appendix D
Monod kinetics	Kinetics defining biomass growth and decay during biotreatment
Nitrification	Biochemical oxidation of ammonia to nitrate
(Organic) loading rate	Rate at which (organic) matter is introduced into the reactor
Panel	FS membrane element
Percolation theory	Theory defining probability of water flowing through a medium containing three-dimensional network of interconnected pores
Permeability	Ease of flow through membrane, represented by flux:pressure ratio
Permeate	Water or fluid which has passed through the membrane
Perm-selectivity	Permeation of some components in preference to others
Plate-and-frame	Synonymous with "flat sheet"
Pleated filter cartridge	Type of flat sheet module
Plug flow	Flow in which no back-mixing or dispersion occurs along the length of the pipe or reactor
Pore plugging	Type of membrane fouling (by blocking of pores)
Porous membrane	Membrane of low selectivity operating by physical straining alome
Psychrophilic	Thriving at relatively low temperatures (0–20°C)
Rack	See Appendix D
Recovery or conversion	Fraction of feed water converted to permeate product
Reduction-oxidation (Redox) conditions	Conditions defined by the presence of either dissolved oxygen or some other species capable of providing oxygen for bioactivity

Relaxation	Ceasing permeation whilst continuing to scour the membrane with air bubbles
Resistance	Resistance to flow, proportional to flow rate:pressure ratio
Retentate	Water or fluid which is rejected by the membrane
Reversible or temporary fouling	Gross solids attached to the membrane surface and which can be removed by cleaning relatively easily
Septum	Coarse membrane filter
Shear (stress)	Force applied to a body which tends to produce a change in its shape, but not its volume
Shell-side	Outside the membrane fibre/filament/lumen
Side-stream	Stream outside the bioreactor
Slug flow	Air/liquid two-phase flow at moderate air/liquid ratios
Stack	See Appendix D
Struvite	Magnesium ammonium phosphate salt
Substrate	Surface or medium on which an organism grows or is attached
Supernatant	Liquid clarified by sedimentation
Isoporosity	Property reflecting narrowness of pore size distribution
Surface porosity	Percentage of the surface area occupied by the pores
Sustainable flux	The flux for which the TMP increases gradually at an acceptable rate, such that chemical cleaning is not necessary
Thermophilic	Thriving at relatively high temperatures (49–57°C, 45°C optimum)
TMP jump	Sudden increase TMP when operating under sub-critical flux conditions
TMP-step method	Critical flux identification method where TMP is incremen tally increased and the flux or permeability response recorded
Tortuosity	Ratio of pore length to membrane thickness
Upflow clarification	Dynamic clarification by sedimentation
Volumetric mass transfer coefficient	A combination of (i) the overall liquid mass transfer coefficient and (ii) the specific surface area for mass transfer. The term measures the mass transfer of oxygen into the liquid via air bubbles
Zeta potential	Potential (in mV) at the shear plane of a solid:liquid interface
α or β factor	Factors applied to correct biological aeration demand for dissolved and suspended solids content of biomass

Index